FIGHTING TECHNIQUES OF THE ORIENTAL WORLD

AD 1200 ~ 1860

FIGHTING TECHNIQUES OF THE ORIENTAL WORLD

AD 1200 ~ 1860

Equipment, Combat Skills, and Tactics

Michael E. Haskew Christer Jörgensen Chris McNab Eric Niderost Rob S. Rice

METRO BOOKS
New York

METRO BOOKS
New York

An Imprint of Sterling Publishing
387 Park Avenue South
New York, NY 10016

This 2013 edition published by Metro Books by arrangement with Amber Books Ltd.

Editorial and design by
Amber Books Ltd
74–77 White Lion Street
London N1 9PF
www.amberbooks.co.uk

Project Editor: Michael Spilling
Design: Zöe Mellors
Picture Research: Terry Forshaw

ISBN 978-1-4351-4532-0

For information about custom editions, special sales, and premium and corporate purchases, please contact Sterling Special Sales at 800-805-5489 or specialsales@sterlingpublishing.com.

Manufactured in China

2 4 6 8 10 9 7 5 3 1

www.sterlingpublishing.com

PICTURE CREDITS:
AKG-Images: 98 (Francois Guenet), 148; **Board of Trustees of the Armouries**: 11, 38, 47 (c&b), 51 (t&b), 56, 64, 90 (t), 95, 99, 106, 110, 111, 114, 116, 125, 154 (b); **Bridgeman Art Library**: 70/71 (Bibliotheque Nationale), 123 (The British Library); **Corbis**: 21 (Bettmann), 43 (Asian Art & Archaeology), 54 (Asian Art & Archaeology), 74/75, 102/103 (Reuters), 130/131, 135, 138 (The Gallery Collection), 154 (t), 163 (Asian Art & Archaeology), 194, 213 (Hulton-Deutsch Collection), 221, 249 (Bettmann); **Dorling Kindersley**: 42, 47 (t), 82, 90 (b); **Eric Niderost**: 175, 182, 203; **Getty Images**: 13 (Hulton Archive), 39, 45 (Hulton Archive), 62 (Time Life Pictures), 100 (Bridgeman), 128/129 (Hulton Archive), 139 (National Geographic), 195 (Hulton Archive); **Heritage Image Partnership**: 6/7, 10, 170/171, 209 (The Print Collector), 216/217 (The British Museum), 223 (The British Library); **Photos.com**: 115; **Photos12.com**: 122 (Panorama Stock);**Thomas Conlan**: 55 (t&b).

All line art and maps courtesy of JB Illustrations.

CONTENTS

CHAPTER 1

The Role of Infantry

The story of the development of oriental warfare during the late Medieval and Early Modern periods is one characterized by a classic military tension between the old and the new. On the one hand, there is the force of tradition embodied by elite, frequently aristocratic, warriors, typically cavalry troops and royal guards. These bodies of men emphasized individual excellence on the battlefield, with hard-learnt skills in handling the sword, bow or spear often backed by strong codes of martial and ethical behaviour. On the other hand, there is the infantry – mass as opposed to individualism.

As we shall see, the infantry was not always divorced from higher martial traditions or the claim that they belonged to an elite themselves. Nonetheless, the power of the infantry over the individual warrior grew with time,

THE TERRACOTTA ARMY *provides us with a striking insight into the composition and uniform of third-century Chinese infantry, although ancient looting has deprived us of many of the real weapons carried by the warriors.*

particularly with emergent forms of gunpowder weaponry. As in the European tradition, there was an inexorable movement towards infantry firepower becoming a decisive, if not dominant, power on the battlefield.

Grist for the Mill – Infantry Soldiers

Infantry recruitment in oriental armies was never a simple matter of scooping up the requisite amount of men and herding them towards the battlefield. Indeed, recruitment of infantry had numerous social and political implications that a ruler was wise to recognize. The first was that in societies dominated by subsistence agriculture, with all the seasonal commitments that implied, it was difficult and frequently inadvisable to strip men from the land. Desertions would often rise steeply during the seasons of planting and harvest, the men simply drifting away to tend to their fields, despite potentially fatal punishments for so doing. Keeping men away from life-critical responsibilities could also foster anger and might encourage rebellions and defections.

> *'Now the resources of those skilled in the use of extraordinary forces are as infinite as the heavens and earth; as inexhaustible as the flow of the great rivers.'*
>
> — *Sun Tzu,* The Art of War

Looking at China, the picture of infantry recruitment is complicated by the geographical extent involved, the length of our time period and the number of competing powers within the territory. Under the Mongol leadership of the thirteenth century, the Chinese armies faced a difficult recruitment juggling act as the previously nomadic Mongol rulers had to adjust to the realities of the agriculturally static Chinese society. Furthermore, the Mongol warriors themselves often had to give up their nomadic ways and become farmers because of a lack of suitable pasture for their mounts.

Infantry recruitment was by conscription from two bodies of citizens. First, there were warrior families who had provided military personnel for generations and who brought with them some body of martial understanding. Second, there was the great mass of civilians: an amorphous pool into which military leaders could dip into in times of emergency. Essentially, in times of military need a political leader or commander would simply levy the requisite number of troops from a geographical area, usually one over which he had an administrative jurisdiction, and herd them into a rough military structure.

The system had an arbitrary quality to it that led to abuses of power and also to frequent indiscipline amongst the ranks of recruits; the ad hoc nature of recruitment was hardly conducive to breeding either loyalty or enthusiasm. This changed, albeit temporarily, at the beginning of the Ming Dynasty in the fourteenth century when Mongol rule was overthrown. Under the Hongwu emperor (1328–98), the system of both recruitment and recruit management received a much-needed injection of professionalism.

One of the phenomena that Hongwu aimed to break was the way that individual leaders could build up their own private armies that owed more allegiance to them than the state. Hongwu established military districts called *so* or *wei* (the latter differed from the first by covering two prefectures instead of one) across China and each district was obligated to produce a set amount of soldiers. The registers of soldiers and recruitment were organized centrally by government staff rather than by local power players, and all commanders were appointed by merit rather than position.

An important point was that the commanders only took charge of their men for the duration of a particular campaign; thus the men serving under them would never identify their leaders as a permanent. Another very important reform under Hongwu was made in recognition that most of the soldiers recruited were subsistence farmers. The system kept up to 80 per cent of soldiers

Yuan Chinese Infantryman (c. 1260)

This halberdier of the Yuan Dynasty is dressed with extreme simplicity, his 'uniform' being a simple coloured tunic (some infantry soldiers might have motifs or words patterned onto the fabric to aid with identification). Atop his head he wears a plain civilian hat, although various styles of helmet were another option, and the long hair protruding out from the hat served as a symbol of courage. The design of his halberd has one particularly interesting feature. Just beneath the main chopping blade is a brightly coloured tassel that swung freely on a ring. The purpose of the tassel was to distract a defending enemy – it drew the eye away from the swinging blade and thereby made the blow more difficult to avoid or parry. For close-quarters fighting, the infantryman wears a short sword on his belt.

employed in grain production, and these soldiers were given enough land to grow food for both themselves and their active service counterparts.

Although the *wei-so* system was not to last throughout the Ming Dynasty – familiar corruption crept in – it demonstrated how large-scale infantry recruitment could be maintained without affecting the social order or political power. The number of wars that passed through Chinese territory meant that, generally speaking, it was not difficult to find experienced infantrymen, although finding competent officers to lead them could be much more problematic. Moreover, the reliance on levied troops was strong – at the battle of Dateng Xia in 1466, for example, a Ming army commanded by Han Yung totalled 190,000 but only 30,000 of that number were actually regular soldiers. As time went on, however, the numbers of infantry available declined dramatically, often because officers kept the names of dead or deserted men on their registers so that they could claim the men's wages. Indiscipline crept into the ranks, and the infantry became known as much for fighting civilians and each other as the enemy.

The social status of soldiering consequently declined greatly during the sixteenth century, fuelled by the heavy use of low-class mercenaries and foreign troops. Use of foreigners, particularly Mongols, could cause problems when the Chinese came up against enemy tribes to whom the foreigners were affiliated. Soldiers were also apt to rebel if a campaign was not going their commander's way.

Incentives and Corruption

Chinese soldiers were often given financial incentives to serve, such as a share in booty. The

THIS MASS PRESENTATION *of the Terracotta Army gives the illusion that oriental infantry were cohesive bodies of battlefield soldiers. All but the closest of formations were often lost once battle was joined.*

rewards could be given on the basis of the number of enemy heads presented after a victorious battle, but this seemingly transparent process hid some grim realities, such as the heads of executed prisoners and civilians being added in amongst the military dead. Yet however strong the financial incentives, potential defeat could quickly undo any loyalty between soldiers and their commanders.

The imperfections of Chinese recruitment continued until well into the nineteenth century and beyond. During the Qing Dynasty, which ran from 1644 to the early 1900s, the Manchu leaders increased the numbers of individuals who could be levied into the infantry, rising to as high as one in three of the male population during the 1640s. The numbers of men drawn into military service were probably necessary to compensate for the general deterioration in standards of training and military expertise amongst the recruits. On the other hand, the sheer numbers of men available to the Chinese did result in their military being one of the world's most potent by the end of the Qing Dynasty. Between 1650 and 1800, the Chinese population doubled (150 million to 300 million) and the consequent rise in infantry numbers fuelled a huge programme of Chinese territorial expansion. This expansion began with Outer Mongolia in the 1690s, and during the next 100 years it incorporated Tibet, Bhutan and Kashgar. Less success was achieved in expanding the empire into Southeast Asia. The complicated and frequently dense jungle terrain of the region was much less forgiving for a large army to travel through than the open plains to the west of China, and the armies that went southwards also found themselves blighted by guerrilla warfare.

> *'When one or two meet it is fine to fight individually, but when spears are employed they must be coordinated and well timed. On this there must to no exception.'*
>
> – *The* Zôhyô Monogatari

Regardless of the period involved, logistics were always a huge challenge for Chinese armies on the move. One basic system of food provision was known as *tuntian* (military habitation), and had roots stretching back into the Qin (221–206 BC) and Han (206 BC–AD 220) Dynasties. Under this system border wasteland areas were converted for the growing of military-stock grain. This system was practically and administratively refined over the centuries, and was heavily supplemented by taxes on the civilian population. At one point in the eleventh century, up to 75 per cent of all taxes were used to finance military operations.

During the Ming Dynasty (1368–1644), the proportion of military revenue provided from taxes rose and the production from *tuntian* fell. This trend continued during the Qing Dynasty (1644–1912), but increased government investment did not mean improvements in the living conditions of the regular soldier. Indeed, corruption amongst the officer class meant that funds were often siphoned off before they reached the lowly infantryman, resulting in desertion, demoralization and hunger.

Only in the late Qing period, as western powers began to flex their muscles over China, did military authorities begin a partial wakeup to the failings of their military system, but inefficiencies and corruption would blight the Chinese military until well into the twentieth century. An illustration of how logistics could affect the outcome of a

A THIRTEENTH-CENTURY *Chinese helmet, constructed as a one-piece steel bowl. The tube extending from the crown of the helmet could have been used to display a plume or standard.*

SIXTEENTH-CENTURY CHINESE INFANTRY *(actually an anti-pirate militia from China's southern coast) are here seen making a charge. The bamboo branches were genuinely useful battlefield tools – they could be used to pin an opponent to the ground for killing.*

battle comes from the beginning of our period, the battle of Tumu.

Battle of Tumu, 1449

The characteristics of the Chinese soldiery and their commanders rendered Chinese forces vulnerable to major defeats by more professional forces. A fine example is that which occurred at Tumu, China, in 1449. It also illustrates how an under-appreciation of logistics can expose even the most powerful infantry force to disintegration.

Tumu happened during the frontier wars between the Mongols and the Chinese Ming Dynasty. In July 1449, a 20,000-strong Mongol force (principally cavalry) invaded China, led by the capable commander Esen Tayishi (d.1454), of the Oirad Choros tribe. The invasion followed three lines of advance, the objectives being Liaodong, Xuanfu and Datong. Initial resistance by local Chinese armies was ruthlessly crushed, and panic started to spread through the Ming court of the Zhengtong emperor, Zhu Qizhen (1427–64). In response, the emperor, encouraged by the influential court official Wang Cheng, ordered the assembly of one of the largest armies gathered during the Ming period. In total, the army may have swelled to some 500,000 troops, principally composed of masses of infantry conscripted specifically to meet the gathering emergency. The confidence behind such an enormous body of men was naturally high, and the army set off on the march to meet the Mongols.

The Mongols were no novices when it came to battlefield tactics and recognized the enormity of the forces marshalled against them. They also correctly understood the huge logistical problems faced by half a million men under forced march. In response, Esen pulled his forces back into the steppe regions, stretching out the distance for the Chinese troops to travel and maximizing their resupply problems. His efforts at subverting the Chinese counter-offensive would be aided by dangerous inadequacies in planning by his enemy. Emperor Zhu himself commanded the expedition, with Wang Chen as his most senior officer.

Nepotism throughout the army ensured that talent was not always the arbiter of rank, while the staff also contained large numbers of civilian officials who would add their own ideas to the mix without the benefit of battlefield experience. Furthermore, only a month's provisions were distributed for a round journey of over 500km (310 miles). The Chinese plan was to make a westerly march from Beijing to Datong, then use Datong as a base from which attacks could be made against the Mongols.

Once the Mongols were defeated, the army would move back to Beijing via an alternative, southerly route – thus avoiding territory they had already scoured for provisions. The southerly route also provided more geographical protection for the army in the form of features such as the Tzu-ching Pass.

THIS REMARKABLE PICTURE *of Chinese infantry was actually taken in the nineteenth century, and illustrates how slow military progression was in China when compared to the West. Matchlock muskets form the unit's main projectile weapon.*

Logistical Setbacks

The march began on 4 August 1449 and immediately ran into trouble. Torrential rains transformed the ground into a boot-sucking morass for the foot soldiers, slowing the pace of advance and in turn rendering the long-term supply situation perilous. Relations amongst the commanders were becoming fraught, particularly between the military officers and the accompanying civilian officials. Seven days of marching also took the army through the battlefield of Yanghe, where the troops were faced with the sobering sight of Chinese corpses from the earlier Mongol victory.

Two weeks after the army had set out, it arrived at Datong. Yet the mass of soldiery was now physically wracked from the journey and a rebellious spirit was rippling through the ranks. Water and food were also running short. Faced with these facts, Wang Chen decided that a further offensive out from Datong was unnecessary. He therefore declared the non-combative march a triumph, and turned the body of men around for home. Noting the condition of the infantry, however, he decided to take them back along the

Battle of Tumu

1449

The battle of Tumu is an object lesson in how important logistics are to any infantry force, as well as the pitfalls of employing a huge number (possibly as many as 500,000) of Chinese troops to meet the Mongolians in 1449. Although the Chinese force was well able to meet the threat of the far smaller Mongol invasion army, poor planning meant that its logistics were stretched to breaking point and then collapsed altogether during the disastrous retreat phase of the campaign. At the same time, the Mongols used their far greater mobility to draw the Chinese infantry deeper and deeper into the countryside while keeping themselves out of serious threat, then turning to make attacks on the Chinese rear and flanks when the Chinese were at a disadvantage. The Chinese infantry were unable to put up a coherent defence: mobile defence was not a traditional skill of the Chinese army in the fifteenth century.

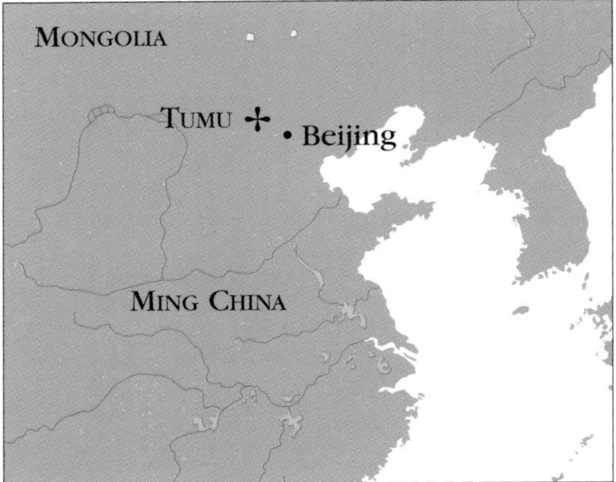

The battlefield of Tumu was located near the walled town of Huai-Lai and near the town of Hsuan-Fu. The advance and retreat took the Chinese army through very barren landscapes.

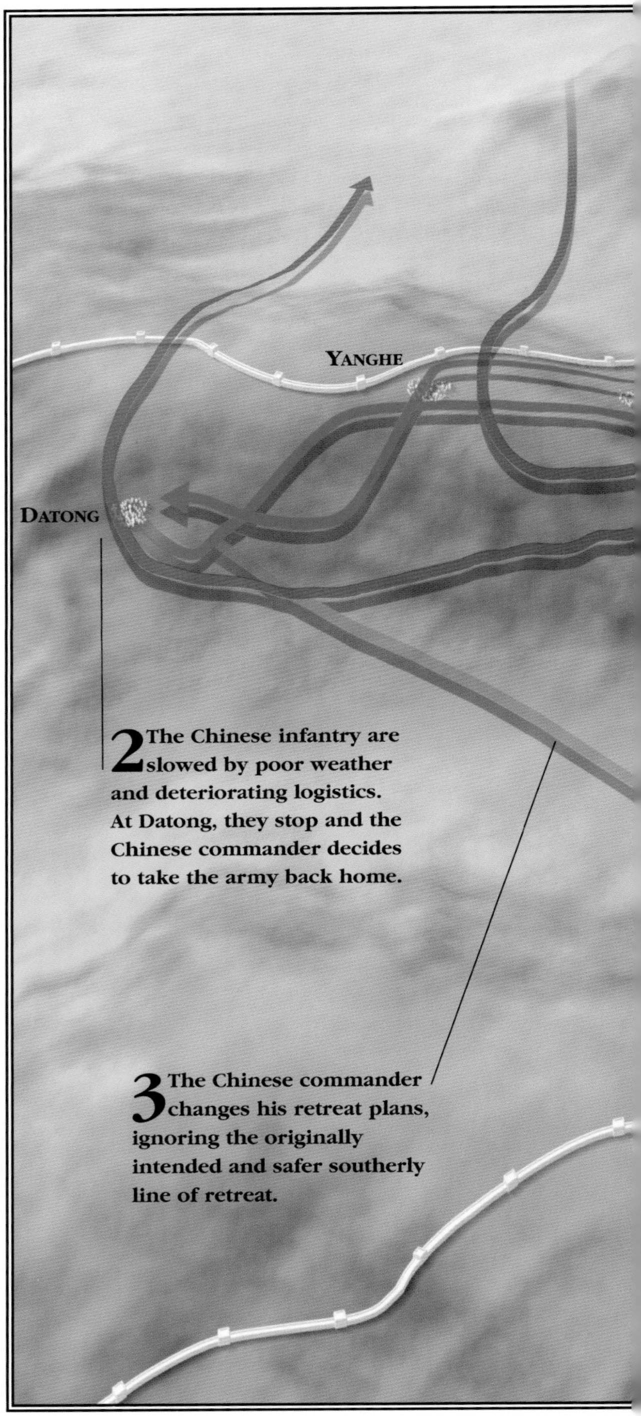

YANGHE

DATONG

2 The Chinese infantry are slowed by poor weather and deteriorating logistics. At Datong, they stop and the Chinese commander decides to take the army back home.

3 The Chinese commander changes his retreat plans, ignoring the originally intended and safer southerly line of retreat.

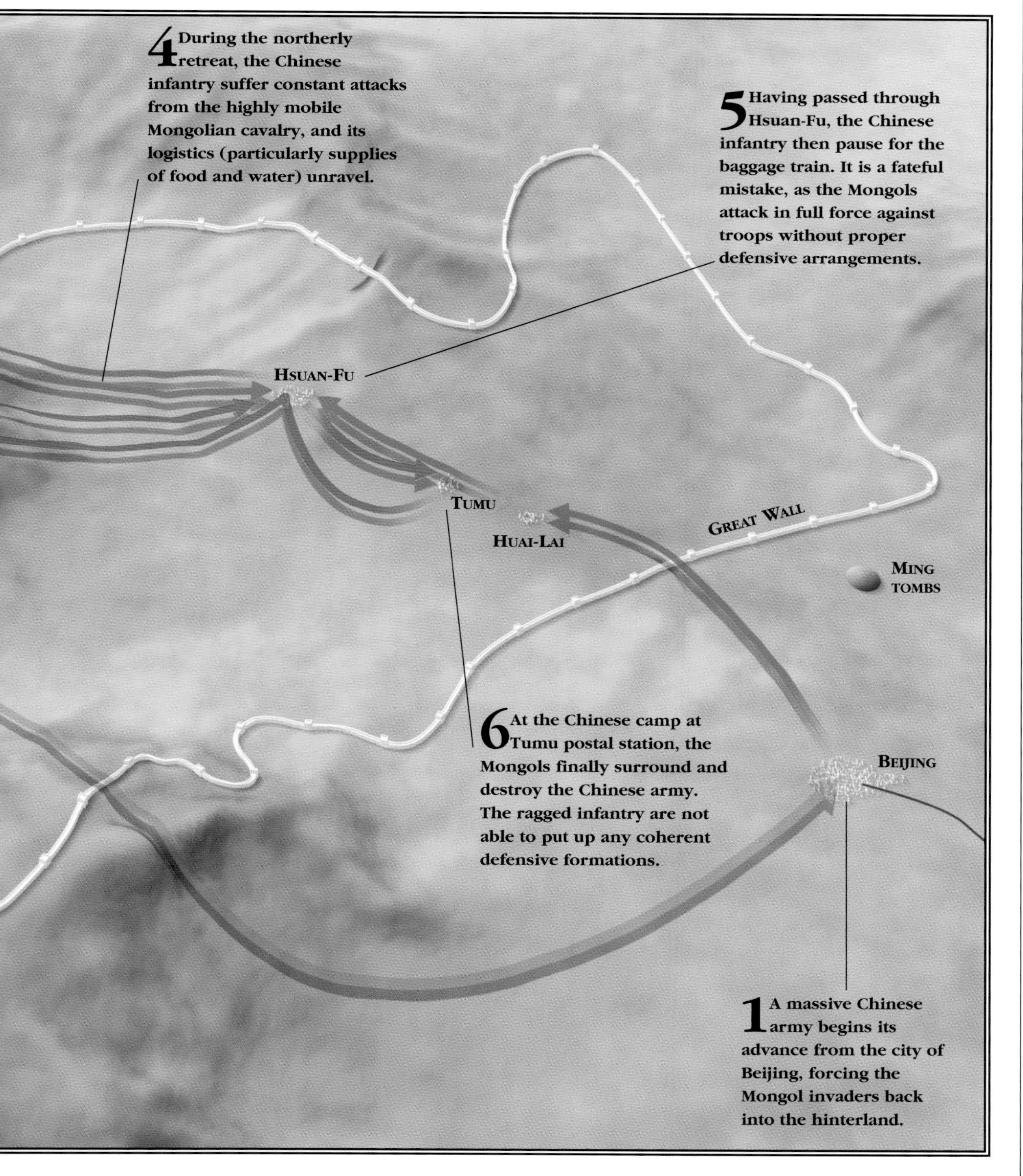

4 During the northerly retreat, the Chinese infantry suffer constant attacks from the highly mobile Mongolian cavalry, and its logistics (particularly supplies of food and water) unravel.
5 Having passed through Hsuan-Fu, the Chinese infantry then pause for the baggage train. It is a fateful mistake, as the Mongols attack in full force against troops without proper defensive arrangements.
Hsuan-Fu
Tumu
Huai-Lai
Great Wall
Ming Tombs
Beijing
6 At the Chinese camp at Tumu postal station, the Mongols finally surround and destroy the Chinese army. The ragged infantry are not able to put up any coherent defensive formations.
1 A massive Chinese army begins its advance from the city of Beijing, forcing the Mongol invaders back into the hinterland.

Ming Guardsman (Fifteenth Century)

This Ming Dynasty guardsman is elaborately kiited out in terms of uniform, armour and weaponry, indicating a higher status when compared to the general ranks of many Chinese armies, who were much more roughly clad. His torso and thighs are well protected behind an armoured panel made from fabric-covered riveted iron plates, fastened by a belt around the waist. Similar armoured panels are seen around the shoulders and upper arms, to protect them from downward blows. The sheet steel helmet has a high dome to shield the top of the skull, and its deep skirt guards the back of the neck and sides of the throat. The soldier carriers a long halberd, fitted with the customary tassle to act as a distraction for the enemy during combat. Halberds had particularly ancient origins in the Chinese army, with examples dating back to the fourth century BC. The advantages of the halberd in action were multiple: it allowed the soldier to attack the enemy while imposing (if only temporarily) a respectably safe distance; the length of the shaft meant that the halberdier could fight cavalry who were sat up high on their mounts; the blade could be used for both hacking and slashing attacks; multiple halberds could make a defensive wall against mass cavalry/infantry attack. On the soldier's left hip is carried a short sword to be used for close-quarters work.

exposed northern route rather than the safer southern course. The decision was primarily based upon his desire to protect his own estates from the effects of an ill-disciplined army passing through them.

It was to be a disastrous decision. As the Chinese retreated, the Mongols turned around to attack their rear, wiping out several major rearguard forces. A critical juncture was reached when the Chinese soldiers camped at the Tumu postal station on 31 August. The choice of camp was a poor one to start with. The infantry were in an exposed position and would have been far better protected in the walled town of Huailai only 12.8km (8 miles) away. Furthermore, lack of water was now driving men and horses to the point of desperation – effective, controlled command was now no longer possible. As the Chinese soldiers camped in chaos, the Mongols steadily surrounded them.

On 1 September, the Mongol forces attacked. In spite of the huge Chinese superiority in numbers, the infantry and the cavalry were unable to organize themselves into any sort of coherent response. The marshalled ranks of infantry so central to a coherent defence crumbled, or were not formed up in the first place, leaving huge gaps of opportunity for the enemy to exploit. Up to 250,000 Chinese troops were either killed, wounded or captured. Wang Chen was killed and two days later even the emperor himself fell into Mongol captivity. Here was one situation where mobility triumphed over infantry mass.

'In all cases you must march quickly to get away from places such as mountains, forests, valleys and marshes. You cannot march sedately ... Carefully observe the state of the enemy's array, and if he falls into disorder, attack him without any hesitation.'

– *Sun Tzu,* The Art of War

Japan's Infantry

Infantry were also a fundamental component of Japanese military power right through the late medieval and Early Modern periods. They are known by the collective term *ashigaru,* which translates as 'light feet' and refers to almost any dismounted soldier who could not be classified as a samurai. The historian Stephen Turnbull has observed how the *ashigaru* had precedents reaching back to the seventh century:

'The antecedents of the *ashigaru* may be traced to one of the earliest attempts by a Japanese emperor to control and systematize the owning and use of military force. To this end, Emperor Tenmu (reigned 673–86) envisaged a national army that was to consist largely of conscripted footsoldiers, but as they often absconded from duty the programme was eventually abandoned.'

As the medieval period progressed, the model of a conscripted army was replaced by armies composed primarily of the mounted samurai, although they operated with foot-soldier support; the lower orders of soldier often being men who worked on the noble's land. Yet it was the rise of mass archers during the second half of the thirteenth century that slowly brought the *ashigaru* back to the fore. During this time, the value of volume fire from hundreds of archers, even if inaccurate, proved itself of greater tactical value than individual but accurate arrows fired by the mounted samurai.

Many lessons were learnt by the Japanese while repelling the Mongol invasions of 1274 and 1281, not least the strength of massed Mongol footsoldiers working in combined arms cooperation with cavalry and missile troops. In fact, had the Mongol fleet not been wrecked by the *kamikaze* (divine wind) storms in August 1281, Japan would have probably sunk under the weight of foreign infantry.

War by numbers rather than quality, particularly with the advent of basic firearms, meant that the Japanese *daimyô* warlords had to find effective methods of bulking out their numbers of foot soldiers. The *daimyô,* by the very fact that they were a landowning class, had one very obvious recruitment pool in the indentured labour employed on their own estates and land. Another pool, by contrast, was the mass of landless peasants in the search for gainful employment. Here was an important point in general about oriental infantry. Joining an army was for many a much better state of affairs than having to hack out a desperate living from often unforgiving soil. From the army's point of view, however, the simple desire to escape from poverty was hardly sound motivation for a professional soldiery.

The fifteenth and sixteenth centuries were periods of transformation in Japanese military organization and saw the importance of large-scale infantry-dominated armies rise exponentially. Between 1467 and 1477 ran the Ônin War, a dispute between the government official Hosokawa Katsumoto (1430–73) and the *daimyô* Yamana Sûzen (1404–73) that spread to become a general war between the shogunate and numerous other *daimyô,* even after the death of the two principals in 1473.

The wars of the thirteenth and fourteenth centuries had often involved armies travelling some distances to fight, say, the Mongols and Koreans, and hence the balance of power leaned towards mobile, mounted troops. The Ônin War, by contrast, was fought at a local level and therefore foot soldiery was a far more important and accessible ingredient in the military mix. This trend continued into the subsequent period known as the 'Age of Warring States' (*Sengoku Jidai*): almost two centuries of civil conflict only brought to a conclusion by the rise of the Tokugawa shogunate established in the seventeenth century. Infantry, although commanded by samurai, rose to become the arbiters of the battlefield.

Age of Warring States

The Age of Warring States not only saw a change in the importance of infantry, it also brought about changes in the social status of the *ashigaru.* The earliest *ashigaru* wore their low class openly – they often had little in the way of armour while their weaponry might consist of nothing more than a sharpened bamboo spear and whatever agricultural implements they could bring to the battlefield. Yet, as time progressed, we see the *ashigaru* donning ever-more

A THIRTEENTH CENTURY ASHIGARU *is here seen charging with a* naginata, *consisting of a long curved blade affixed to a laquered wooden shaft. A* katana *sword in inserted through his belt.*

sophisticated armour and equipment, much of it provided directly by the *daimyô* - an indication that the military commanders no longer treated the footsoldiers as expendable auxiliaries. Furthermore, *ashigaru* status increased as they took over the bulk of responsibility for missile weapons. The bow had been the traditional province of the samurai warrior, but gradually that weapon moved into *ashigaru* hands. At the same time, the samurai themselves showed an increasing tendency to fight dismounted, thereby reducing one of the outward signs of social distinction - the mounted warrior's elevation upon a horse.

Soldiers and Civilians

The social and cultural division between *ashigaru* and samurai closed to an even greater extent with the Separation Edict of 1591, one of the most momentous decisions in sixteenth-century Japan's social history. The context of the Separation Edict is the rise to power of Toyotomi Hideyoshi (1536–98), himself the son of a former *ashigaru* and a rising star amongst the military commanders of Oda Nobunaga (1534–82). Hideyoshi rose to became conqueror and regent of Japan, and in 1588 he began moves to alter the balance of power amongst the Japanese people. His 'sword hunt' order instructed the Japanese peasantry to hand over any weapons in their possession. An important consequence was to massively reduce the numbers of men from which *daimyô* could draw their armies, thereby minimizing their potential powerbase to professional soldiers already under their command. A victorious *daimyô* could, however, draw defeated *ashigaru* into his own ranks: an option that ensured a *daimyô* became more powerful with each victory.

For *ashigaru* already in service, the sword hunt instantly raised their status, which improved even more with the Separation Edict. In this edict, Hideyoshi proclaimed that in the future a farmer could not aspire to any other profession, including that of an *ashigaru*. Similarly, an *ashigaru* or samurai could never return to life as a farmer. Severe punishments were put in place for anyone who tried to jump out of the newly established social order. The Separation Edict ensured that the age-old problem of who would tend the fields during times of war was mostly solved. The only military service open for a peasant was a logistical one - acting as a human packhorse by carrying soldiers' equipment. Conversely, *ashigaru* were by default moved into same type of military order as the samurai.

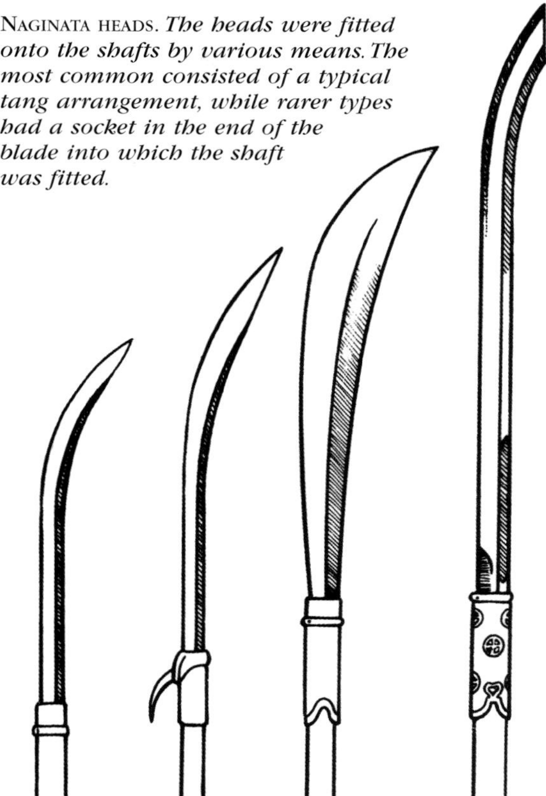

NAGINATA HEADS. *The heads were fitted onto the shafts by various means. The most common consisted of a typical tang arrangement, while rarer types had a socket in the end of the blade into which the shaft was fitted.*

The Separation Edict aided the creation of a full-time soldiery within Japanese society. Prior to that time, a *daimyô* would simply call up men to arms by mobilizing his samurai retainers, who in turn would bring with them numbers of servicemen according to the samurai's classification of wealth. Stephen Turnbull explains the system:

'The number of troops supplied and their equipment depended upon the samurai's recorded wealth, which was expressed in terms of the assessed yield of rice fields he possessed. Such assets were traditionally measured in 'koku', one koku being the amount of rice thought necessary to feed one man for one year... The samurai knew exactly how many men he was required to take with him on campaign. Some would be other samurai who were usually related to him. The rest would be ji-samurai [part-time samurai] or farmers who may not have had long family connections, but as the years went by and casual recruitment became less common, a family tradition of service to a particular samurai family would develop.'

Such is not to say that there were no permanent *ashigaru* units in service prior to the Separation Edict. Many *daimyô* had standing personal bodyguard units, consisting of men who were respected warriors with considerable combat experience. Some of these units could reach considerable size, numbering several thousand troops for the most important leaders. Conversely, there were rapid-response units of *ashigaru* that could be raised in a matter of hours to respond to sudden crises, such as an enemy invasion. Takeda Shingen (1521–73) created an alarm system of fire beacons, arranged in a chain between easily visible landmarks. When a beacon was lit, neighbouring beacon attendants would see the flames and light their own beacons, the signals notifying soldier-farmers that they had to mobilize immediately and report to their respective commanders.

Organization and Manoeuvre

A sound introduction to the world of oriental infantry manoeuvre comes from one of China's greatest military thinkers, Sun Tzu (c.544–496 BC). Although composed in the fifth and sixth centuries BC, Sun Tzu's *The Art of War* was still holding enormous influence over military thinking in Asia at least 1000 years after it was first penned. During the Ônin War in Japan, for example, samurai would sometimes gather for a public reading of passages from the book, adding to their martial education. Of course, the tenets of Sun Tzu would be unlikely to have percolated downwards into the largely illiterate masses of the infantry, but the principles mapped out would have influenced their deployment on the battlefield.

ASHIGARU ARCHERS, SIXTEENTH CENTURY. *Bows were the principal projectile weapons of the Japanese until the widespread introduction of firearms. Archers were often mixed with spearmen for protection.*

Structure and Control

One element of the successful army clearly recognized in *The Art of War* is the necessity for good organization and proper control of the various elements within a unit or formation. The need to divide a mass of men into respective groups is strictly delineated:

'One man is a single; two, a pair; three, a trio. A pair and a trio make a five, which is a squad; two squads make a section; five sections, a platoon; two

PORTRAIT OF TOYOTOMI HIDEYOSHI. *Born in 1536, Hideyoshi would later become the most powerful* daimyô *in Japan, conquering almost the whole of Japan by the later 1580s. He was himself from* ashigaru *stock. While in office, he dramatically changed the martial structure of Japan by removing weapons from the hands of peasants.*

platoons, a company; two companies, a battalion; two battalions, a regiment; two regiments, a group; two groups, a brigade; two brigades, an army. Each is subordinate to the superior and controls the inferior. Each is properly trained. Thus one may manage a host of a million men just as he would a few.'

The structuring given here is little different from that of used by armies up to the present day. While today's armies are faced with a challenge of coordinating technological assets as well as manpower spread over large distances, a challenge for Asian armies throughout much of history was that of controlling large numbers of human assets often within relatively small geographic areas. Above all, the commander had to bring some measure of order to the chaos that constantly hung over a fighting army – the fighting force that maintained some measure of control and discipline was likely to be the one that won the day.

Control of the mass of infantry forces depended very much on each individual soldier having a clear place within the command structure of the army and being able to orient himself on the battlefield in relation to that command structure. Regarding the first point, the principle of subdivision outlined by Sun Tzu applied to almost all armies, even if they incorporated their own particular cultural and regional variations. The weapons with which the soldier was armed with would often dictate his position within the army and on the battlefield. Hence the armies of, say, eighteenth century China would have some structural similarities with those of the fifteenth century, albeit with a heightened emphasis on the placement of troops armed with gunpowder weaponry (muskets and artillery).

C. J. Peers has noted how the Chinese medieval Song Dynasty (960–1279) apparently based its system of military organization directly on the recommendations of Sun Tzu:

'A late Song commentator on the *Sun Tzu* describes a rather idealized table of organization according to which an "army" of 3200 men is made up of two sub-units, each progressively divided into two, down to a "platoon" of five sections of ten men each. A section consists of two squads of five, each made up of a "pair" and a "trio". This suggests that a squad may have contained men with different weapons, perhaps three spearmen and two archers, but another Song source

Chinese Infantryman (1500)

This Chinese infantryman of the fifteenth century is a rocketeer. As with almost all gunpowder weapons of this period, the rocketeer's purpose was not accuracy but the general imposition of casualties and disorder upon the enemy. The wooden launcher contains multiple rockets, the entire battery fired by the lighting of a single fuse at the rear. The soldier would close to within the limited effective range of the rockets – roughly 100m (328ft) – whereupon he would launch his rockets into the mass of enemy infantry. The rockets were often splayed in the launcher to create as wide a pattern as possible, and their flames, smoke, bangs and shrieks were designed to create maximum confusion amongst the enemy personnel, and hopefully break the integrity of their defensive or offensive arrangements. For body protection (launching the rockets was a fiery and dangerous job), this soldier wears a leather lamellae coat. Launching a battery of unpredictable rockets from the waist must have been an alarming prospect for the soldier.

condemns the practice of mixing close-combat and missile weapons.

The word 'idealized' is an important one here. There is implicit suggestion that the availability of weapons would to a significant extent dictate the type of formations that were possible within any army. As some infantry conscripts brought no weapons other than crude farming implements, finding a rational place for them within the ranks of soldiers may have been something of a challenge. A further level of complication, particularly for some of the culturally complex Chinese armies, was how to incorporate infantry of different ethnic and linguistic backgrounds. This was not just a matter of acknowledging language issues; different ethnic groups might have little affection for one another even if they were fighting within the same army, and the placement of units had to reflect this reality by, say, separating antagonistic groups on opposite flanks.

In Chinese armies, an added issue was the frequent discrepancy between the on-paper strength of the forces and those that could actually be mustered. During the 1560s, for example, a commander in the Hsuan-Fu district had to gather his forces rapidly to face a Mongol invasion. His registers indicated that a total of 120,000 men were drawing salary as soldiers, yet when the actual headcount came to be made only 30,000 soldiers were available to serve. In lining their pockets with salaries to fictitious soldiers, commanders could ultimately pay a heavy price on the battlefield.

Divisions and Subdivisions

Systems of structuring Chinese armies went through subtle variations throughout our period, but the general principle of subdivision according to numbers and weapon specialisms holds true. For example, at the beginning of the 1400s the Miao army was divided into 5000-strong *wei,* equivalent to modern brigades, which were each then subdivided into five *chien-hu so* (battalions), and these in turn consisted of 10 *po-hu so.* During the seventeenth century, the Manchu powers organized their forces according to the *pa-ch'i* (eight banners) system. This literally involved marshalling the units under eight banners, each banner having its own distinct colour scheme. The troops gathered under each banner were divided into five *jalans* (regiments) of roughly 1500 men, with each of these further divided into five *nirus* ('arrows') of about 300 men apiece.

Each subdivision of a unit would bring with it its own leadership structure, in the same way that modern military ranks are used. For the Japanese *ashigaru,* the highest command rank was the *ashigaru taishô,* probably equating to the rank of colonel today. Note that this individual was not an *ashigaru* himself, but would belong to the samurai class. An *ashigaru* company, however, would be the responsibility of an *ashigaru kashira,* equivalent to a captain, and beneath him was the 'lieutenant', the *ashigaru ko gashira,* who would probably have control of around 30 men.

Japanese armies were also horizontally structured according to weapons specialization: spearmen, arquebusiers and archers all having their own units and lieutenant-rank commanders. For example, arquebusiers were led by a *teppô ko gashira,* who probably controlled two or three groups, each of five arquebusiers and one archer. Several of these groups would then fall under the command of an *ashigaru ko gashira.* The point of organizing according to weapons speciality was critical, especially regarding infantry. Because of the inaccuracy of early firearms, arquebuses had to be concentrated in mass volley fire, and the same applied to archery. Spearmen had to maintain proximity to one another to successfully repel cavalry attacks – any gaps in the line would quickly be exploited.

Battle Formations

Today, the effective control of a large army still remains a challenge, and it was even more so amongst the oriental infantry armies of the past. Through much of the medieval period the mounted cavalry (particularly Japanese cavalry) considered themselves as the elite, hence infantry were often relegated to a reserve role so that they did not steal the battlefield glory. With the introduction of improved missile weaponry and the development of large armies this emphasis changed, and the infantry steadily gravitated to the forefront of combat. In Japan, figures such as Oda Nobunaga began to insist that the *ashigaru* were properly drilled in their combat skills and also decked out in appropriate armour. In China, the size of the armies meant that the infantry were frequently less skilled than those found in Japan, although the sheer mass of manpower available still gave them critical importance.

Infantry forces could be ordered most coherently in defensive or pre-attack formations, before the chaos of battle had taken over. In fact, many formations were often abandoned once actual battle had begun. The guiding considerations behind most oriental battle formations were as follows:

1 The flanks had to be protected against attacks from enemy infantry and cavalry.
2 The army should present a good defensive front, usually consisting of a solid wall of spears.
3 If possible, the formation should attract the enemy into controlled avenues of assault, inviting them into traps and ambushes.
4 Missile troops should occupy protected frontal positions, giving them clear fields of fire while offering security behind spear troops.

The Japanese developed some 22 pre-battle formations, and there were many such similar formations in use in Chinese armies. (Indeed, the Japanese borrowed some of their formations directly from the Chinese.) Space does not allow us to consider all of them here, but a selection will illustrate some of the thinking behind infantry deployment in particular.

Hoshi (arrowhead): In this formation the missile troops were arranged in ranks at the front of the army, set in staggered lines. Their purpose was to deliver rippling fire using artillery, arquebuses and archery, opening up gaps that could then be exploited by an arrowhead formation charge by the samurai cavalry backed by infantry spearmen. Note that *ashigaru* troops, including arquebusiers and archers, were

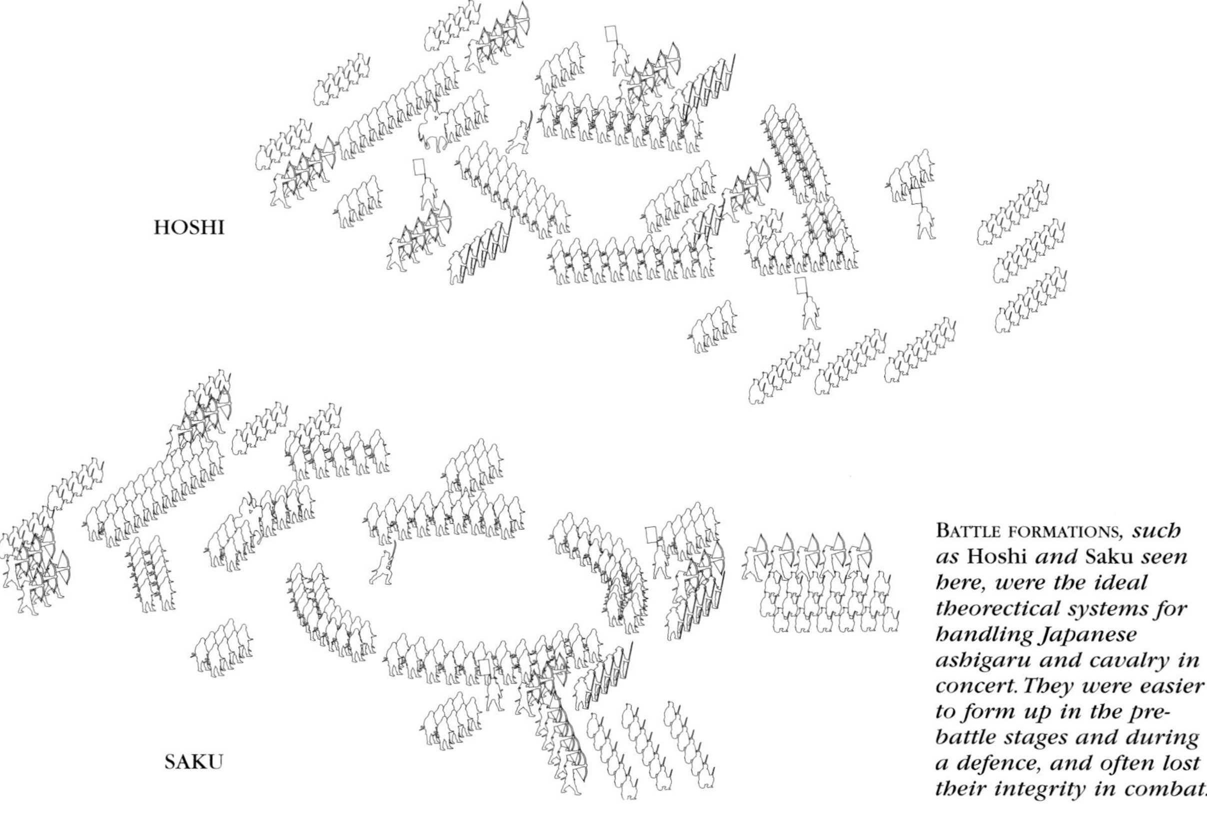

BATTLE FORMATIONS, *such as* Hoshi *and* Saku *seen here, were the ideal theorectical systems for handling Japanese ashigaru and cavalry in concert. They were easier to form up in the pre-battle stages and during a defence, and often lost their integrity in combat.*

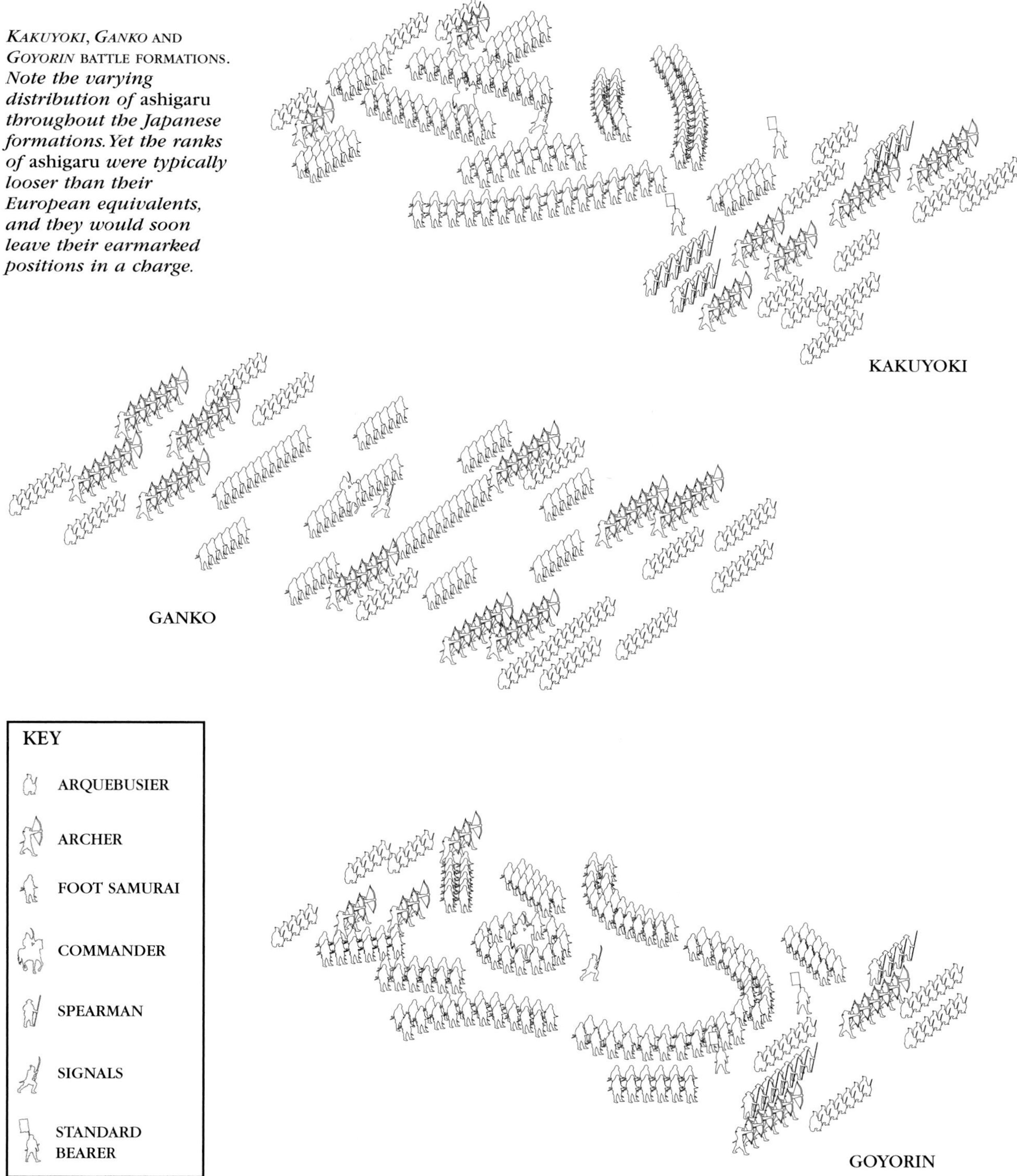

KAKUYOKI, *GANKO* AND *GOYORIN* BATTLE FORMATIONS. *Note the varying distribution of* ashigaru *throughout the Japanese formations. Yet the ranks of* ashigaru *were typically looser than their European equivalents, and they would soon leave their earmarked positions in a charge.*

deployed also on the flanks and at the rear to provide some measure of all-round protection.

Saku (keyhole):The *saku* is the counterpart to the *hoshi,* and was designed to crush an arrowhead attack by the enemy. To receive the 'point' of the enemy attack, the infantry were arranged in a V shape, the opening of the V extending out towards the attack. This formation would allow the arquebusiers and archers to pour fire into the enemy flanks, breaking up the attack so that *ashigaru* spearmen and samurai could then close in to finish off the enemy. Note that the heavy concentration of infantry at the front of the *saku* could leave the flanks quite unprotected, although a wise commander would also have strong infantry contingents at the rear who could move forward in an emergency.

Kakuyoku (crane's wing): The title of this formation refers to the curved front described by the cavalry elements of the force. By adopting bowed ranks, the cavalry could spread outwards rapidly and envelop the flanks of an unwary enemy. Note, however, that the frontal ranks of the force were still composed of regular lines of infantry missile troops and spearmen. Their job was to 'fix' the enemy in position, locking him into a frontal battle that would take his attention off the cavalry curving round to attack the flanks. An example of the *kakuyoku* formation in use is the battle of Kawanakajima (1561) described later.

Ganko (birds in flight): The *ganko* formation allowed a commander to hedge his bets by creating a force with good protection to front, flanks and rear. The front and rear of the formation would be heavily protected by staggered ranks of arquebusiers, archers and spearmen, between which would be the samurai cavalry. Some *ashigaru* would also take up position on the flanks. The overall advantage of this formation was not only its robust defensive options but also its

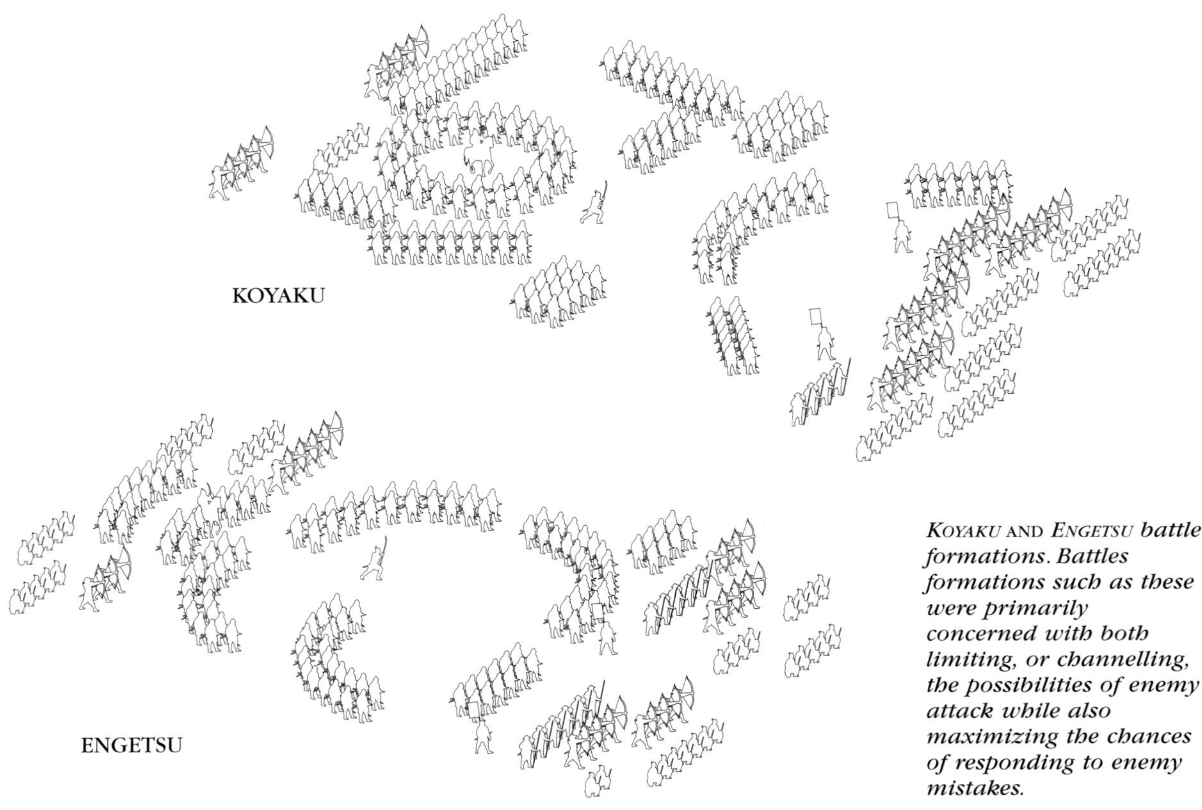

KOYAKU AND ENGETSU battle formations. Battles formations such as these were primarily concerned with both limiting, or channelling, the possibilities of enemy attack while also maximizing the chances of responding to enemy mistakes.

capacity to respond to emergencies, such as flanking attacks.

Gyorin (fish scales): Here the infantry again form a strong frontal defence against the enemy, while the samurai cavalry behind them adopt various semicircular patterns that provide them with flexible defensive positions or offensive responses. The infantry also provide rear defence.

Koyaku (yoke): The *koyaku* was essentially a defensive formation. Once again, the infantry presented itself in a strong frontal rank, but the samurai behind it adopted a wide curve. If it looked as if the infantry defence might crumble, the *koyaku* could be converted into a *saku* formation in the hope of trapping and defeating the enemy attack.

One notable point in most oriental military formations is that the role of the infantry was fairly consistent. The first, and arguably principal, role of infantry is to provide frontal firepower and defence. There is the implicit recognition here that the cavalry is dependent upon infantry either breaking up enemy ranks with firepower or a spear charge, or at least reducing enemy numbers by defending against their initial attacks. In a sense, the cavalry simply wait for the opportunities created by the battles of the infantry. This becomes increasingly true as we advance through the gunpowder era, where firepower becomes much more influential.

A CHINESE INFANTRYMAN, *seen here armed with nothing more than a dao broadsword and a circular shield. The dao was often large enough to be used with two hands for powerful armour-splitting attacks.*

Battle of Maymyo, 1767

The battle of Maymyo is an interesting study in how the conduct and manoeuvre of a military campaign could be undone by an enemy that used terrain and unorthodox tactics to his advantage. It took part during one of China's regular incursions into Burmese territory. The military picture of Southeast Asia from the fifteenth to the nineteenth centuries is a mixed one. As they did with the Chinese and Japanese, Portuguese travellers introduced gunpowder weapons into the region during the sixteenth century, and Portuguese mercenaries often served as both elite soldiers (principally bodyguards) and as artillery instructors. Regarding infantry, the situation was much like that in China and Japan, with the bulk of the armies made up of men pulled temporarily from the fields holding whatever weapons they could find, or sometimes equipped with officially issued swords, shields, spears, battleaxes and daggers. Indian influence was strong over military thinking, with the infantry occupying the traditional role of inflicting attrition upon the enemy, while the cavalry (mounted on horses or elephants) provided the dash and manoeuvre. And there was much for the soldiery to do. Between the fifteenth and seventeenth centuries, Siam (modernday Thailand) and Burma were in a regular state of conflict, principally from the expansionist ambitions of various Burmese rulers.

In the eighteenth century, however, Burma attracted the attention of the Chinese, in particular the Ch'ien-lung emperor, who launched four invasions into the country between 1767 and 1771, all of which were repelled. The 1767 invasion led to an especially grievous defeat at the battle of Maymyo.

The Chinese emperor had put together an army of 50,000 soldiers, the vast majority being infantry. The mountains and thick jungle covering much of Burma made the employment of cavalry forces problematic, so the bulk of the work would have to be done by foot soldiers. The commander of the Chinese army was Ming Jui, the son-in-law of the emperor. He divided his massive army into two main columns for the southeastern advance - one would strike through Bhamo on the upper Irrawaddy river while the other would move through Hsien-wi further south in Shan country. The ultimate objective was for both columns of troops to clamp themselves in a pincer action on the Burmese capital of Ava (near what is today Mandalay). Ming himself led the southern fork of the advance.

Burmese forces had the advantages of easy access to local supplies and familiarity with the terrain. The Chinese troops, by contrast, soon found themselves with stretched supply lines and having to tackle physically draining obstacles such as dense jungle foliage and thick swamps. In such a climate, diseases like dysentery and malaria began to cut down the Chinese ranks even before battle was joined. Medical care for the Chinese infantry was utterly rudimentary, principally consisting of herbal remedies brought from home, and those who became seriously ill effectively either marched or died. Furthermore, the military situation was soon not going their way.

At Kaughton, a fortified position in the way of the northern column's advance, a vigorous defence by the Burmese stopped the Chinese column dead in its tracks. Repeated attacks by the infantry simply resulted in more casualties, so the Chinese commander in the north decided to proceed no further and retreat back towards China. This was against the express orders of Ming, and left his southern column isolated. Nevertheless, Ming drove his army hard and managed to fight through to within 48km (30 miles) of Ava. The situation was looking desperate for the Burmese King Hsinbyushin (1736–76). With large numbers of his troops already committed to fighting in Siam, and with the Chinese on his doorstep, he prepared to abandon his capital. However, the battle was soon to shift in his favour.

Encirclement and Annihilation

The fact that there was now only one Chinese column instead of two offered Hsinbyushin the opportunity for an encirclement operation. A division of 12,000 men was despatched to meet Ming head on. The composition of this force reflects how central infantry fighting power was to the Burmese army - the infantry numbered 10,000 as opposed to only 2000 cavalry. The Burmese force locked horns with the Chinese troops, but ultimately the Chinese numbers won out and pushed them back.

Yet the victory was not all that it seemed to be. Instead of simply retreating, many Burmese soldiers peeled off into the jungle and from there began a steady campaign of harassment of the Chinese flanks. Archers and gunners proved particularly invaluable, as they were able to pour fire into the Chinese column before disappearing back into the vegetation.

More seriously for the Chinese, Burmese infantry also managed to work their way around the rear of the column, and began to disrupt or cut the Chinese supply lines. When Chinese infantry and cavalry ventured out on foraging missions, foot soldiers would be waiting in ambush. Before long the Chinese supply situation was becoming critical, and the massive army was beginning to go hungry.

It was at this point that the Burmese forces, under the command of Maha Thiha Thura, played their masterstroke. While Ming struggled with the irritations of the first Burmese division, another much larger division had advanced through a mountainous route to emerge directly behind the Chinese. Through careful manoeuvring they managed to achieve complete encirclement of the Chinese column, and over the course of three days of bloody fighting the noose tightened. The Burmese infantry proved themselves just as adept in defence as in attack, and Chinese numbers were steadily whittled away through death and injury. Eventually the surviving Chinese troops had no option but to surrender. Some 2500 of them passed into captivity where they faced a life of brutal slavery. Of the remaining troops, probably numbering in the region of 30,000 men, all were killed, either on the battlefield, through disease or through execution after their surrender.

The battle of Maymyo was a cataclysmic defeat for the Chinese, and Ming Jui himself committed suicide in despair. Here was an engagement that clearly demonstrated the superiority of manoeuvre over sheer numbers of men. Moreover, despite the fact that both sides were reliant upon foot soldiers, the Burmese used familiarity with local terrain to enable the soldiers to gain a position of advantage. By avoiding pitched battle when necessary, and embracing it when the moment was right, the Burmese troops illustrated that open battlefield clashes was not the only way to win a war.

Infantry During Combat

It is hard to overstate the sheer chaos of infantry engagements amongst many Asian armies. Chinese infantry armies grew so large that the most the command structure could aspire to was to point fairly large units in roughly the right direction and launch them into attack at the right time. Once committed to an offensive action, the infantry would often charge in a chaotic fashion, with each man essentially fighting for himself. This martial individualism was encouraged by a reward system based on enemy heads taken, a situation that was not really conducive to fostering team thinking. Individual aggression was promoted and was assisted by huge amounts of noise from percussive

ASHIGARU MAKE A CHARGE. *Note the uniformity of their armour and outfits and the motifs on their helmets. Such standardization is more common during later post–medieval periods of Japanese history, when military formations became more unified.*

Battle of Maymyo

1767

The battle of Maymyo illustrates the disastrous consequences for infantry when a strategy unravels. At Maymyo, the Chinese pincer action against the Burmese capital of Ava failed after one force became isolated and cut off. The Burmese infantry had the major advantage of being familiar with their country's convoluted terrain, and they put up such an impressive defence at Kaughton, a fortified position, that the northern arm of the Chinese pincer was stopped dead in its tracks, and then put into retreat. The Chinese infantry of the southern column, therefore, were left isolated against the increasing Burmese guerrilla-style attacks, and then against a Burmese encirclement manoeuvre. The Chinese suffered large casualties from Burmese archers and gunners, who had ample covered positions from behind which they could fire on the disordered Chinese ranks.

MANCHU CHINA
INDIA
MAYMYO
Ava
BURMA
SIAM

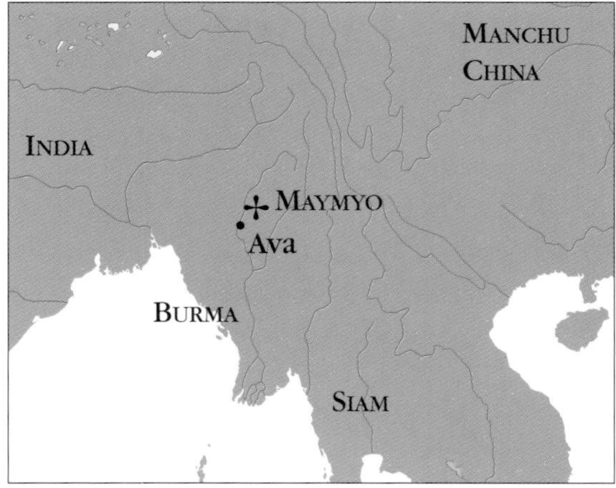

Terrain was an important factor in the Chinese defeat at Maymyo. The jungle and mountains of Burma made the Chinese advance difficult and aided Burmese guerrilla-style tactics.

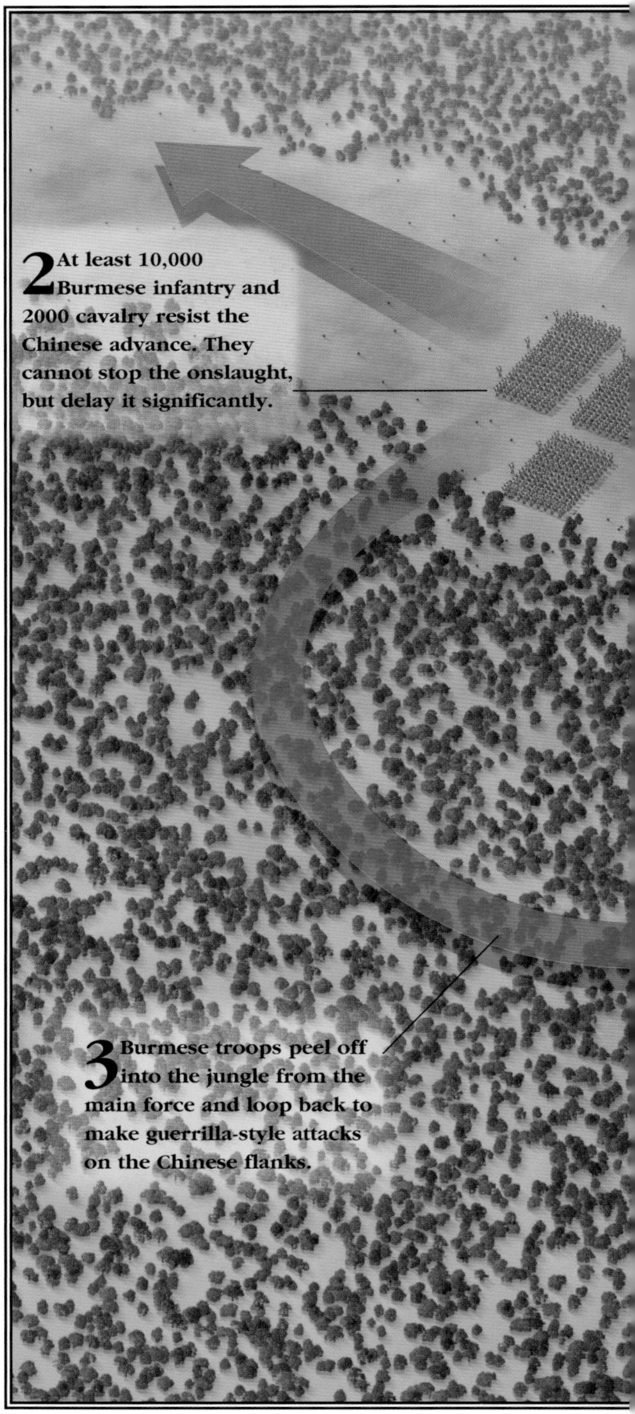

2 At least 10,000 Burmese infantry and 2000 cavalry resist the Chinese advance. They cannot stop the onslaught, but delay it significantly.
3 Burmese troops peel off into the jungle from the main force and loop back to make guerrilla-style attacks on the Chinese flanks.

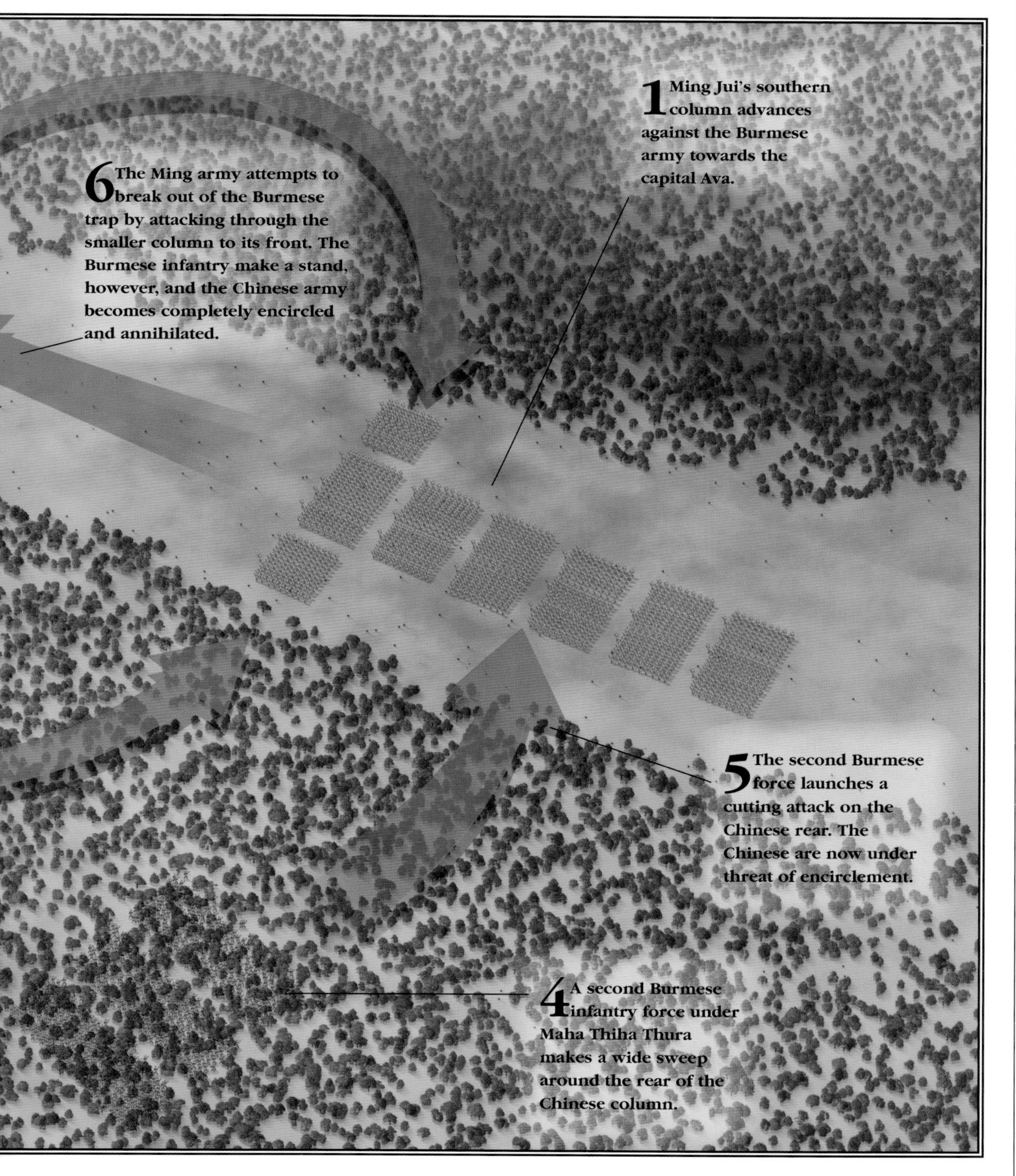
1 Ming Jui's southern column advances against the Burmese army towards the capital Ava.
6 The Ming army attempts to break out of the Burmese trap by attacking through the smaller column to its front. The Burmese infantry make a stand, however, and the Chinese army becomes completely encircled and annihilated.
5 The second Burmese force launches a cutting attack on the Chinese rear. The Chinese are now under threat of encirclement.
4 A second Burmese infantry force under Maha Thiha Thura makes a wide sweep around the rear of the Chinese column.

and brass instruments, adding to the intimidation of the enemy. Infantry soldiers did have some form of orientation, however. Their commanders would have individual flags hoisted high that gave the troops something to follow across the battlefield, while signal flags gave instructions such as advance, retreat or adopt defensive positions.

In all oriental armies, the ability of infantry to identify their unit and commanders within the chaos of battle was always a problem. Banner systems provided the most ready mechanism for giving some form of battlefield coherence. In the Japanese forces, not only did samurai carry the *nobori* flag attached to the back of their armour – giving the attendant *ashigaru* some prominent form of bearing – but the *ashigaru* themselves would often transport various insignia around the battlefield. Some would carry the *hata-jirushi* (flag streamer), the base of the flagpole sitting in a holder fixed to the waistband. This flag would denote the nearby presence of a particular samurai. Some soldiers were esteemed enough to carry the *uma-jirushi* flag attached to the back of their armour, thus indicating that a general or *daimyô* was present. While these positions obviously carried a certain honour, they could dramatically shorten the life of the flag carrier – enemy soldiers were attracted to such insignia as they were obvious targets.

> *'Regarding spear bearers … there is no place amongst their ranks for cowardly spirits. They serve with great devotion and perform a great service ….'*
>
> – *The* Zôhyô Monogatari

In many instances attempts were made at a standardized uniform, as much to create a psychological impact as for unit identification. For example, the *daimyô* Ii Naomasa clad all his troops in a strident red-lacquered armour, regardless of whether they were samurai or infantry. (The effect was also no doubt part of the *daimyô's* attempt to attract personal glory to himself.) Other systems by which the infantry could identify themselves included the *mon* of their samurai stencilled onto their helmet or armour, or the wearing of various small flags with Japanese insignia painted on them.

The sophistication of these systems was elementary and, in the gunpowder age, following such motifs could be problematic amongst clouds of smoke. Auditory command instructions were provided by various forms of musical instrument, principally percussion instruments but also trumpets and horns. The Japanese had dedicated infantry drummers who played out instructive rhythms on a small drum carried on the back of their comrades. Larger drums were carried by two men, the drum hanging from a pole or even suspended from a large wooden frame. Chinese troops of the Ming Dynasty could be seen with large drums measuring over 1m (3.2ft) across fixed in wooden frameworks on the ground and being played by two men each holding sticks the size of small clubs. Bells and gongs were also used; the former could be especially useful, as higher pitched sounds would often cut more successfully through the rumbles of battle than the sound of a bass drum.

Much as in the European tradition of military drummers, oriental percussionists would learn a fixed set of beats, each one of which would give a different instruction to the elements of the army.

Battle of Kawanakajima, 1561

The battle of Kawanakajima provides an ideal example of how infantry forces were controlled around a battlefield. It was one of five battles fought on the plain of Kawanakajima between 1553 and 1564. The participants in these battles remained constant – the armies of the *daimyô* Takeda Shingen and Uesugi Kenshin (1530–78). The battleground sits in northern Shinano province, near the city of Nagano. Shinano separated the provinces of the warring parties – Uesugi's Echigo province in the north and Takeda's Kai province in the south-east – and became one of many battlegrounds during the *Sengoku Jidai* (Age of the Warring States). The three battles at Kawanakajima that preceded the 1561

Ashigaru Spearman (1561)

Spearmen were critical to the effectiveness of Japanese armies. Although Japanese firearms have rightly captured much interest from historians, spearmen were still central to any final victory or defeat. In massed ranks the spearmen would provide both a defensive body against enemy cavalry or infantry attack, while on the offensive they provided the main force for overwhelming an opponent and consolidating the battlefield. The straw raincoat worn here provides another reason why spearman were never entirely supplanted by firearms – they could fight even in the wettest of conditions. The spearman seen to the right carries the mochi-yari, *a spear with a relatively short shaft when compared to the* nagae-yari *that was introduced during the sixteenth century. Here the blade tip is exposed, but when not in use the blade was often covered with a short scabbard, both for safety and to protect the metal against the elements. On his belt he carries the two blades typically wielded by* ashigaru *– the long* katana *sword and the shorter* wakizashi *dagger. Headgear is provded by the classic conical-shaped* jingasa *war helmet with the identifying* mon *badge displayed at the front.*

Battle of Kawanakajima

1561

The battle of Kawanakajima in 1561 was, despite the heavy reliance on footslogging *ashigaru*, manoeuvre warfare in every sense. Speed, secrecy of deployment and the use of formal battle formations characterized the engagement, as did the personal fighting skills of the *ashigaru*, who found themselves locked into numerous melee actions. Spearmen provided much of the offensive impetus during the battle, serving to pin the enemy forces while the cavalry sought weaknesses in the enemy ranks. The key to the outcome of the battle was the fast deployment of the 12,000 Takeda troops from the Saijosan onto the Kawanakajima plain, a deployment that threw the Uesugi attack against the Takeda forces already down on the plain. Note in this regard how the battle demonstrates that infantry, not just the cavalry, could be used in rapid tactical manoeuvres. The Uesugi also used the system of rotating fresh troops in and tired troops out of the battle to maintain the momentum of the assault.

Kawanakajima was located in Shinano province in central Japan, a particularly troubled region during the sixteenth century, with various local warlords vying for power.

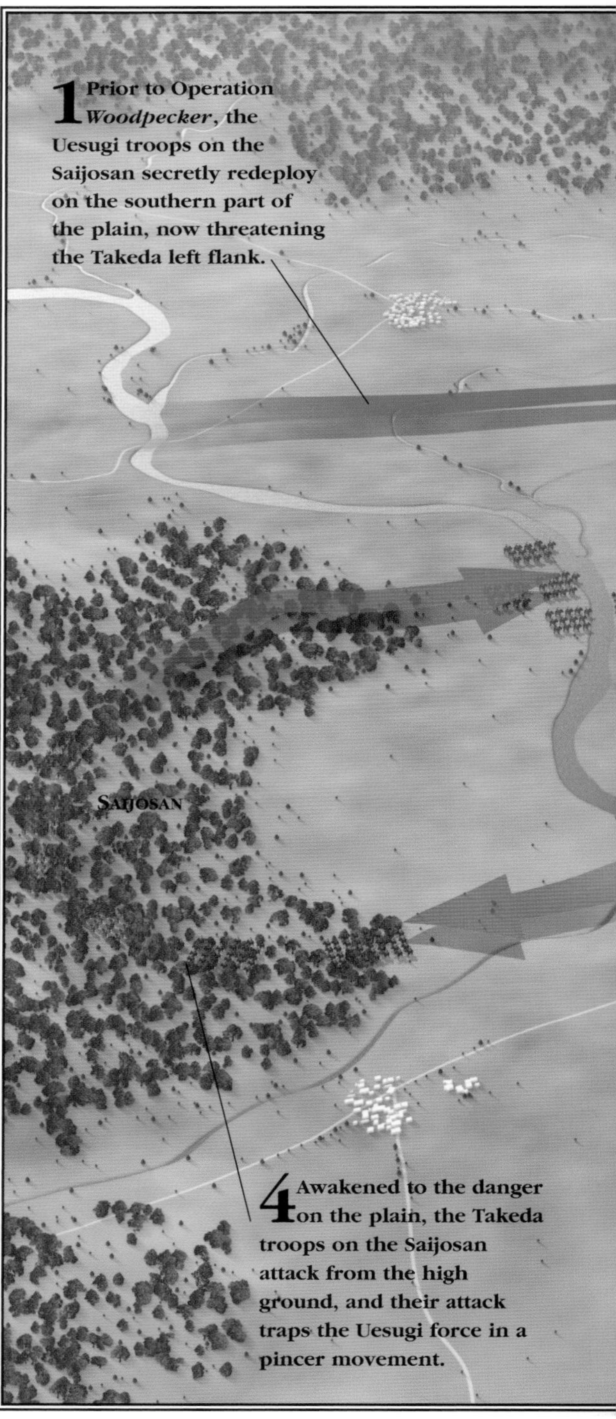

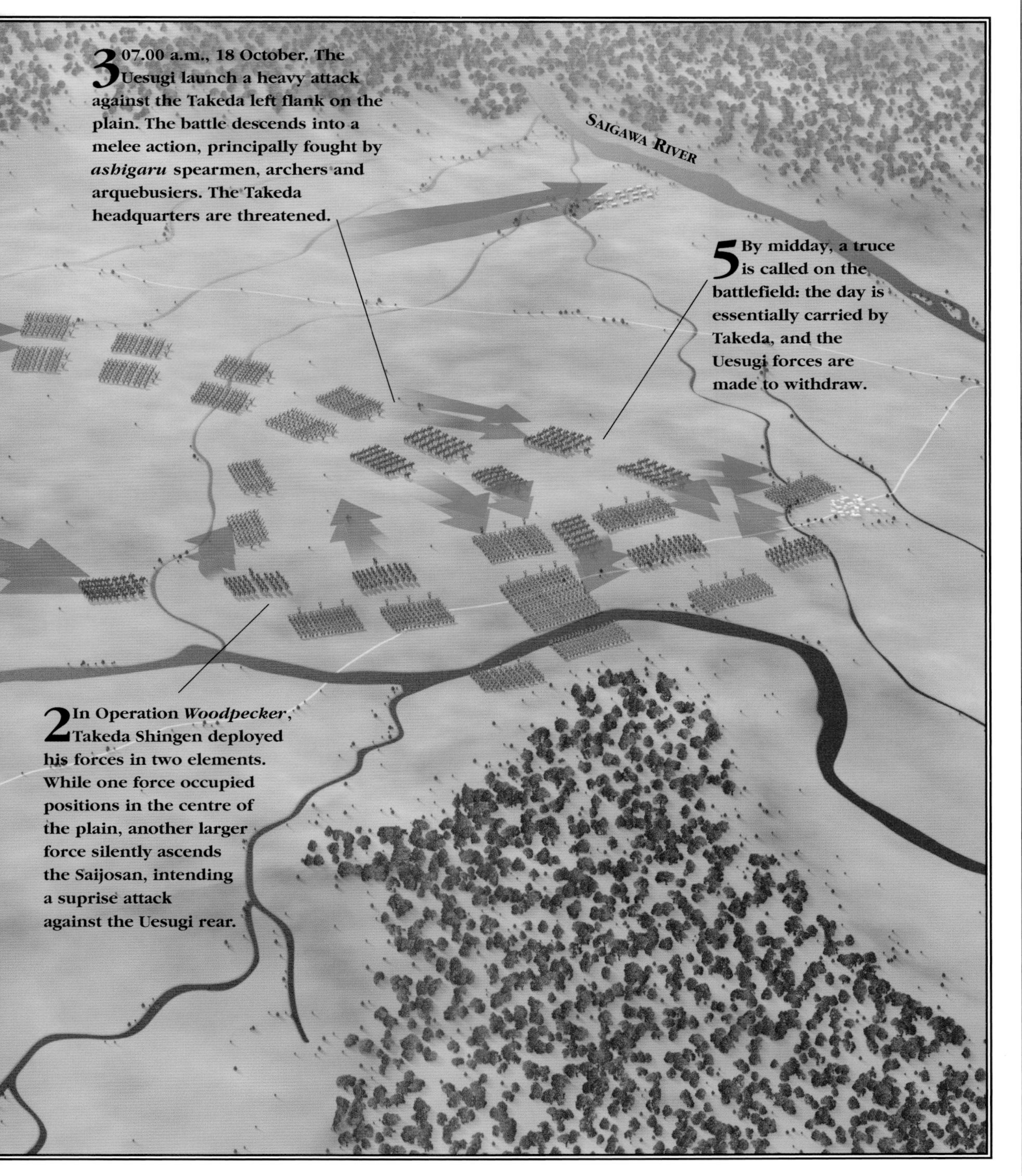
3 07.00 a.m., 18 October. The Uesugi launch a heavy attack against the Takeda left flank on the plain. The battle descends into a melee action, principally fought by *ashigaru* spearmen, archers and arquebusiers. The Takeda headquarters are threatened.
Saigawa River
5 By midday, a truce is called on the battlefield: the day is essentially carried by Takeda, and the Uesugi forces are made to withdraw.
2 In Operation *Woodpecker*, Takeda Shingen deployed his forces in two elements. While one force occupied positions in the centre of the plain, another larger force silently ascends the Saijosan, intending a suprise attack against the Uesugi rear.

Samurai with Standard (Seventeenth Century)

Battlefield command was always a problem for oriental armies, particularly once gunpowder weapons added their literal smoke to the metaphorical fog of war. Auditory methods of identifying units and receiving instructions included various types of musical apparatus, principally percussion instruments such as drums and gongs. Visual means included banners; here a Japanese warrior displays a banner from a socket attached to his back. The flag itself bears the name of a particular noble: with multiple individuals wearing such banners, it was possible to achieve a degree of unit cohesion on the battlefield. This warrior also wears a suit of laquered-steel armour, and the colour of such armour could also serve to give an identity to some units.

engagement, precipitated by Takeda's incursions into Shinano from the 1530s, were fought in 1553, 1555 and 1557, but the violent clash that came in 1561 eclipsed the earlier battles in scale and import. In August/September 1561, Uesugi began a large deployment of forces towards Kawanakajima based on the expectation that Takeda was building up to an invasion of Echigo itself. He mustered a force of 18,000 troops and began his advance, leaving a unit of 5000 men further north at Zenkoji to act as a rearguard.

The opposing armies were still very much governed by the traditions of the samurai and organized into samurai warriors, mounted attendants and *ashigaru*. *Ashigaru* comprised some 70–80 per cent of the total numbers and would have a major influence over the battle. They included arquebusiers, but by 1561 the full potential of these weapons had yet to be realized and volley-fire tactics had not been as developed as they would be later. Instead, the main ashigaru influence came from archery (the arquebusiers were mixed in amongst the archers to provide additional firepower) and the *nagae yari* spears.

> *'Some 3600 or 3700 soldiers, friends and enemies alike, joined in a melee, stabbing and being stabbed, slashing and being slashed...'*
>
> – *THE* KOYO GUNKAN

Tactical positioning was going to be crucial for the forthcoming battle. The Kawanakajima plain is cut along a north–south line by the River Chikumagawa, which runs down the western portion of the plain. The plain itself is framed by mountains and hills on all sides, and control of such high ground was vital for any army attempting to secure tactical advantage. By 1561, Takeda controlled the southern and eastern sides of Kawanakajima, with a fortress (known as Kaizu) built in the southeastern corner providing a convenient launchpad to deploy forces rapidly into the plain. Uesugi's immediate objectives were to safely cross the Chikumagawa, at the Amenomiya Ford, just to the southwest of Kaizu, and to dominate Kaizu from the wooded Saijosan high ground. All these objectives were achieved by the end of September, putting Uesugi in a position of strength to threaten Takeda's flanks.

Manoeuvre and Countermanoeuvre

The next two weeks involved a series of manoeuvres that eventually placed the Takeda army in a position of tactical advantage. Takeda had marched out from Kai with an army of around 16,000 troops on 27 September, reaching Kawanakajima in about six days. At first, he deployed the bulk of his army on the Chausuyama heights to the west of the plain and effectively blocked the return path for the Uesugi army. Then, on 8 October, Takeda took his force down from the Chausuyama, crossed the Chikumagawa and occupied Kaizu, all under the noses of the Uesugi force on the Saijosan. The Kaizu did not have the capacity to host the Takeda army, which had by now swelled to 20,000 in number.

Moreover, it would surely only be a matter of time before the Uesugi began their assault. In a pre-emptive move, Takeda launched Operation *Woodpecker.* The plan ran as follows. Takeda would take out 8000 men from Kaizu during the night, moving them silently across the Chikumagawa to the Hachimanbara area in the centre of the Kawanakajima plain. At the same time, a force of 12,000 soldiers was to climb the Saijosan and, when the time was right, attack the Uesugi troops there from the rear. Theoretically, the Saijosan assault would force the Uesugi men down from the mountain on to the plain, where they would be driven onto a crushing flank attack launched from the Hachimanbara.

Ashigaru were fundamental to the success of this plan. While cavalry could undoubtedly inject the fast manoeuvre element necessary to cause chaos amongst the enemy ranks, the *ashigaru* would be the ones to apply the wall of pressure against the Uesugi infantry that would drive their mass onto the Hachimanbara attack. The spear troops of the *ashigaru* at this time were used in a slightly different manner to the equivalent pikemen of Western warfare. Although the spearmen would present a united front in defence, their spears forming a vicious barrier to penetration by horseman, on the attack the lines could be much more fluid and individualistic, each soldier charging with his own intent.

Operation *Woodpecker* was launched at midnight on 18 October, although by this time there was already some significant movement in the Uesugi camp. At 10.00 p.m. on 17 October, Uesugi had actually taken his troops down from the Saijosan in secret (even the horses' hooves were wrapped in cloth to dampen the noise), alerted by his scouts to the intended movements out from Kaizu. He took his troops across the Amenomiya Ford and assembled them into two columns on the west side of the Chikumagawa, ready to launch their own attack. Takeda's troops later ascended the Saijosan, unaware that their enemy had moved. But the tactical advantage did not lay entirely with Uesugi. Takeda's night-time movement across the Hirose Ford into attack positions went entirely unobserved. Here the 8000 troops deployed into a westward-facing *kakuyoku* (crane's wing) formation and readied themselves to make a flanking attack on the enemy force as it was driven northwards.

The Attack Begins

For the Takeda army, the battle would not play out as planned. At 7.00 a.m. on 18 October the left flank of the unit on the Chikumagawa suddenly faced a fast, violent assault from out of the morning mist by a surge of Uesugi *ashigaru* and mounted samurai. While the cavalry cut into the Takeda ranks from horseback, the *ashigaru* drove forward with their spears, and the battlefield descended into a melee.

A Japanese cuirass (*yukinoshita do*) *from the Edo period. Such pieces of armour gave respectable protection against sword blows and arrows, but provided less assurance against close-range volleys of musket fire.*

Ashigaru arquebusiers engaged targets of opportunity, particularly samurai commanders. During this stage of the battle, both the Uesugi samurai and *ashigaru* were used in a *kuruma gakari* ('winding wheel') attack system, in which (as far as we can tell) fresh troops were rotated into the attack as battle-tired troops moved out, thus maintaining the momentum of the assault. The attack was bearing fruit, and at around 9.00 a.m. the Uesugi vanguard even closed on Takeda's headquarters, resulting in an individual combat between the two principal commanders. Uesugi received two injuries in the battle, but his bodyguard force managed to drive away Takeda and his retainers. Yet despite fearful losses the Takeda defence held out. Furthermore, by this point Kasaka Danjo's Operation *Woodpecker* detachment up on the Saijosan had discovered that their enemy was elsewhere. Hearing the sound of a battle on the plain below them, they moved at speed down the mountainside and locked in battle with the Uesugi rearguard out of the Amenomiya Ford. They cut through resistance at the ford, and began to assault the rear lines of the main Uesugi force.

Only at this point was the original Takeda plan starting to bear fruit – the Uesugi troops were now caught in a classic pincer movement. By midday the fighting was at a close, a truce having been called. Ascertaining who to declare the victor was not easy, as both sides had suffered terrible casualties, estimated in the region of 60–70 per

A JAPANESE WARRIOR, *armed with a glaive, displays a full suit of body armour, the* tosei gusoku. *The armour is constructed from laquered metal plates riveted or laced together. The armour includes protection for the forearms and shoulders and a* kabuto *helmet that shielded the back of the neck.*

cent. Nevertheless, the Uesugi force subsequently withdrew from the Kawanakajima battlefield.

The 1561 battle at Kawanakajima illustrates how the battle tactics of the infantry were in a transitional phase. Infantry were absolutely critical to maintaining mass momentum in the attack and also the consolidation of the battlefield – cavalry are often good at smashing into enemy ranks, but finalizing a defeat is typically the role of the foot soldier. There were also some emblematic moments that illustrated how the elite samurai would one day be humbled by infantry firepower.

Yamamoto Kansuke, the Takeda commander who had been responsible for planning Operation *Woodpecker,* was killed after charging into the enemy's midst. (He did so before Kosaka Danjo made his counter-attack, and so felt that his plans had only brought disaster upon the Takeda army.) Rather than dying from the slashes of opposing samurai swords, he fell to a hail of arquebus and archery fire, his wounds leading to him retreat and commit *seppuku* (ritual suicide).

> *'If the enemy moves to close quarters ... cross swords with them. Aim at the helmet, but if you have loan swords that are blunt, aim at the hands and legs...'*
>
> — *The* Zôhyô Monogatari

Armour and Uniform

The standard of oriental infantry equipment and dress varied tremendously depending upon the time period, the status of the soldier and the conflict in question. Broadly speaking, there was improvement in armour distribution during our period, this being based upon the growing awareness that keeping the infantry alive was a militarily beneficial practice. Such a policy was no doubt encouraging to the infantry themselves. Yet there were many exceptions to this general rule. Soldiers who were drawn from peasant stock on a temporary basis were the least likely to have any sort of armour or specialized equipment. Armour was a considerable financial investment, and one beyond the means of most subsistence farmers. Illustrations in military manuals used by Ch'i Chi-kuang depicts Chinese peasant soldiers as wearing nothing more than their civilian clothes – a soft, belted tunic, cotton trousers bound to just below the knee with puttees, and simple slip-on sandals. The only items that distinguish them as soldiers are their weapons. Some illustrations depict infantry carrying bamboo trees stripped of their leaves. These were actually a form of weapon, used to pin down an enemy soldier so that an allied spearmen could kill him with greater ease.

Japanese artworks depict peasant soldiers in similar states of dress, and worse. One painting from the late fifteenth century shows an *ashigaru* without footwear, his work clothes hanging ragged from his body. Yet another artworks from the same period show *ashigaru* elaborately dressed in full samurai-type armour. The fact that they are *ashigaru* rather than samurai is indicated by a generally unkempt state of facial hair, an artistic motif used to denote lower-class individuals in Japanese art. Most of the armour in this instance will have come from looting the bodies of samurai killed on the battlefield. While looting could be performed for simple material gain, there was also an element of necessity to it. An *ashigaru* with some degree of armour was much more likely to survive an arrow strike, spear thrust, gunshot or sword slash than his unarmoured comrade.

Once the Japanese *daimyô* also recognized this fact, they began the issue of *okashi gusoku* (loan armour). This consisted of a basic body armour system called a *dô*. The *dô* consisted of strips of metal laced together and formed into a protective bodice for the torso, or was made out of plate armour in the familiar European mould. Another variety of armour popular with the *ashigaru,* principally on account of its cheapness, was made from metal strips or rectangles attached to mail armour. This was relatively easy to produce once the mail coat was available, and it also gave a good degree of mobility to the wearer. In all armour types, the *ashigaru* would often have the emblem of their *daimyô* stencilled on his front to assist with battlefield identification.

Hanging down from the *dô* was a *kusazuri,* a short skirt made of strips of iron that provided some measure of protection to the groin and upper thighs, key target areas for enemy spearmen. Early varieties of *dô* hung from the soldier's shoulders, but more advanced types shifted the weight to the waist, giving the soldier greater upper-body mobility. The armour was lacquered to give it some form of rust proofing, and the colour of the lacquer could also be a means of identifying the soldier's unit.

Helmets for Japanese infantry consisted of the *jingasa* (war hat). This was a simple, conical-shaped helmet made from metal or tough leather. When helmets were not available, *ashigaru* created a simple form of protection called the *hachimaki.* This was nothing more elaborate than a section of mail fitted to a headband. It would provide some measure of resistance against a blade slash, but gave little protection from powerful impact injuries.

Because of the size of Chinese armies, armour tended to be less widely distributed. Typical depictions of an armoured infantryman show an individual with a basic protective breastplate, this being made up of either leather or metal jointed panels. On better-armoured individuals, this type of armour could extend over almost all exposed parts of the body, including the arms and lower legs. As with Japanese armour, it could be brightly

BELOW: THREE TYPES OF JINGASA *war helmet.* Jingasa *were made from a wide variety of materials, including wood (1), bamboo (2), metal (3) and paper, and consequently offered varying degrees of protection. The key problem with such helmets was the lack of adequate neck protection.*

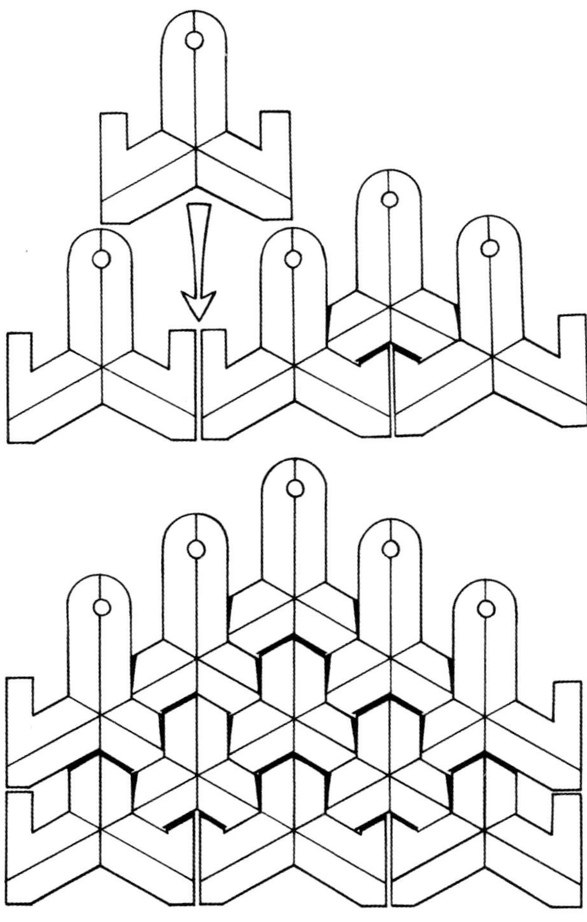

RIGHT: CHINESE SCALE ARMOUR AND STYLE OF CONSTRUCTION. *Each section locks into the others to form a regular pattern with no obvious gaps.*

decorated for show, but such decoration was more common amongst the wealthier mounted troops, than amongst foot soldiers.

Weapons and Tactics

The weapons of oriental warfare break down into two categories: edged and missile, although this simple categorization contains some internal complexities. What is important to note is how the weapons contained within an army dictated the way it could fight, as demonstrated by the battle of Nagashino in 1575 (see below).

Armies composed of hastily levied workers would display all manner of weaponry, including simple agricultural tools such as scythes and hoes converted to military use. Excerpts from a recruitment document issued by Hyjy Ujimasa (1538–90) in 1560 illustrate how samurai commanders were realistic about what the *ashigaru* might bring to the battlefield:

'All men, including those of the samurai class in this country district, are ordered to come and be registered on the 20th day of this month. They are to bring with them a gun, spear, or any kind of weapon, if they happen to possess one, without fearing to get into trouble. [...] Men must arrive at the appointed place properly armed with anything they happen to possess, and those who do not possess a bow, a spear or any sort of regular weapon are to bring hoes and sickles.'

The tactical expectations for those armed with 'hoes and sickles' must have been particularly low, and they can have been seen as little more than a hacking and slashing mass of men used to cause general chaos amongst the enemy ranks (who may well have been similarly armed). Note also that Ujimasa acknowledges that some men might join his ranks without any form of weapon at all. Artworks from the time of the Ônin War (1467–77) depict *ashigaru* figures armed with nothing more than sharpened bamboo poles. For such poorly-armed soldiers, looted weaponry and armour became much more important, hence the mixed composition of armament seen in many military artworks.

Spears

Of course, not all medieval Asian armies were so haphazardly armed, and as battlefield tactics were refined during the Early Modern period the standardization of infantry weapons increased. Large portions of most armies were made up of spearmen, the term 'spear' here referring to any stabbing or slashing weapon mounted on a shaft. Chinese armies had a long history of using halberds stretching back to the fourth century BC, and these were used alongside pikes and spears ranging in length from about 2.8m (9ft) to a prodigious 5.5m (18ft). Japanese armies were equally wedded to the use of spearmen or pikemen; the proportion of men armed with spears in a typical Japanese army rarely went below 25 per cent but could climb as high as nearly 70 per cent. Classic forms of Japanese pole arm include the *nagamaki* and *naginata,* consisting of long shafts terminating in straight or slightly curved blades respectively. Use of such weapons is associated with mounted samurai, who found them practical for unseating enemy horsemen at a safe distance. They were wielded by *ashigaru* hands as well, for the same purpose. Cavalry shin guards known as *sune-ate* were introduced to provide some measure of protection from these weapons.

A more common *ashigaru* spear arm was the pike-like *nagae yari,* produced from about 1530, which was longer than earlier spear weapons. Call-

THE NAGINATA *was an extremely effective weapon in trained hands, and its origins date back to the eighth century AD. During the Edo period, however, it became less used on the battlefield and more associated with home defence, particularly in the hands of women.*

SAMURAI LEADER, painted in *1886 – General Yamanaka Yukimori stands in full armour, a crescent moon as the crest of his helmet. He placed great faith in this symbol, but lost his life fighting the Mori clan in 1579, when he was 34. He is armed with a* yari, *or spear, a weapon which became increasingly used by samurai from the sixteenth century onwards.*

to-arms decrees from the late sixteenth century give instructions to *ashigaru* recruits not to bring along any spear that was less than 4m (13ft) long. By the 1600s it was typical for the *nagae yari* to be 5–6m (16.4–19.7ft) long.

Korean armies also had a variety of spear weapons in their arsenals, and length was again a key ingredient. The shortest of the spears was the *jang chang*, a 1.5m (5ft) weapon made of yew or bamboo, the flexibility of the shaft making it resistant to blows. A far longer version, the *juk jang chang*, by contrast reached up to 6m (20ft) in length and was capped by a 10cm (4in) blade – the great length enabled the weapon to probe deep into enemy ranks while leaving the user at a decent protective distance. Korean armies also created multi-blade spears such as the *nang sun*, which had 9–11 auxiliary blades jutting out from the 4.5m (15ft) shaft, creating an effect like a giant woodsaw when pushed into tightly-packed enemy soldiers.

The Role of the Spearman

The role of the spearmen in oriental warfare was similar to that seen in European warfare of the same period. A massed body of men, all presenting spears of uniform length, could form a powerful phalanx to smash through enemy infantry ranks, while also offering a formidable barrier against penetration by cavalry. However, depending on the particular circumstances, spearmen could also be used in more flexible arrangements. An especially interesting example comes from the Ming era generalship of Qi Jiguang (1528–88), a Chinese general who gained much combat experience in the mid to late 1500s fighting Japanese pirates and Mongol mounted raiders. In response to the latter, he developed units based around two-wheeled carts, each cart having screen walls and containing two light artillery pieces and four arquebusiers. When the unit was attacked, the carts could be pulled together to form a long, solid barrier, from which artillery and musket fire was unleashed at the attacking

OPPOSITE: ASHIGARU WEAPONS. *This useful nineteenth-century photograph shows three principal types of* ashigaru *weaponry (from left to right): the* naginata; *the bow; and the* nagae yari *long-shafted spear.*

BELOW: SPEAR HEADS. *Oriental spears came with a variety of heads, not only simply straight blades. Various configurations gave additional capabilities, such as trapping enemy blades and unseating horsemen.*

1 STRAIGHT BLADE WITH SIDE PICK; 2 AND 3 SPEARS WITH CROSS BLADES; 4 CHINESE CRESCENT BLADE WITH PICK; 5 CHINESE HOOKED BLADE WITH SIDE PICK.

cavalry. Each cart also held another four soldiers armed with spears whose job it was to drive away attackers who made it to close quarters. They would also venture out through special swing doors in the cart 'wall' and attack individual riders. Here was an example of the dismounted spearman functioning as an individual warrior, rather than working as part of a mass of infantry, although there was a strict distance prescribed as to how far the spearman could actually wander away from the carts (about 7.6m/25ft).

In Japan, infantry spearmen were rigorously drilled in tactics and formations. In defence, Japanese *ashigaru* spearmen would adopt close-set kneeling ranks, the spears angled upwards in an even wall. The angle of the spears was designed to attack horses' breasts, either killing the horse or at least causing it to rear and throw its rider. When facing other enemy infantry, the *ashigaru* spearmen would stand and present a similar wall of blades while usually advancing to contact. At this point they would engage in thrusting combat, the spearmen stabbing their spears into the mass of opponents. Such a clash would not only serve to inflict casualties on the enemy but could also 'fix' the enemy centre (spearmen were usually positioned in the centre of a line) while faster cavalry forces manoeuvred around the flanks.

WIELDING THE SWORD. *Although the samurai were traditionally mounted troops, it was perfectly common for them to fight on foot alongside the* ashigaru.

Swords and Daggers

Although swords were often held to be the province of the cavalry, at least in terms of being wielded with any expertise, the fact remained that swords were practical close-quarters weapons for any infantryman. Access to swords varied according to an individual's status: Japan in particular had strict restrictions on weapons distribution under the laws put in place during the sixteenth century.

In Chinese armies, however, swords of many different types were common, even amongst infantry ranks. The most prevalent of these was the *dao,* a single-edged sabre-like blade that was essentially a melee weapon used for powerful slashing attacks.

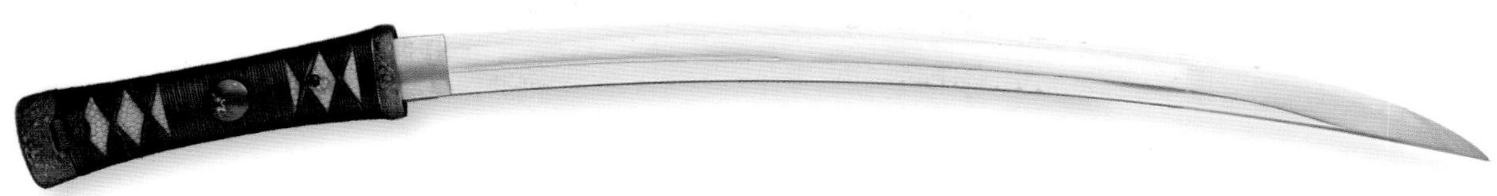

CLOSE-QUARTERS WEAPON. *This Japanese* wakizashi *sword was usually worn alongside the longer* katana *blade, and had various functions: it could be a backup weapon, it could be used to decapitate the enemy, or* in extremis *it could be used as a suicide weapon.*

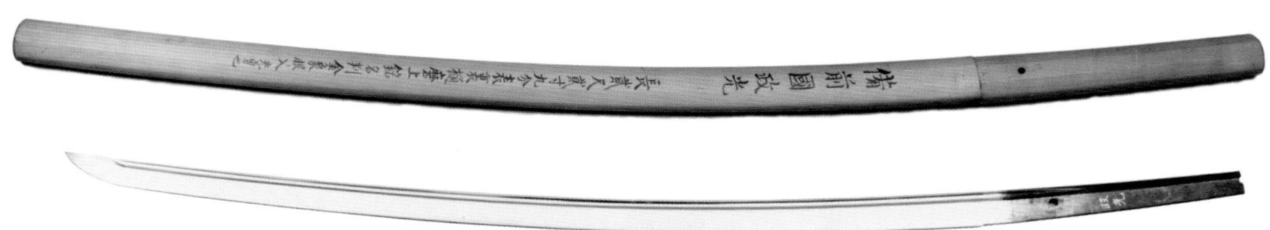

SAMURAI BLADE. *The Japanese* tachi *sword has no clear definition. Some argue that it is longer and more curved than a* katana, *while others say it is the manner of wearing – from the belt with the edge down – that defines it. This example was made c.1370.*

CHINESE DAO AND SCABBARD. *This Chinese sword, probably dating from the seventeenth century, is likely a ceremonial weapon, on account of its well-preserved state and its elaborate hilt and scabbard.*

The other type of sword commonly seen in Chinese armies – the *jian* – was a straight-bladed sword of ancient ancestry, but a weapon that took considerable skill and training to master. During the Ming Dynasty it tended only to be seen in the hands of cavalry and officers and it was the *dao* that became the standard-issue weapon. *Daos* fall into several different categories, separated mainly according to the degree of curvature of the blade. The most widely used was the *liuye dao,* a more moderately curved blade than the slashing *pain dao* or the *niuweido* sabre, the latter being the deep broadsword immortalized in many a martial arts movie.

> *'By victory through fighting with the sword, against individuals or in battle against large enemy forces, we can attain power and fame for ourselves or our lord.'*
>
> – MIYAMOTO MUSASHI

Japanese Swords

Japan, of course, is famous for its superb swords, and most illustrations of professionalized *ashigaru* (as opposed to rough-shod peasant recruits) show the infantryman with one, commonly two, swords held in scabbards on his *uwa-obi* belt. The paired swords would imitate those of the samurai warrior – a long *katana* sword, used as a lethal slashing weapon, and a shorter version known as a *wakizashi.* This latter could be used as a backup or close-quarters weapon, or for other purposes such as the beheading of fallen enemies and the committing of *seppuku* (ritual suicide). Together the pair of swords was known as a *daishô.* The swords of the general run of *ashigaru* reflected the *daishô* of the samurai in terms of design, but frequently not in quality. During the Warring States period, swords had to be manufactured in mass quantities to arm infantry as well as cavalry, and quality consequently suffered. The martial arts manual known as the *Zôhô Monogatari,* composed in 1646 by an experienced commander for his less-than-experienced *ashigaru* troops, noted that in hand-to-hand fighting with swords: 'Aim at the helmet, but if because the loan swords have dull blades you can only chop, aim at the enemy's legs and you can cut at them.' The context of this passage is that the enemy has closed to such as distance that using spears and arquebuses is no longer viable. There is no suggestion of sophisticated swordplay here, only the brutal slash and hack of close-range killing by hardened men. Indeed, rarely are *ashigaru* depicted armed only with swords – spears, bows and arquebuses formed their main weapons, with the swords acting as backup.

Another type of emergency killing tool carried by many oriental warriors was a simple dagger. In Japan the classic style of dagger was the *tanto.* This dagger had a cross guard and a blade with a maximum length of 30.5cm (12in). Amongst higher ranking individuals *tanto* blades tended to have a more ceremonial significance, although *in extremis* they could be used for *seppuku.*

Within the infantry, however, any blade was a potentially useful weapon, especially as a last-ditch fighting tool or as a general implement of self-defence outside formal combat situations.

Korea had its own diverse range of swords, many inspired directly by Japanese models. Korean martial authorities frankly accepted the superiority of Japanese swordsmanship over their own and, during the sixteenth century, there was a concerted attempt to codify Japanese sword fighting techniques in Korean martial manuals. The variety of different sword types used by the Koreans is too broad to allow any detailed examination here, but some key examples are illustrative. *Ssang gun* referred to twin swords used for fighting, although the term did not specify the blade types. The two-sword technique emphasized attack with one sword and defence with the other. Of similar principle was *deung pae,* which used a short sword as the attacking instrument and a 0.9m (3ft) diameter shield (made of wisteria or willow branches wrapped in leather) as a defensive tool. Single sword types included the *ssang soo do,* a powerful cutting sword used for close-range fighting, and the *ye do* short sword, as well as training tools such as the crescent *wol do* and *hyup do* spear sword.

Other Hand-held Weapons

While Asian armies became increasingly standardized in their armaments as time went on, there were still one or two hand-held oddities that cropped up in battle. One weapon that appeared in Japanese, Korean and Chinese hands was the war-flail: a simple military conversion from the humble rice flail. The Korean *pyun gon,* for example, consisted of a staff that was 2.4m (8ft) in length, with a 0.6m (2ft) impact section connected to one end by a length of chain. This weapon, although unwieldy, would be capable of delivering an enormously powerful blow to a seated rider, and they were also used by the defenders of fortifications to slap away wall-scaling attacks.

More specialist and unconventional forces also used a range of flail weapons, although often combined with edged instruments. Chinese martial artists, for example, used iron chain whips featuring a blade connected to a long shaft by a chain, while the Japanese *ninja* combined a sickle and chain, the chain featuring a heavy weight at one end. The idea was that the weight was swung to unseat a rider or knock a soldier off his feet, whereupon the sickle would be used as a killing tool.

Missile Troops

Missile troops were principally a mix of artillerymen, arquebusiers, archers and crossbowmen, the relative mix of these elements depending upon the period and the nature of the army involved. As a general rule, we see a steady increase in the proportion of arquebusiers to archers with the passage of time, although the expense and sometimes unavailibility of firearms, plus some of the advantages of sprung weapons over gunpowder arms, means that archery was never completely displaced. Furthermore, a lack of investment in firearms, or suspicion about their value, led to a reduction in their numbers in some armies.

In Japanese culture, the bow was initially associated with the mounted samurai, but gradually shifted into *ashigaru* hands as the

SAMURAI ARCHER, SIXTEENTH CENTURY, *here armed with a* yumi *bow. The Japanese archer was never entirely replaced by the arquebusier, primarily because his rate of fire was faster and his accuracy was generally better.*

samurai switched more to the use of couched lances and blade weapons at the same time as the Japanese armies sought ways to deliver massed firepower.The classic Japanese bow was the *yumi.* It was a composite bow with great elasticity and power, formed from a strip of hardwood, such as sumac, sandwiched between strips of bamboo. Rattan binding was used as a reinforcement for the glue then the whole weapon was further covered in a waterproof lacquer. The overall length of the *yumi* varied considerably, some bows being a short as 0.6m (2ft) while the longbow varieties could measure up to 2.4m (8ft).The smallest of the bows were adapted to close-range mounted warfare, and their size often meant that they were made out of horn or whalebone instead of wood. The main objectives of *ashigaru* archery were range and weight of fire, not the precision shots at individual targets sought by the samurai warriors, hence the longbow was the most common infantry bow.

Bows and arrows retained enough advantages to keep them relevant until well into the nineteenth century. They had a faster reload time than matchlock firearms and a decently-trained archer was able to achieve at the least double the rate of arquebus fire. For this reason, the *Zyhy Monogatari* recommended that one archer be stood between two matchlockmen to provide covering fire while the gunners were reloading. This configuration not only ensured that the matchlockmen had some protection during their most vulnerable moments, but also maintained a continuity of fire between volleys. A further advantage of the bow and arrow was simply that they could be used in all weathers – many arquebuses were reduced to little more than iron and wood clubs by rain. Although the penetration of an arrow was generally less than that of a matchlock ball, particularly under 100m (328ft), range was still decent. Typically, archers would engage these distances, but if need be they could make long-range attacks with some measure of accuracy out to 400m (1312ft). Note that samurai armour was quite effective in stopping individual arrows, so weight of fire was necessary to improve the chances of a fatal blow. Often it was cumulative injuries that brought down a well-armoured samurai, unless he took an arrow to the small exposed portion of his face.

Crossbows

Crossbows were a lesser presence on oriental battlefields when compared to traditional bows

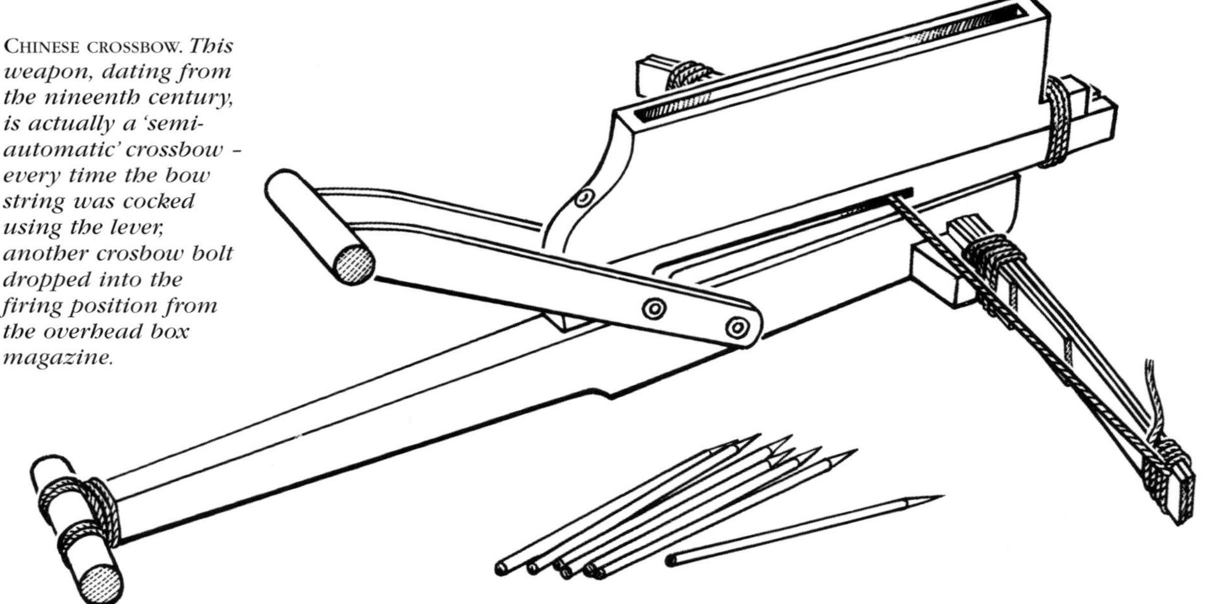

CHINESE CROSSBOW. *This weapon, dating from the nineenth century, is actually a 'semi-automatic' crossbow – every time the bow string was cocked using the lever, another crosbow bolt dropped into the firing position from the overhead box magazine.*

and arrows, although they were still seen in action late in the Early Modern period, particularly in China. Crossbow bolts had excellent powers of penetration against armour at short ranges. The bows were fitted with basic sight and could be extremely accurate, although these advantages were offset by a much lower rate of fire, closer to that of a matchlock gun rather than a conventional bow.

For this reason, crossbowmen were sometimes deployed in different ways to those of standard archers. One Chinese doctrinal manual, the *Wu Ching Tsung Yao,* advocated that crossbowmen reload their weapons behind a protective shield of infantry spears and swords, then venture out into the open to engage individual enemy warriors. They should quickly retreat behind their comrades to reload after each shot. How this potentially exhausting and dangerous routine might have worked is unclear, but there is no doubt that crossbowmen did pose a serious threat to mounted cavalry because of their weapons' twin virtues of accuracy and penetration.

One method of attempting to improve the crossbow's rate of fire was the repeating crossbow, which entered service with Chinese forces during the Ming Dynasty era. At its most basic, the repeating crossbow consisted of a standard crossbow above which was mounted a wooden box magazine containing multiple arrows. When the crossbow was cocked, the first bolt in the box dropped into place and was then fired, the recocking allowing the next bolt to load and so on. Spanning the bow was performed by a lever set-up at the rear of the magazine, and the rates of fire for the time could be extremely good – a bolt could be unleashed about every second.

There were downsides to this, however. The automatic loading system demanded that the bolts be featherless to ensure a smooth feed, dramatically reducing both accuracy and penetration. Also, because spanning had to cycle the magazine as well as the bowstring (the magazine was moved backwards and forwards to perform the correct loading action) there was a consequent reduction in the power of the bowstring, further reducing range. Nevertheless, as a close-range tool, the repeating crossbow could pose an additional problem for an enemy.

EARLY FIREPOWER. *A three-barrelled hand cannon from Korea, dating from the seventeenth century. It would have been mounted on a wooden haft to provide stability, but its accurate range was extremely short.*

Firearms

The development and application of firearms in oriental warfare is a curious mix of revolution and regression. In terms of revolution, the introduction of firearms would eventually alter the entire balance of power on the battlefield, finally unseating the mounted cavalryman as the supreme arbiter of victory or defeat. Although cavalry would still retain its central position in any form of manoeuvre warfare. The central point is

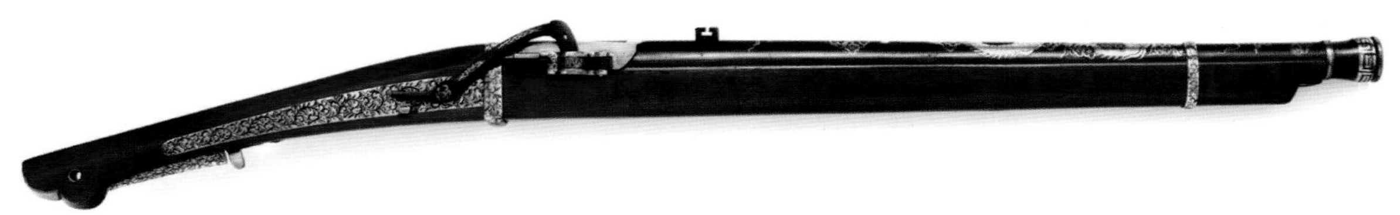

MATCHLOCK TEPPO (MUSKET). *This Japanese matchlock weapon dates from the nineteenth century, indicating just how far Japanese firearms lagged behind the flintlock weapons universally adopted by European armies from the eighteenth century.*

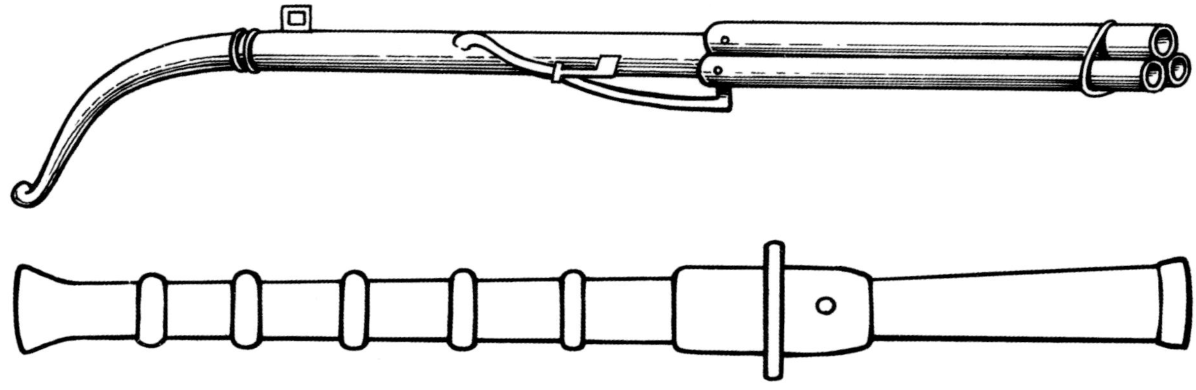

TOP: EARLY TRIPLE-BARRELLED CHINESE GUNPOWDER WEAPON. *Accuracy from this firearm must have been extremely poor – note the stock, which would have been clamped under the armpit. Each barrel has a separate vent hole for firing.*

LOWER: SEVENTEENTH CENTURY 'SILK' GUN. *The name of this Chinese gun refers to the way that the iron barrel was covered with raffia and then wrapped in silk, making for an extremely attractive light artillery piece.*

that other forms of weapons technology – particularly bows and swords – demanded decent levels of training to handle well, even if that training did not reach the heights of the samurai master. Firearms, by contrast, could give even the most sketchily-trained infantryman lethal killing power, although discipline and training in reloading were important to achieve consistent volley fire. Moreover, the powerful penetration of a musket ball meant that a peasant infantryman with a few weeks of gun training was now quite capable of downing a samurai warrior who embodied years of personal training and centuries of martial ancestry. In short, firearms were central to the rise of the infantry and the democratization of the battlefield.

Then comes the regression side as a counterbalance. In Europe, firearms development from the fourteenth to the nineteenth century went through steady advances – matchlock, wheellock, flintlock and percussion (cap to cartridge). In many ways oriental forces remained stuck at the matchlock stage, partly through choice and partly through necessity. Japan's enforced isolation from the international world, implemented by Tokogawu Ieyasu (1543–1616), froze firearms development at that particular moment in history, and western adventurers re-entering Japan in the nineteenth century found matchlock firearms still central to infantry units.

China's experience was similar, although probably for different reasons. Being derived from a hugely populous but poor collection of states, Chinese armies often struggled to equip small portions of the infantry with a relatively inexpensive matchlock guns, let alone the more expensive European flintlocks. Furthermore, the mechanically simple matchlock gave less room for misuse and malfunction in the hands of ill-trained conscripts than did the more complex successor weapons. A closer look at the experience of Japan and firearms gives a good feel for both the capabilities of the weapons and how they were integrated into infantry forces. The arquebus was introduced into Japan in the 1540s by Portuguese traders, when the new technology was fairly enthusiastically embraced by the Japanese. In contrast, China had firearms in service far earlier than the Japanese, with handguns seen in action by the early fourteenth century, indicating an almost parallel development with early European firearms. The arquebus was a muzzle-loading weapon, the barrel being charged with powder and ball before a small quantity of powder was tipped (from a powder horn) into a pan adjacent to a vent hole connecting the pan to the main charge in the gun's chamber. Ignition was performed by a smouldering slowmatch gripped in the jaws of a hinged metal arm, known as a serpentine. Pulling on the lower end of the arm – effectively the trigger – swung the match down into the pan, igniting the priming powder, the flame of which flashed down the vent hole to ignite the main charge and fire the gun.

By the time the arquebus reached Japan, the simple hinged trigger arrangement had generally been replaced by a spring-loaded 'snapping' matchlock system, the match arm being released under spring tension for much faster lock times.

Firearms Technology

The arquebus had a number of problems despite its rapidly gained popularity. The first issue was its reloading time. In well-trained hands the arquebus could only average about one shot every 15–20 seconds. Even this would decrease in the hands of a less expert infantryman or during a battle as the gun suffered from increased fouling. After a few minutes' firing, reload times would increase and volleys would become extremely ragged instead of in concert. In bad cases the fouling could reduce the rate of fire to almost one shot per minute.

In dry weather, the reliability of the arquebus was good, especially if good quality powder was used. In damp weather, however, misfires became common, and if the powder or slowmatch became wet under rain then the gun would become inert, at which point the arquebusier would be forced to rely on more traditional backup weapons.

There were also safety issues for the infantryman. Keeping a burning slowmatch and explosive propellant in close proximity is a recipe for potential disaster: many infantryman were injured by premature detonations during loading, or by the accidental ignition of a powder bag or horn.

Maximum theoretical range for an arquebus could be up to 500m (1640ft), but in reality any ball still flying at over 200m (656ft) would have little penetrative force. Furthermore, at ranges beyond 50m (164ft) the arquebus was woefully inaccurate. The inaccuracy was a product of several factors – a smoothbore barrel (meaning that the ball had no stabilization), a frequently excessive windage (the degree of movement between the ball and bore), unwieldy dimensions making the gun hard to handle, crude sighting systems, and a relatively slow lock time when compared with modern propellants and percussion systems. Turnbull (*Warriors*) notes a modern test of an arquebus in which five bullets were fired at a samurai-sized target at 30m (98ft), then again at 50m (164ft). At the shorter range all the bullets hit the target in the chest but at the longer range only one of the five bullets struck home. With this formidable list in the arquebus' deficit column, why, then, did the gun come to have such a major impact upon warfare? An initial reason is shown by Turnbull's analysis of arquebus range. Even though the arquebus ball fired at 50m (164ft) was inaccurate, it still carried enough power to penetrate 1mm (0.03in) of steel plate, and as Turnbull notes: 'The scales of a typical lacquered *dô maru* were of similar thickness.' Factor in the large calibre of many of these

TWO TYPES OF JAPANESE MATCHLOCK, SEVENTEENTH CENTURY. *These represent the summit of the matchlock gun. Both have snapping matchlock mechanisms, the serpentine being dropped by spring power to speed up the time between trigger pull and ignition.*

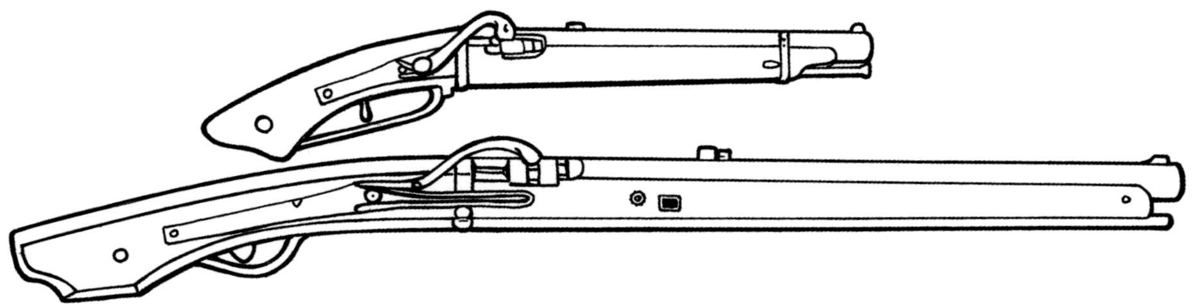

arquebus weapons - some of which could approach 2.5cm (1in) - then a single strike from one of these balls would likely incapacitate even the best-armoured samurai.

Furthermore, the fact that years of training were not necessary to develop proficient use of the arquebus meant that massed volley fire became possible, and amidst a huge ripple of gunfire individual accuracy was not important. Guided by a *teppô ko kashira* (firearms officer), the arquebusiers would be grouped into units of five gunners and one archer, and several of these units gathered into a larger formation. The gunners would be split so that rotating volleys could be delivered (one group fires while the other reloads), a process that required good command skills, rigorous discipline amongst the gunners and a smooth proficiency in reloading. If the arquebusiers could achieve all these standards during the confusion of battle, the dividends could be exceptional. One Japanese account, the *Kirin Gunki,* explains how force of 100 *ashigaru* armed with arquebuses produced an enemy death toll of 600–700 during one battle.

Of course, firearms would not be the only factor to decide the outcome of any battle. Although the arquebusiers did have a critical effect there was still much would be done by other forms of *ashigaru* and also by samurai cavalry. However, infantry firepower became an even more central element in the considerations of generalship. War was no longer an enterprise of the elite.

Battle of Nagashino, 1575

Few engagements in the history of oriental warfare illustrate better how new weapons changed the balance of power on the battlefield than Nagashino. Fought in 1575 in southern Japan, the engagement was a struggle for power between

KATSUTAKA, *besieged in Nagashino castle, broke out and sought help from Tokugawa Ieyasu. He was captured when he tried to return, but in an act of heroism he shouted to the castle garrison that they would soon be relieved. He was immediately killed.*

PLAIN OF SHIDARAHARA TODAY. *The fence represents the barricades behind which the Nobunaga arquebusiers deployed. The barricade imposed enough delay on the enemy cavalry to present them as clear targets for the Japanese volleys, which were delivered in ranked fire.*

cavalry and the arquebusiers. Although there is the temptation when analyzing Nagashino to focus on the arquebusiers to the exclusion of the many other forces on the battlefield that day, firearms undoubtedly tipped the balance and proved how integral the *ashigaru* were to the outcome of future conflict.

By the end of the 1560s, the great samurai general Oda Nobunaga had emerged as a salient figure from the great struggles between Japan's *daimyô,* controlling much of central Japan from his powerbase in Kyoto, which he had taken in 1568. In 1575, however, the *daimyô* Takeda Katsuyori (1546–82), ruler of many of Japan's eastern provinces, invaded the provinces of Mikawa and Tôtômi. These belonged to the Nobunaga ally Tokugawa Ieyasu, and they

JAPANESE ARQUEBUEIERS. *This rank of arquebusiers, decked out in straw rain coats, has fitted boxes over the locks of their guns in an attempt to keep out the damp during the firing process. Such measures had limited success.*

bordered Oda's own territories in the south. The message was clear – Takeda was heading for Kyoto, much as his father, Takeda Shingen, had unsuccessfully attempted in earlier years.

Takeda's initial military forays against Okazaki Castle and Yoshida Castle, two principal Tokugawa fortifications, were thwarted by Tokugawa's intelligence services and military strength. Takeda therefore turned his attention to a new target, Nagashino Castle in Mikawa province. The castle nestled between the confluence of the Takigawa and Onogawa rivers, where they joined to form the Toyokawa. It was a well-protected fortification: the banks of the rivers were up to 50m (164ft) high and formed natural barriers to an assault, while the castle itself was guarded by multiple ranks of walls and outer defences. Nevertheless, the castle's garrison numbered only 500 men and by mid-June it was surrounded by 15,000 enemy soldiers. After initial Takeda attacks were beaten off by remarkable resistance, Takeda eventually opted to place the castle under siege and starve the defenders into submission.

The besieging forces were soon to face a challenge far more severe than a recalcitrant castle. By the end of June, a relief force of 38,000 men – 30,000 belonging to Oda and the remaining 8000 to Tokugawa – was marching for Nagashino, intent on smashing both the siege and Takeda's pretensions to power. *Ashigaru* infantry formed a significant part of both armies. We know little about the composition of the Oda–Tokugawa forces, but what we do know is that they contained some 3500 *ashigaru* arquebusiers. In 1570 Oda had been on the receiving end of a similar number of arquebuses when he fought against Buddhist Ikko-Ikki rebels around Ishiyama Hongan-ji. On an earlier occasion Oda himself had been shot twice in combat. Only luck and generous layers of body armour saved him from fatal injuries. Nevertheless, his experience led him to give arquebus-armed troops a central role in the forthcoming battle at Nagashino, having ensured that they were fully trained and disciplined in volley tactics.

The Takeda army also contained a decent proportion of *ashigaru,* including arquebus troops, but as we know more about the composition of his army it is clear that he favoured a traditional weighting towards cavalry. Of the 33,000 troops at his disposal (only half this number was assigned to Nagashino), over 27,000 consisted of horsemen and their followers, leaving only about 5500 regular *ashigaru* (there were some 800 bodyguard troops) and just 650 arquebusiers. In essence, the battle that was gathering steam would be a test between cavalry and infantry; between sword and gun.

The Plain of Shidarahara

By 27 June, the Oda and Tokugawa troops had assembled on the plain of Shidarahara, roughly 5km (3 miles) from the castle, behind the banks of the River Rengogawa. Here would be site of the main

TAKING AIM. *A Japanese matchlockman takes rudimentary aim along the barrel of his matchlock. Separate bags for powder and ball hang from his belt, while ration packs are slung diagonally across his chest.*

battle. Their first action was to construct a series of defences. The shallow river that sat 100m (328ft) in front of the Oda lines was a natural barrier to retard a cavalry attack, but the troops also constructed a three-tiered palisade of wooden stakes just behind the river. Not only would this palisade break up the charge of the horsemen but it would also present slower-moving targets to the arquebusiers, 3000 of whom sat in massed ranks behind the palisade. They were arranged in three ranks, and each rank would fire in turn. Openings in the wooden structure provided the routes through which Oda-Tokugawa soldiers could make counter-attacks. Natural flank protection on the right came from the Toyokawa river, while on the left flank there was forested and hilly ground. The critical point here is that the *ashigaru* were deployed at the forefront of the battle. For Oda, gone were the days when the cavalry led the charge: here defence and firepower formed the core of his tactics.

Despite warnings from other officers not to be drawn into open battle, it seems that Takeda could not resist the potential glory of the challenge waiting for him to the west. Consequently he ordered 12,000 of his men away from the siege lines and headed out to fight. He ordered his troops into four 3000-strong divisions, arranged as right, centre and left wings and a headquarters reserve. Because of the nature of the terrain, which had been judiciously chosen by Oda, Takeda decided to make a fast frontal attack with his cavalry. In many ways this made sense. The open ground between the frontlines was only 200m (656ft), a distance that could be covered extremely quickly by horses at a gallop. The speed would also limit the number of volleys the enemy arquebusiers could deliver – Takeda was fully aware of their presence – and hence would restrict their influence.

These judgements, however, had several faults. First, although the distance to cover was short, the ground was not conducive to speed. It had been raining, and the weather had not only caused the Rengogawa to swell, but it had made the ground boggy, and unsuitable for galloping. Takeda also thought that the wet weather would cause problems for the defender's arquebusiers, but in this case they had managed to keep their powder dry. Takeda had further underestimated the effect

Battle of Nagashino

1575

There is a danger of overstating the role played by firearms at the battle of Nagashino. Much about the battle hung on traditional close-quarters, hand-to-hand fighting as much as missile weapons, and tactical decisions (and a selection of critical command mistakes) were equally important for the eventual outcome of the battle. Nevertheless, Nagashino was significant for offering a glimpse of the future of infantry warfare in the Far East. By delivering effective mass volleys of arquebus fire, the Japanese troops of Oda Nobunaga and Tokugawa Ieyasu effectively altered the social order of warfare - the less skilled arquebusiers could achieve superiority over proudly martial samurai warriors backed by the best military training. From the sixteenth century onwards, firepower would become an increasingly important part of oriental warfare, although not on the scale seen on European battlefields.

HONSHU
Kyoto
Mt Fuji
NAGASHINO

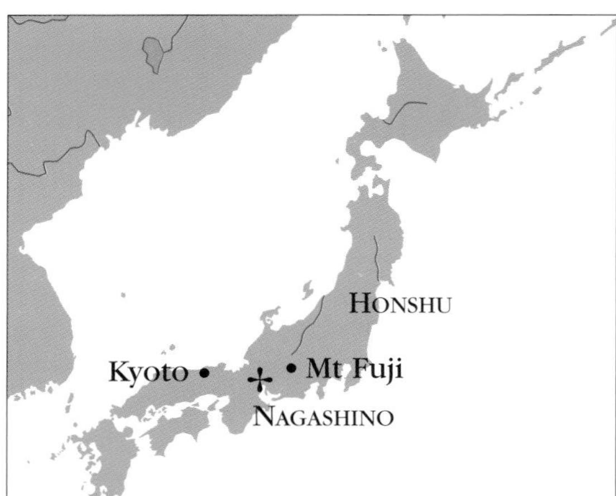

The cross marks the location of Nagashino Castle. The struggle for political power in sixteenth- and seventeeth-century Japan focused upon central Honshu.

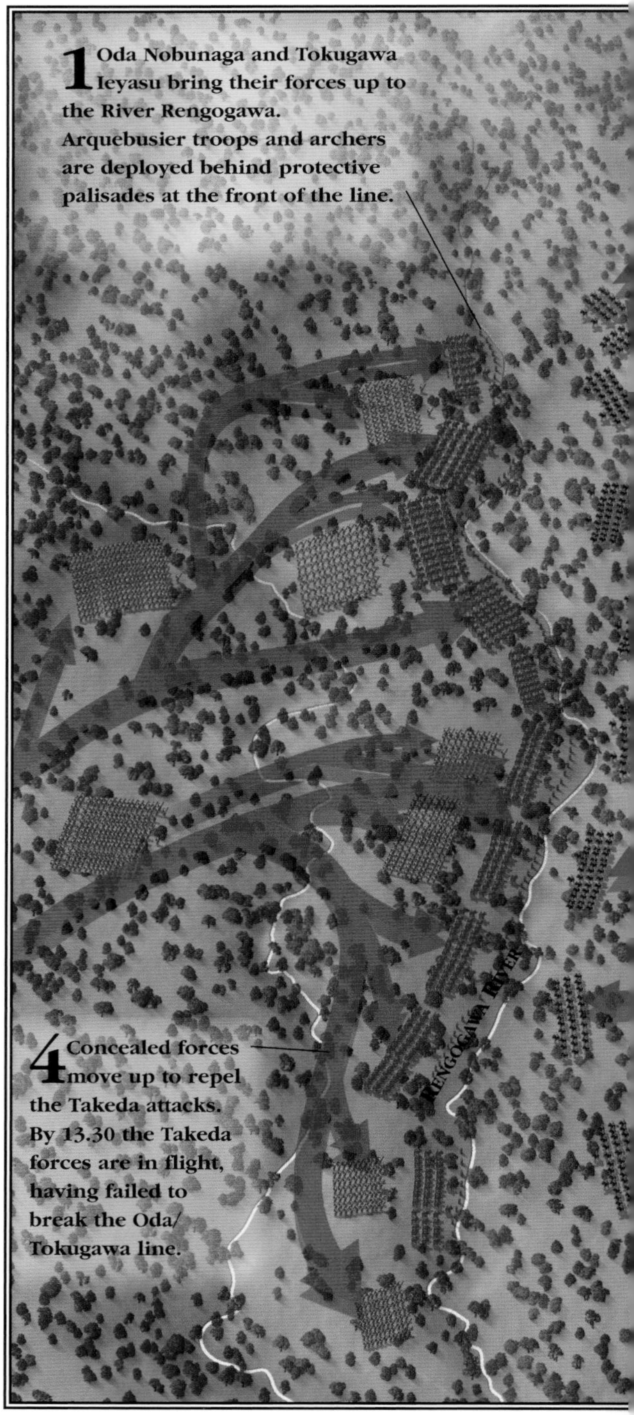

1 Oda Nobunaga and Tokugawa Ieyasu bring their forces up to the River Rengogawa. Arquebusier troops and archers are deployed behind protective palisades at the front of the line.
4 Concealed forces move up to repel the Takeda attacks. By 13.30 the Takeda forces are in flight, having failed to break the Oda/ Tokugawa line.
RENGOGAWA RIVER

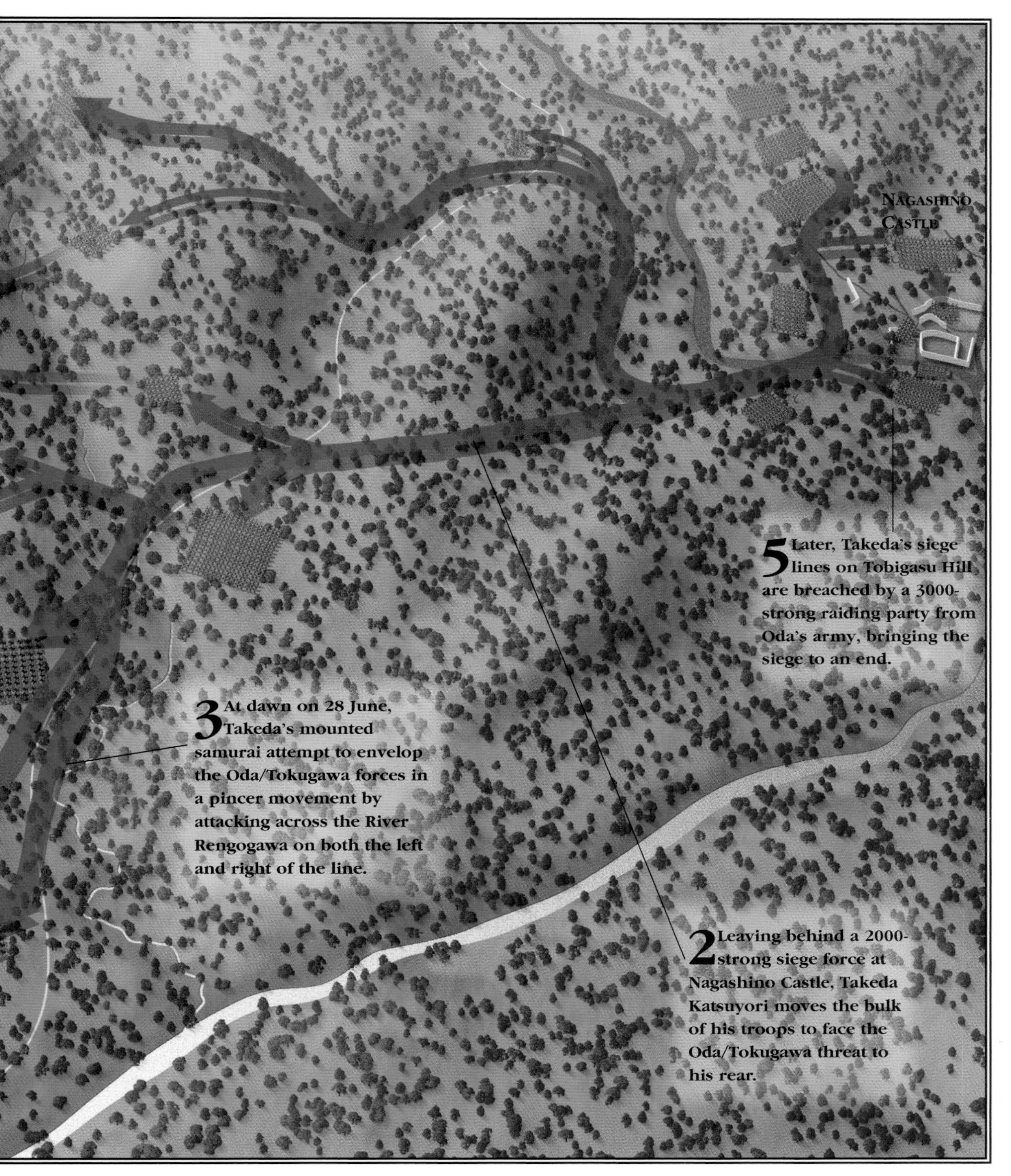
NAGASHINO CASTLE
5 Later, Takeda's siege lines on Tobigasu Hill are breached by a 3000-strong raiding party from Oda's army, bringing the siege to an end.
3 At dawn on 28 June, Takeda's mounted samurai attempt to envelop the Oda/Tokugawa forces in a pincer movement by attacking across the River Rengogawa on both the left and right of the line.
2 Leaving behind a 2000-strong siege force at Nagashino Castle, Takeda Katsuyori moves the bulk of his troops to face the Oda/Tokugawa threat to his rear.

ASHIGARU FIRING TEAM. *Rates of fire for arqubusiers varied throughout a battle and slowed as the gun became increasingly fouled with powder residues. After prolonged shooting, the soldier might be reduced to firing once a minute.*

of the enemy palisade, namely that his cavalry would have to slow to an almost dead stop right under the muzzles of the enemy guns.

The Attack Begins

On 28 June at 06.00, Takeda's cavalry began their attack. The charge rolled out at as brisk a pace as the ground conditions allowed, then slowed up considerably for the crossing of the Rengogawa. The cavalry rode up the far bank and against the palisade, at which point the *ashigaru* gunners opened fire with huge rolling volleys. The three volleys fired in that first instance sent 9000 heavy-calibre balls at the enemy, all within the optimum penetration range for the arquebus. Large numbers of horsemen and mounts were killed or injured within the first few minutes of the battle, including some of Takeda's most important samurai leaders. The arquebusiers also had their ranks interspersed with spearmen. With their long *nagae yari* spears they could attack any surviving enemy cavalry who pushed up against the palisade. The spearmen themselves were protected from the mounted soldiers' weapons by the palisade. Any cavalry who made it through the palisade were finished off by the waiting samurai and *ashigaru* positioned further back.

Throughout the morning of the 28th, Takeda's cavalry persisted in their assault, but the waves of cavalry simply provided more targets for the *ashigaru* gunners. An attack which was delivered against the right flank, around the palisade, did result in more hand-to-hand fighting, with the attacker's samurai skills having an impact here. Nonetheless, rapid redeployments of *ashigaru* troops helped to quench the assault. One of Takeda's most senior commanders, the 60-year-old Yamagata Masakage, was downed by a group of arquebusiers.

As the Takeda cavalry finally weakened, the Oda-Tokugawa troops finally surged forward from behind the palisade, creating an intense hand-to-hand melee that last until early afternoon. At one point it even looked as if the Oda-Tokugawa troops might have to make a retreat, but then the Takeda troops finally broke and scattered in a frenzied run for survival.

It was not only on the Shidarahara battlefield, as it became known, that *ashigaru* gunners and spearmen were deployed. A group of around 3000 men, including 500 arquebusiers, had formed a raiding party who made an attack against the

Japanese Monk (1600)

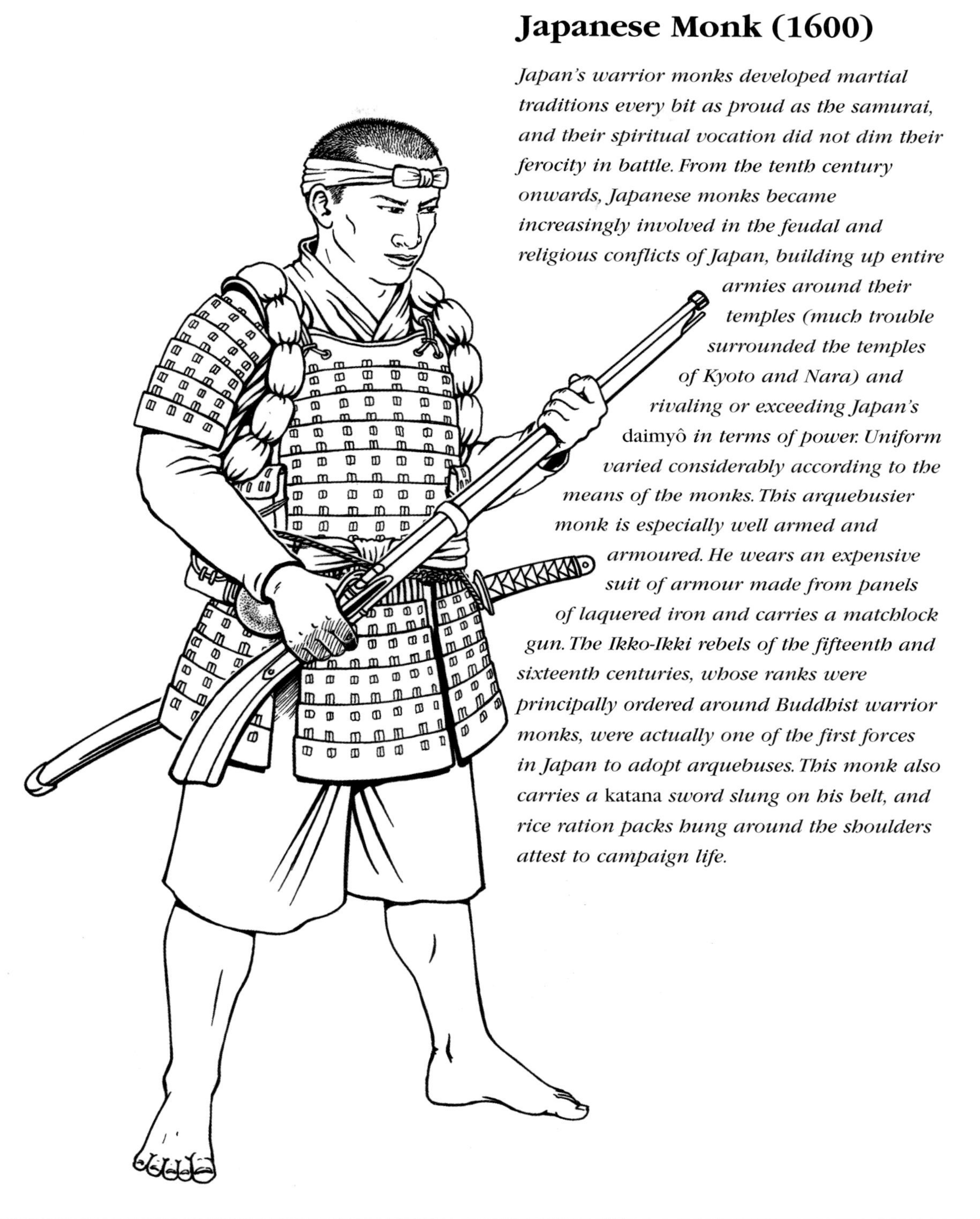

Japan's warrior monks developed martial traditions every bit as proud as the samurai, and their spiritual vocation did not dim their ferocity in battle. From the tenth century onwards, Japanese monks became increasingly involved in the feudal and religious conflicts of Japan, building up entire armies around their temples (much trouble surrounded the temples of Kyoto and Nara) and rivaling or exceeding Japan's daimyô *in terms of power. Uniform varied considerably according to the means of the monks. This arquebusier monk is especially well armed and armoured. He wears an expensive suit of armour made from panels of laquered iron and carries a matchlock gun. The Ikko-Ikki rebels of the fifteenth and sixteenth centuries, whose ranks were principally ordered around Buddhist warrior monks, were actually one of the first forces in Japan to adopt arquebuses. This monk also carries a* katana *sword slung on his belt, and rice ration packs hung around the shoulders attest to campaign life.*

Takeda siege lines around Nagashino castle. The attack was a brutal surprise to those besieging the castle, and such was its effect that the castle's defenders emerged from their fortifications to help finish off their besiegers. The battle of Nagashino lasted for around eight hours, and killed 10,000 of Takeda's troops and over 50 per cent of his samurai commanders.

Militarily, the battle had proved several key points. The first was that the *ashigaru* now truly occupied the front ranks of the battle. Although the cavalry still provided mobility and speed in the attack, volume missile fire was demonstrated to be superior to cavalry dash if it was combined with good defensive positions. The Oda-Tokugawa cavalry therefore acted as shock troops to take advantages of weaknesses opened by the *ashigaru*. By relying on his cavalry in a traditional way, Takeda had made a mistake that military commanders would continue to make well into the twentieth century, that of thinking that willpower would always triumph over firepower.

Artillery and Bombs

Beyond hand-held weapons, gunpowder also fuelled the rise of artillery as a major force on the oriental battlefield. Before considering gunpowder artillery, however, it is worth acknowledging some of the other forms of 'artillery' in use on oriental battlefields, particularly at the end of the Medieval period and the beginning of the Early Modern period. Torsion-powered stone or missile throwers were used in many battles, particularly in siege actions but also against gathered ranks of enemy infantry. These weapons could be trebuchets in the familiar European mould, or other large catapult devices. Both Chinese and Japanese infantry made

PITCHED BATTLE. *This nineteenth-century Chinese woodcut shows a pitched battle from the same era. Despite the advent of gunpowder weapons, swords, spears and shields were still part of the Chinese weapons inventory into the early twentieth century.*

use of these devices, and the missiles they hurled varied considerably – stones and balls of terracotta and metal could be thrown, and even blocks of ice. By the eleventh century the Chinese were also hurling explosive devices made from gunpowder wrapped in paper or a bamboo casing, and by the fifteenth century such 'shells' had also entered Japanese use, both in naval warfare and on land.

Of all the Asian nations, it seems fitting that it was the Chinese who seem to have made the most creative use of gunpowder. Metal-encased fragmentation bombs capable of impressive explosive effects were in practical use by the thirteenth century, and flamethrowers of various descriptions had been terrorizing enemies since the ninth century. Skills in producing bombs led to the development of hand grenades and even defensive mines, probably during the fourteenth century. Some mines could even be tripwire detonated by connecting the wire to a mechanism similar to that of a flintlock gun.

An early form of light field artillery was provided by Chinese mobile rocket launchers. These devices consisted of hundreds of slender rockets stacked in boxes, which were in turn mounted on a simple wheelbarrow structure that was easy for two men to wheel into a firing position. Each rocket was little thicker than an arrow, and the accuracy was negligible – in fact, *inaccuracy* was partly intended, as the hundreds of rockets were arranged to fire in many directions to cause maximum confusion and casualties. Individually the rockets posed a minimal threat to an enemy soldier, but when several batteries fired simultaneously the effect must have been harrowing, with injuries compounded by the psychological disruption of shrieking rockets and choking levels of smoke. Their main disadvantage was limited range; in fact, enemy archers would have a range advantage over rocket troops and the rocketeers would have to survive a deluge of arrows to deliver their barrages.

Cannon

Like the arquebus, it was cannon that helped to revolutionize infantry warfare. The primary application of gunpowder artillery was in the conduct of siege operations, where they were used to batter down the stone or wooden walls of the defenders. That role is studied in depth in the later chapter dedicated to siege warfare. Nonetheless, artillery was used in a battlefield anti-personnel role as well, although the frequency of this usage changes with the country and time period – Japan, for instance, used less anti-personnel artillery than China.

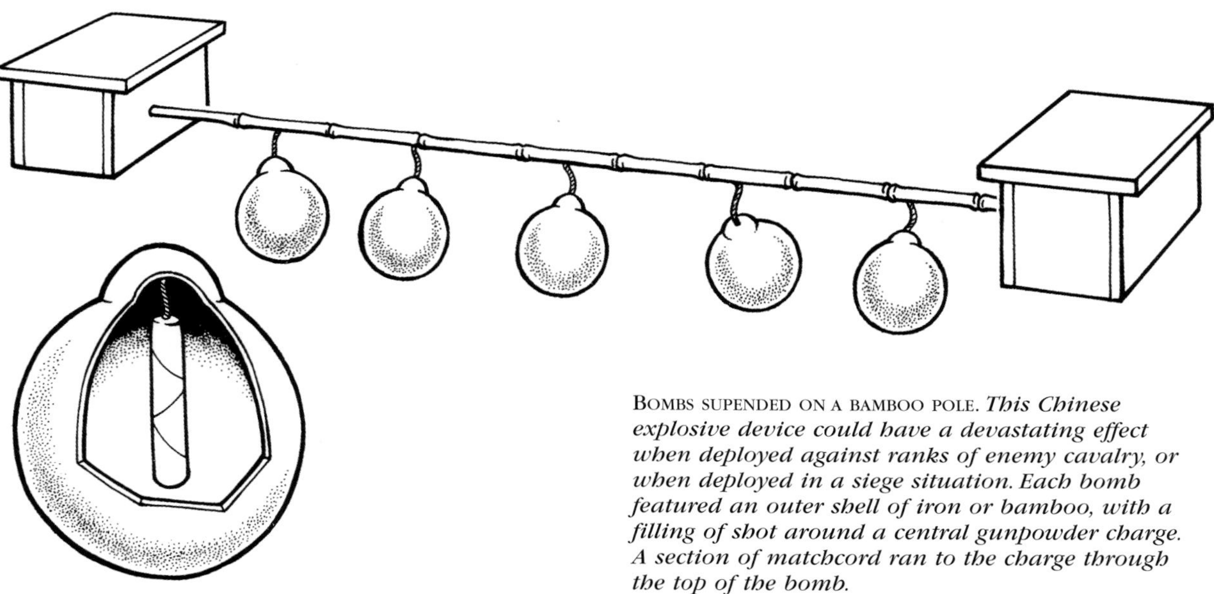

BOMBS SUPENDED ON A BAMBOO POLE. *This Chinese explosive device could have a devastating effect when deployed against ranks of enemy cavalry, or when deployed in a siege situation. Each bomb featured an outer shell of iron or bamboo, with a filling of shot around a central gunpowder charge. A section of matchcord ran to the charge through the top of the bomb.*

JAPANESE FIELD CANNON. *The cannon here depicted, probably from the fifteenth or sixteenth century, is little more than an enormous arquebus mounted on a simple two-wheel field carriage. The implications of recoil for the gunner here look a little worrying.*

One artwork from Qing Dynasty China, for example, shows two armies opposing one another across a narrow river. The river is preventing hand-to-hand combat, but an exchange of fire is being maintained by arquebusiers and also by a battery of artillery positioned centrally and frontally amidst one of the armies. The artillery itself is crude and small – simple cast iron barrels around 1m (3.2ft) in length, with elevation provided by blocks of wood positioned beneath the muzzles. Nor does the artillery seem to be applied as a consistent barrage – of the eight guns displayed, only two are firing, the other six being either reloaded, repositioned or inactive.

We must always be cautious about extracting general principles from contemporary artworks, as the artist himself may not have personally witnessed battlefield action. Yet we do know that there were distinct limitations to oriental battlefield artillery. The Chinese were producing cast-iron cannon and mortars by the mid-fourteenth century, but it was only during the sixteenth century that such artillery reached an adequate degree of sophistication, as a result of Portuguese importation of breechloading technology. The western world had more than overtaken China in artillery technology and professionalism, and it was often left to Portuguese mercenary gunners to give Chinese armies some degree of artillery competence.

Battlefield mobility could also be a problem, and many of the drawings of Chinese artillery show crude carriages that cannot have promoted easy combat use. Some cannon, such as the *hudun pao* ('crouching tiger') had just two frontal legs for support, the cannon being physically lifted into position. Others, however, were formed into field artillery pieces by being strapped to wheeled carts, although the crudeness of some of these casts many questions over their handling of recoil. Nevertheless, a decent level of mobility could be maintained by the artillery – during the Ninghsia campaign (1592) Chinese troops were observed to move 400 field artillery guns over 483km (300 miles) in several weeks.

Outside of naval use and siege warfare, oriental use of artillery in infantry warfare seems to be somewhat limited. Partly this is due to conservatism in military outlook, possibly stemming from the fear that gunpowder technologies could be socially disruptive by reducing the importance of the traditional warrior classes. Even in 1640s China, for example, several important court officials advocated restricting the use of artillery in warfare and even recommended that China stop producing its own guns.

MING CHINESE CANNON, SEVENTEENTH CENTURY. *Artillery became increasingly common amongst Chinese ranks from the sixteenth century onwards. This cannon was made of cast iron, with thick reinforcing rings around the barrel being necessary to prevent barrel ruptures during firing.*

The other factor was often a lack of professionalism. Artillery is a sophisticated arm to operate and to apply tactically, and in mass conscript armies training often fell woefully short of what was required. The results could at times be almost laughably horrific. In 1605, for example, a group of Chinese soldiers set to remedy the problem of a congealed gunpowder store by attempting to hack up the gunpowder with axes. The resulting explosion killed several hundred soldiers.

Specialist Infantry

We have already seen in that infantry in the oriental world were typically organized according to the basic division system of the army and their weapon specialism. Yet there were other forms of infantry, including some that we might describe today as special forces.

While not strictly specialist, one important category of 'infantry' did not fight at all. These were the masses of labouring troops who physically transported supplies and war materials to and from the battlefield. Labouring troops were particularly important in campaigns conducted a long way from the home base, or during siege operations that demanded regular supply lines to the besieging army. In Japanese armies, supplies such as rice, armour, arrows, cooking utensils and clothing would be carried either on horseback, or by soldiers using backpacks or special chests slung from poles and carried by two men.

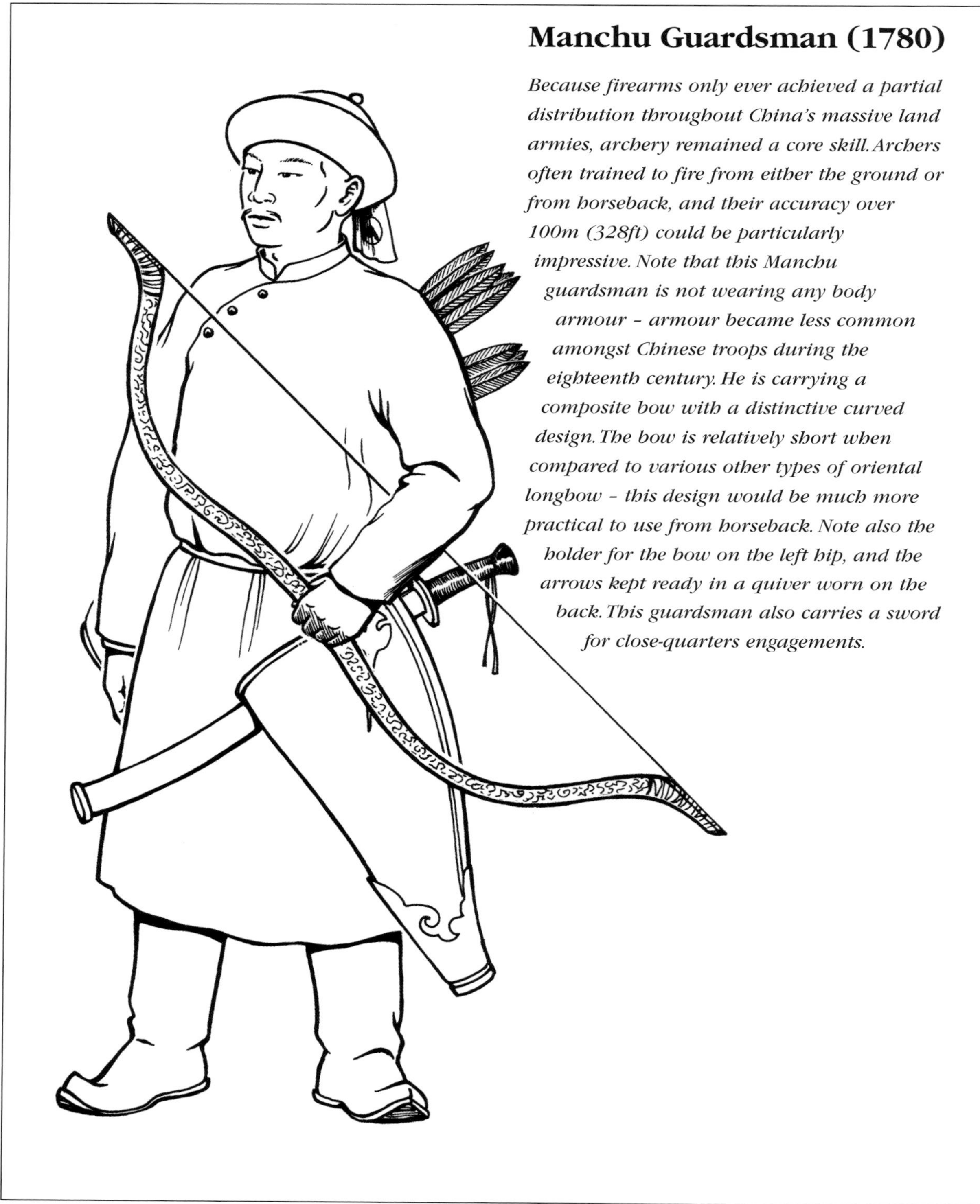

Manchu Guardsman (1780)

Because firearms only ever achieved a partial distribution throughout China's massive land armies, archery remained a core skill. Archers often trained to fire from either the ground or from horseback, and their accuracy over 100m (328ft) could be particularly impressive. Note that this Manchu guardsman is not wearing any body armour – armour became less common amongst Chinese troops during the eighteenth century. He is carrying a composite bow with a distinctive curved design. The bow is relatively short when compared to various other types of oriental longbow – this design would be much more practical to use from horseback. Note also the holder for the bow on the left hip, and the arrows kept ready in a quiver worn on the back. This guardsman also carries a sword for close-quarters engagements.

Non-combatants were not only found amongst the lower echelons of the *ashigaru*. Commanders and nobles of both sides would have their own retinues of troops who acted as everything from bodyguards to personal assistants. In Japan the origins of this form of infantry lie in the *genin* (warrior's attendant) of the medieval period. Although the *genin* would fight if necessary to save the life of his master, generally he served as an equipment carrier and a groom.

By the sixteenth century, the varieties of attendant serving a samurai's needs had expanded impressively. They included the *zori tori* (sandal bearer), a man whose main purpose in life was to carry spare footwear for the samurai, although given the short lifespan of most contemporary footwear, this was not necessarily a trivial job. The *zori tori* took his place amongst several general assistants, but he would not have enjoyed the prestige of another retinue member, the *mochiyari gumi* (spear bearer). This man not only carried the samurai's pole-arm, but also acted in a bodyguard capacity when necessary.

'When your weapons are dull and spirits wane, your strength exhausted and your wealth gone, neighbouring rulers will use the moment to turn against you.'

– Sun Tzu, The Art of War

On the battlefield, the standard of a samurai or *daimyô* was borne by an *uma jirushi*, the word denoting 'horse insignia'. Because the insignia represented not only a means for command and control but also indicated the nearby presence of a samurai, the lifespan of an *uma jirushi* could be a short one. Therefore, while not a combat role, it was still held in wide respect as requiring someone of exceptional courage and a willingness to head into the thick of the fighting.

Bodyguards and Mercenaries

Bodyguard units attached to a ruler were to be found throughout the oriental world, and emperors or commanders would have their own personal guards, some of which were on the scale of a small army in themselves. For example, in China between 1260 and 1352, the imperial guard force rose from 6500 to more than 100,000. Remember that this army was personally devoted to the emperor himself and was, therefore, separate from the millions pulled into general military service. On the battlefield, a sizeable bodyguard force could greatly increase the chances of a commander achieving survival or personal glory. The Chinese commander Liu Ting, for example, took a personal bodyguard of over 700 men onto the battlefield during the 1600s, and such men established a strong reputation for being vigorous fighters.

Mercenaries were a special category of troops, although the military skills they brought could be extremely mixed in terms of quality. (Japan rarely used mercenaries – with one important exception discussed below – while the Chinese used mercenaries extensively.)

On some occasions the mercenaries were little more than opportunistic individuals from conquered territories who contributed nothing much apart from an addition to the overall numbers. On other occasions they brought their own specialist skills. Mongol mercenaries, for example, were frequently skilled with bow and crossbow, although most Mongol troops demonstrated these skills best from horseback rather than on foot. The problem with mercenary soldiers was quite simply their suspect loyalty and their general lack of discipline, particularly in the aftermath of a victory or defeat.

The Chinese Yuan army experienced severe problems with the *gantaolu* profiteers who accompanied them into battle during the thirteenth century. Once the battle was over and victory was secured, these individuals then tended to wander off to exploit the civilian population for money and goods. In this way the Yuan lost any support or goodwill they might have gained and civilian populations were much more likely to resist them than if they had behaved with discipline in victory.

Silent Warriors

Another category of foot soldier worth mentioning, albeit one that strains against the category of 'infantry', is the *ninja* of Japan. Although armies since the beginnings of recorded time have used spies and assassins to achieve their goals, the *ninja* are one of the clearest examples of a special forces type of grouping in the oriental world. The *ninja* essentially sat outside the formal Japanese military order, as their multifarious roles - which included assassination, sabotage, intelligence-gathering and espionage - were often considered disreputable by conventional society. This did not mean that *ninja* were excluded from participation in conventional military campaigns. They would indeed often accompany an army into a siege or battle, but their role in it was likely to be collecting information on enemy movements or attempting to assassinate enemy commanders and key personnel. On the whole, regular *ashigaru* would probably have no awareness of the presence of the *ninja* as part of their operation.

The origins of *ninja* are uncertain and stretch back well into the medieval period but, by the fifteenth century, they had taken shape as a specific force that could be hired to perform

NINJA IN ACTION. *The ninja warrior to the right has snared the samurai's sword with a chain and sickle combination (*shinobi gama*), and is preparing to deliver the fatal blow.*

duties outside of the samurai's code of honour. Their rate of employment was accelerated as Japan descended into its long period of civil unrest, when assassinations and spying became central to the competition amongst Japan's *daimyô*. *Ninja* soldiers tended to be attached to certain families or villages, these having passed down martial skills for generations. *Daimyô* would support the development of their own *ninja* armies, and these would often train in the isolation of the *daimyô's* estate, never mixing with regular infantry soldiers.

Ninja training bore no comparison to that received by the *ashigaru*. While a typical *ashigaru* foot soldier might be taught nothing more than the thrusting of a spear, the *ninja* would be instructed in handling a huge variety of close-quarters weaponry, from the conventional *katana* sword through to weird and wonderful weapons such as *shuriken* (throwing stars), *hokode* (hand claws) and the *shinobi gama*, a combined flail and sickle weapon. They would also be trained in explosives and poisons, and in the arts of delivering an effective arson attack. Other specialist equipment used by the ninja included listening devices and implements for scaling or cutting through walls.

'They journeyed in disguise to other regions to spy on the enemy, and they would work their way into the midst of the enemy to spot weaknesses, and enter enemy castles…'

— The Buke Meimokusho

Accounts of *ninja* operations are often wreathed in myth-making and exaggeration, but there is no doubt that they played a part in several major Japanese battles. Even at the great battle of Sekigahara in 1600 we find references to *ninja*-type activities. *Ninja* of the Shimazu clan, fighting against the forces of Tokugawa Ieyasu, left behind sharpshooters lying on the battlefield, disguising themselves as dead bodies. When the main Shimazu troops were put into retreat, the gunners popped up at random to engage targets of opportunity, including the famous commander Ii Naomasa, who was seriously wounded by a gunshot and had to be carried from the battlefield. This example from Sekigahara shows how the *ninja* were not only consigned to covert operations in the shadows of castle walls, or sneaking into a bedroom to carry out an assassination. They did perform such missions, although the resulting improvements in security for powerful figures (including the creation of large bodyguard units) meant that many of these operations ended in failure or the death of the operative. Nevertheless, by the time the *ninja* fell out of employment in the seventeenth century, they had proved how useful elite military units could be as an adjunct to conventional forces, a lesson that is not lost today.

The Decline of an Elite

As we have seen throughout this chapter, as time progressed infantry became more and more central to the strategy and tactics of oriental warfare. The armies of countries such as China and Japan gradually accepted that an individual warrior's skill would eventually be trumped by the sheer momentum of infantry mass. Battles such as Nagashino radically challenged the notions of military elitism that carried centuries of martial tradition.

It is an undeniable fact, however, that European advances in infantry combat (in terms of weapons and tactical sophistication) would exceed those achieved over the same period in Asia. The reasons behind this fact are complex, and reach from economic conditions through to cultural resistance to change, the latter being particularly strong in many Asian societies, which value tradition over innovation.

Probably the greatest example of this reticence is the ambivalence towards the adoption of gunpowder weaponry on the mass scale of European armies of the same period. What is certainly true is that while the mounted warrior gained much of history's attention, Oriental history itself was arguably more shaped by the efforts of farmers, conscripted soldiers and professional infantry.

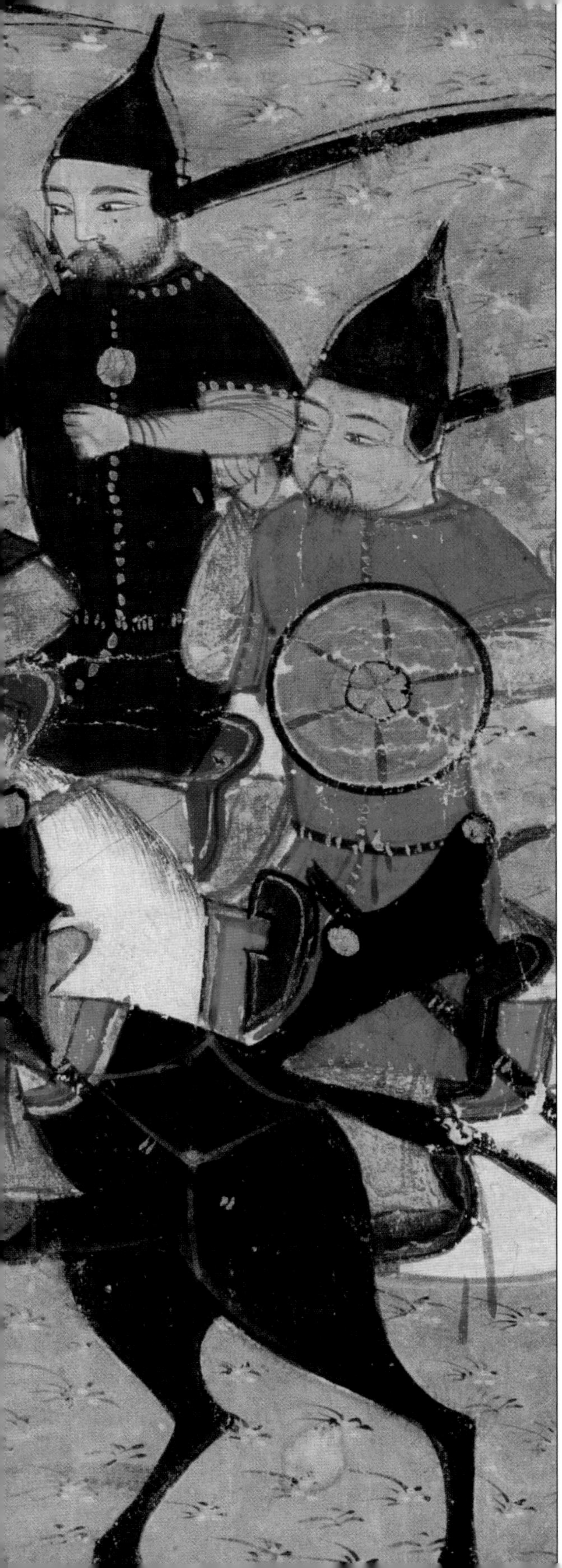

CHAPTER 2

MOUNTED WARFARE

Since the foundation of China over 3000 years ago, the Middle Kingdom has faced an almost constant threat from the north – from the nomadic cavalry of the steppe regions of Mongolia and Manchuria. There were only four ways for the Chinese (Han) nation to deal with these 'barbarians': buy them off, divide them among themselves, shut them out or fight them on their own ground. The Song tried the first, the Jin the second with the Mongols, the Ming tried the second and only the Qing (Manchu) finally conquered the steppes during the eighteenth century and put an end to the threat once and for all.

The Qin and Han dynasties in earlier times had built their versions of the Great Wall but these had failed to put a stop to nomadic raids by the Turks and Xiongu peoples. The

TERROR OF THE STEPPES: *the Mongols attack at close quarters a Muslim enemy. A vivid illustration from a Persian chronicle of the history of the Mongols. The swords shown were probably a bit shorter but the helmets, shields and overcoats the Mongols wore have been portrayed with uncanny accuracy.*

Northern Song (960–1126) trusted to bribes, diplomacy and sheer numerical preponderance with their army of over 1.3 million (mainly infantry) in 1040. None of these methods proved very successful.

The first threat came from the northwest when the warlike Tanguts conquered the strategic bend of the River Yellow (Huang He) known as the Ordos region. This threatened to undermine the western flank of the Song empire and placed the finest horse-breeding region of China in hostile hands. Both were disastrous for the Song empire. The Tanguts relied upon cavalry to invade Song territory and their Royal Guards included 3000 heavy cavalry used as a strike force in battle.

In 1004 the Mongol Qidans invaded China as far south as the River Yellow, taking the complacent and unwarlike Song completely unawares. The Mongol army, however, being composed entirely of cavalry, could not make any impression upon Song fortified cities, including the capital of Shanxi, Ying-chou. Fortunately for the Song the Qidans eventually settled for tribute.

They ended up controlling the increasingly sinicised kingdoms of Liao, Manchuria, Mongolia and Korea. The Koreans were subdued using 400,000 horsemen, while the Tartars who threatened the grazing lands of the Qidans were crushed in a series of cavalry expeditions from 1092 to 1102. For all nomadic empires based on cavalry power, it was vital to keep control over the grazing lands of the north.

The Qidan cavalry army was a model for all subsequent ones. The core of the professional, hard-hitting Qidan army was regular heavy cavalry known as the *Ordo*. This consisted of armoured horsemen armed with a lance, bow, sword or mace mounted atop a steppe pony or horse also covered in armour. Each *Ordo* heavy cavalryman had at least one or more orderlies or pages who were themselves mounted and armed with bows, lances or halberds. These troops functioned as skirmishers, together with Mongol and Manchurian (Jurchen) auxiliaries, in front of the advancing cavalry army.

The Liao army was divided into units based on the simple, effective and logical decimal system that all nomadic cavalry armies used for their command structure. The army was divided into light, medium and heavy cavalry. The light cavalry, operating as mounted archers, would soften up the enemy with a shower of arrows, javelins and other missiles. The medium cavalry would harass and tire the opposition with charges, feints and sudden moves, including burning the grass behind or in front of the enemy. This simple tactic would be used, together with the feigned retreat, by all Mongol and Manchurian cavalry armies thereafter, with devastating effect. The Qidans, like other Mongols, would use a human screen when attacking a fort, city or entrenched enemy position. Finally, once the other cavalry had left the

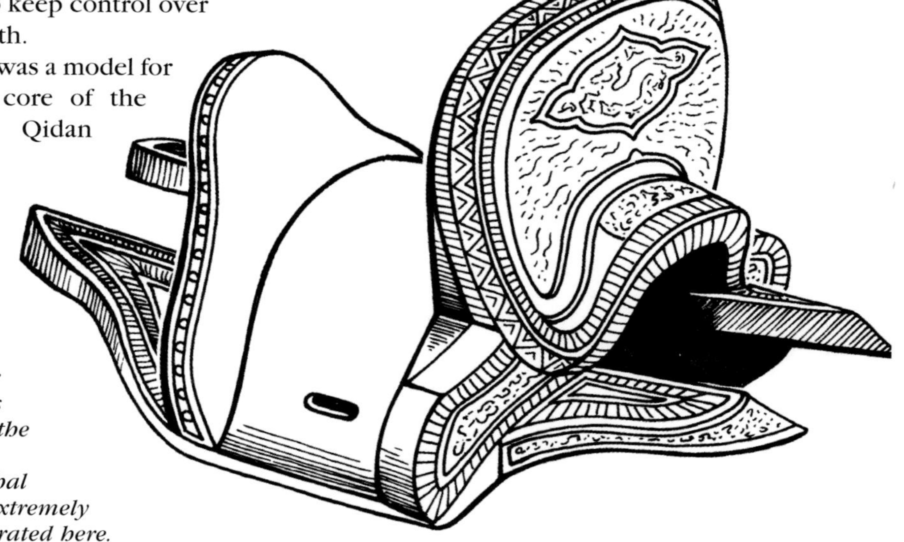

THE MONGOL SADDLE. *A western saddle had equally high pommels at each end of the seat of the saddle as its purpose was to allow the rider to fight at close quarters with sword, mace or lance. The Mongol horseman's saddle was primarily to give the mounted archer support and balance while he fired his lethal volleys of arrows. Hence the extremely high back of the seat as illustrated here.*

enemy tired and confused, the war drums, horns and flapping flags would signal the advance of the *Ordo* regiments (each some 500–700 men) into the attack. The outcome was usually a routed enemy who was pursued with relentless energy and ruthlessness.

The Manchurian Threat

Manchuria, a million square kilometres of green lush forest, mountains and grass steppe lands, was the homeland of the formidable cavalry nation of the Jurchen who created the Jin (1115–1234) and Qing (1644–1912) empires of China.

In 1125, the Jurchen warlord Ukimai toppled the Liao kingdom and proceeded, a year later, to conquer northern China. In 1127, Ukimai seized the Northern Song capital of Kaifeng, almost took Nanjing and made an unsuccessful bid, in 1130, to cross the River Yangtze. The Jurchen's lack of ships and infantry support, minimal experience in siege warfare and their nomadic dislike of the rice fields of southern China saved the now Southern Song from a catastrophe.

Ukimai became the first Golden or Jin Emperor Taizong (1075–1135) and settled for the whole of China north of the River Huai, south Mongolia, Korea and Manchuria. The latter supplied his needs for horses and riders.

When the Jin army entered Song China, it consisted, as did the earlier Qidan, entirely of cavalry. Rough steppe warriors dressed in quilted jackets, skins and leather boots were armed with lances, maces and bows, and rode small Manchurian steeds. The basic unit, a *mou-kou* regiment, was made up of 50 warriors comprising 30 light horsemen and 20 heavy ones. The core of the Jin army was the 10,000 men of the Imperial Guards, of whom a mere 200 formed the Emperor Khan's bodyguard.

'A speedy victory is the main object in war. If this is long in coming, weapons are blunted and morale depressed.... When the army engages in protracted campaigns, the resources of the state will fall short.'

— *Sun Tzu*, The Art of War

The Jin freely borrowed tactics from their former Qidan overlords. A Jin cavalry division (*wan-ku*) would form into five ranks of mounted troops known as a horse team (*kuai-tzu ma*). The first two ranks of heavy cavalry formed the front facing the enemy while the following three were made up of mounted archers who would attack, Qidan-style, in relays, leaving the heavy cavalry – in a tight formation – to crush the enemy through sheer brute force. Movement and orders were directed using drums and flag signals.

That was the theory. In reality as soon as the Jin moved the centre of their vast empire into the densely populated agricultural lands of northern China, the quality of their cavalry, both horses and men, deteriorated quickly. There was a shortage of good, tough riders and an even greater shortage of good horses – those that were to be had came from Manchuria or the unreliable Xia. As for riders these were more or less unreliable and reluctant Uighurs, Mongols, Qidans, Xia and nomadic ('wild') Jurchen.

In 1161, the Jin army, increasingly dependent upon Chinese infantry, showed its mettle in battle. The result was mixed. Against the well-entrenched Song the 400,000 Jin infantry and cavalry crossed the River Huai but made little progress, although the cavalry managed to outflank the Song army which had almost no cavalry at all. The campaign ended with the brutal and unpopular Jin Emperor, Hailingwang (1122–61), being assassinated on 15 December.

Meanwhile, on the steppes Jin cavalry, supported by their Tartar allies, struck deep into Mongolian territory. The allied cavalry army came upon the unprepared Mongols at Buyur Nor where they annihilated the enemy. Before the Mongols had realised what was happening the

Tartar archers showered them with arrows and light javelins as a prelude to the Jin heavy cavalry crushing them completely. For a generation, the Jin were to have peace on the steppe.

Unfortunately, the Mongols were quick to recover. Six years after Buyur Nor, a boy named Temüchin was born into the Borjigit clan. By the early 1190s, Temüchin, now a warrior, united the Mongols, defeated the Tartars (1196) and had by 1206 brought all the greater Mongol tribes – including the Keraits and Tartars – under his rule. Elected Grand Khan (Khagan) of Mongolia, he took the name Genghis Khan or Universal Ruler.

The Mongol Army

Genghis's imperial plans seemed to have been the wild dreams of a megalomaniac. After all he had less than three million Mongol warriors to conquer a huge, sophisticated empire of 53 million, with an army of at least half a million troops backed by a paraphernalia of modern weaponry, including the first primitive handguns.

JUSTIFIABLY PROUD of their ancestors' amazing prowess in the saddle and in battle against vastly superior enemies, present-day Mongols in the Republic of Mongolia stage impressive and, on occasion, frighteningly realistic battle enactments.

Furthermore, Jin Emperor Shìzông (1161–89) had built 5000km (3110 miles) of wooden forts, palisades and ditches well inside Mongol territory and a secondary line of similar length to protect Manchuria. How were a mere 95,000 Mongol cavalrymen to overcome such stupendous defences?

One advantage they had was Genghis's own abilities as a field commander, organizer and disciplinarian, backed by such brilliant Mongol commanders as Subutai (1176–1248) and Mukhali (1170–1223). Genghis would prevail over his enemies because he had to keep winning in order to keep his tribal federation together. Expand, conquer and prevail, or die, was the motto of the Mongol empire.

Then there were the Mongols themselves. Every man from 16 to 60 was available for military service. Inured to cold, paltry food, and riding in all weathers and conditions, the Mongol was a born warrior whose combat skills were tested twice yearly by the Khan's grand hunts (*nerge*). Here animals were hunted in a wide arch, herded together in an even smaller space and then finally trapped in a valley or ravine. There the Khan shot the first arrow as a signal for the slaughter to begin. The *nerge* would serve as a model for all future campaigns.

The Mongol soldier could keep moving on a horse further and longer than any other. He subsisted on a meagre nomadic diet of millet meal, dried milk curd, cured meats and *koumiss* (fermented mare's milk). In dire emergency he would open a vein in his horse's neck and drink its blood. Since childhood, the Mongol had been trained to ride a horse, shoot with a bow and arrow, and most difficult, combine the two.

The nomadic composite bow was the only 'secret' weapon the Mongols possessed. It was made of supple wood with the stress points reinforced by fish glue and the angled 'ears' reinforced with bone for greater effectiveness. The Mongol bow had twice the range and pull (45–68kg/100–150lb) of the famous English longbow, with a flatter trajectory and more regular release of tension. At close range, and in the hands of an experienced horse archer, it was a devastating weapon.

The Mobile Archer

The Mongol horse archer had several functions. He operated as scout, skirmisher, harasser and finally as mobile artillery. Each man carried two or three

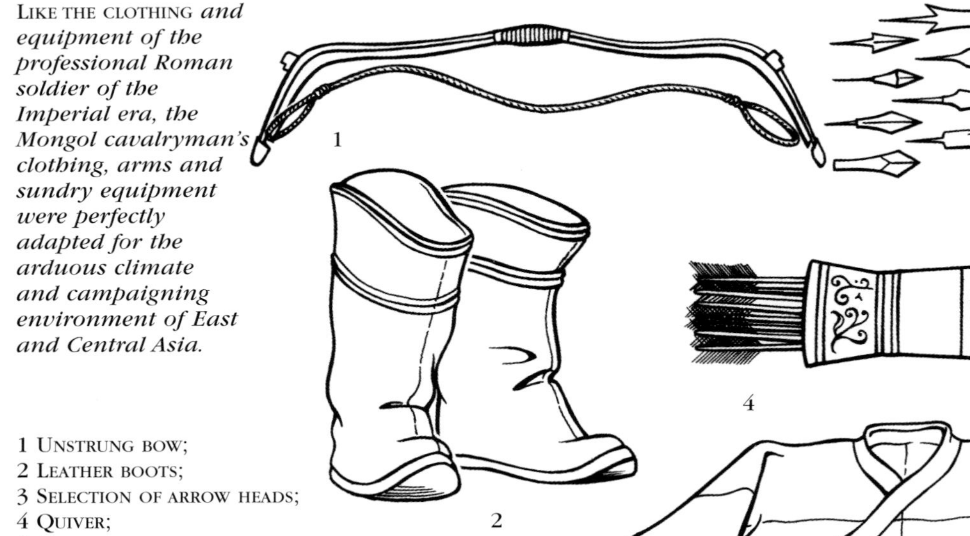

LIKE THE CLOTHING *and equipment of the professional Roman soldier of the Imperial era, the Mongol cavalryman's clothing, arms and sundry equipment were perfectly adapted for the arduous climate and campaigning environment of East and Central Asia.*

1 UNSTRUNG BOW;
2 LEATHER BOOTS;
3 SELECTION OF ARROW HEADS;
4 QUIVER;
5 COAT.

bows for both long- and short-range fire. He was also equipped with three quivers, each with 30 arrows so that he would not run out of projectiles during battle. On the march both he and his heavy cavalry colleagues would have two or three horses so that they always had a rested mount available.

The arrows in the quiver also held one of the keys to the Mongols' success. All had iron tips that had been heated until glowing red hot and then plunged into salt water for hardening. Arrows prepared like this could even pierce armour. At close range, the archer would use arrows with large, broad heads while at a distance short arrows with small pointed heads were most effective. Another improvement was that the Mongol used a metal thumb ring to give him greater pull and tension in the bowstring.

The mounted archer was never supposed to get entangled with the enemy at close quarters but instead keep him pinned down and harassed by a constant barrages of arrows. Close-quarter combat was to be left to the heavy cavalry who wore lamellar armour and were armed with lances. Their armour was made of light metal or hardened leather that was covered with a black pitch 'lacquer'. Their horse too was covered in similar lamellar armour. Each rider had a small metal helmet and was equipped with a lasso, a small, razor-sharp Turkish dagger or sword and a long, steel-tipped lance with a grappling hook to unseat his opponent. He would also use a mace to hit his enemy over the head in close combat.

Their mounts were even more impressive, except in size and appearance, than their riders. Small, only 13–14 hands high, like a modern pony, the steppe or Mongol horse had amazing stamina and range and would subsist on the most meagre diet. With such mounts the Mongol army was able to cover up to 210km (130 miles) in two days and still fight and win at the end of the gruelling march.

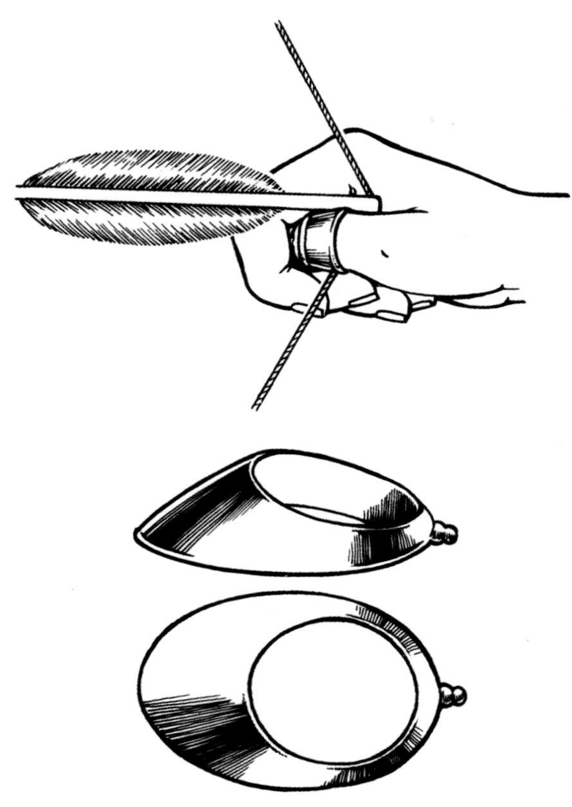

EUROPEAN ARCHERS *would use their fingers to draw the string of their longbows back. The Mongols, as usual, had come up with an originally brilliant technique that greatly improved upon the range, power and penetration power of their composite bow: the thumb grip. The thumb being the hand's strongest digit had greater pull than fingers and this was massively enhanced by the use of a thumb ring. These came in a variety of shapes and materials, ranging from hardened leather to metal.*

The Mongol cavalry army was centred around the Khan's guard, the *Keshik,* that numbered 1000 men in 1206 and had grown to 23,000 men – in line with the general expansion of the army from 95,000 to 135,000 – at the time of Genghis's death in 1227. The Guards' officers were all potential *tumen* and army commanders while the Guards were often responsible for turning the tide of battle. Discipline in the *Keshik* was draconian, with desertion, disobedience or sleeping on duty punishable with immediate execution.

The rest of the army was organized along the decimal system with a unit of ten (*arban*) being the soldier's 'home' in the army. Ten *arban* made a *jegun* (100 men) and 10 *jeguns* made a *minghan* – the basic battle unit. Ten *minghans* made a division or *tumen* and two or three *tumens* made a Mongol army corps.

The Mongols would make ready for battle in ranks of five *minghans*: two ranks of heavy cavalry up front with three units of light cavalry behind. The archers were sent through gaps in the heavy cavalry lines and poured withering fire into the enemy. At the same time the archers would also make a sweep (*tulughma*) on both flanks of the enemy. Controlled by white and black pennants these movements were carried out in complete, eerie silence. Then the massive *naccara* (war drums that were so heavy they had to be carried by camels) were beaten as a signal for attack.

Suddenly the Mongols, baying and screaming like wolves, would attack the confused, bruised and weakened enemy, showering them with arrows and javelins. The enemy would be caught in a trap, as in the *nerge,* although the Mongols would leave an opening at the back – providing a dangerous temptation to flee. This 'escape route' was in fact another deadly Mongol tactical trap. Giving the enemy the chance to flee would enable the Mongols to avoid a pitched battle and allow them to pursue and slaughter the broken enemy for days or even weeks across hundreds of kilometres.

Battle of the Passes, 1211–13

Few armies, before or after, could match the Mongols when it came to tactics, organization or strategy. Given their meagre manpower resources the Mongols could not afford costly frontal assault and would normally attack indirectly or outflank with deadly effect. They would envelop, harass, weaken and finally destroy the enemy.

Genghis decided that he could not attack the Jin without dealing with the Xia first. Not only would this open up a flank route into the soft sides of China but it would deprive the Jin cavalry of a valuable source of mounts.

Battle of Nankuo Pass

1213

In late 1213 Genghis Khan had tried to smash the formidable Jin frontier defences with mixed success. The shortest but most perilous route to Jin Yenching (Beijing) lay via the Nankuo Pass to the northwest of the capital between the fortified village of Huailai and Nankuo village. This village lay at the edge of the 'Plateau of Dragon and Tiger' and a mere 30km (20 miles) from the walls of the capital. But to reach this point the Mongols would have to pass through the narrow, deep, gorge-like pass – a distance of 22km (14 miles) from one end to the other with Jin forts, towers, palisades and troops atop steep hills on either side. Genghis ordered Tolui to take this strategically vital outpost. General Jebe took his cavalry down the Nankuo Pass, making it appear to the enemy that his was only a reconnaissance operation. He advanced half way down pass before he ordered his men to retreat. The Jin took the bait, abandoned their fortified posts, gave chase and only realised when the Mongols halted their 'retreat' that they had walked into a deadly trap. The Jin infantry were slaughtered, their positions were occupied and Jebe marched on Beijing.

MONGOLIA
NANKUO PASS
Jin Yenching
JIN EMPIRE
SONG EMPIRE

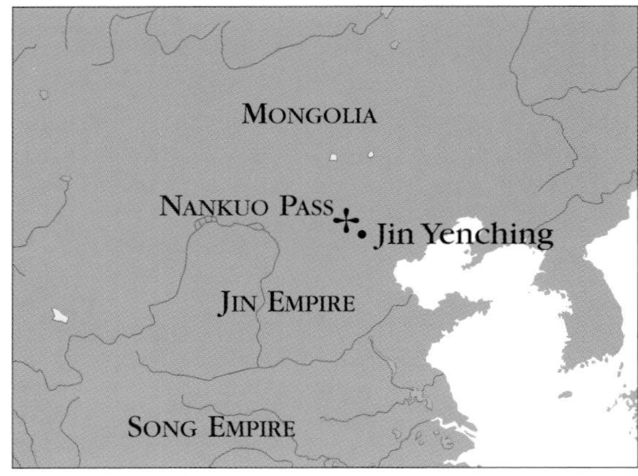

Less than 30km (20 miles) from Beijing (Jin Yenching), Nankuo Pass protected the shortest and fastest road from the north to the Chinese capital.

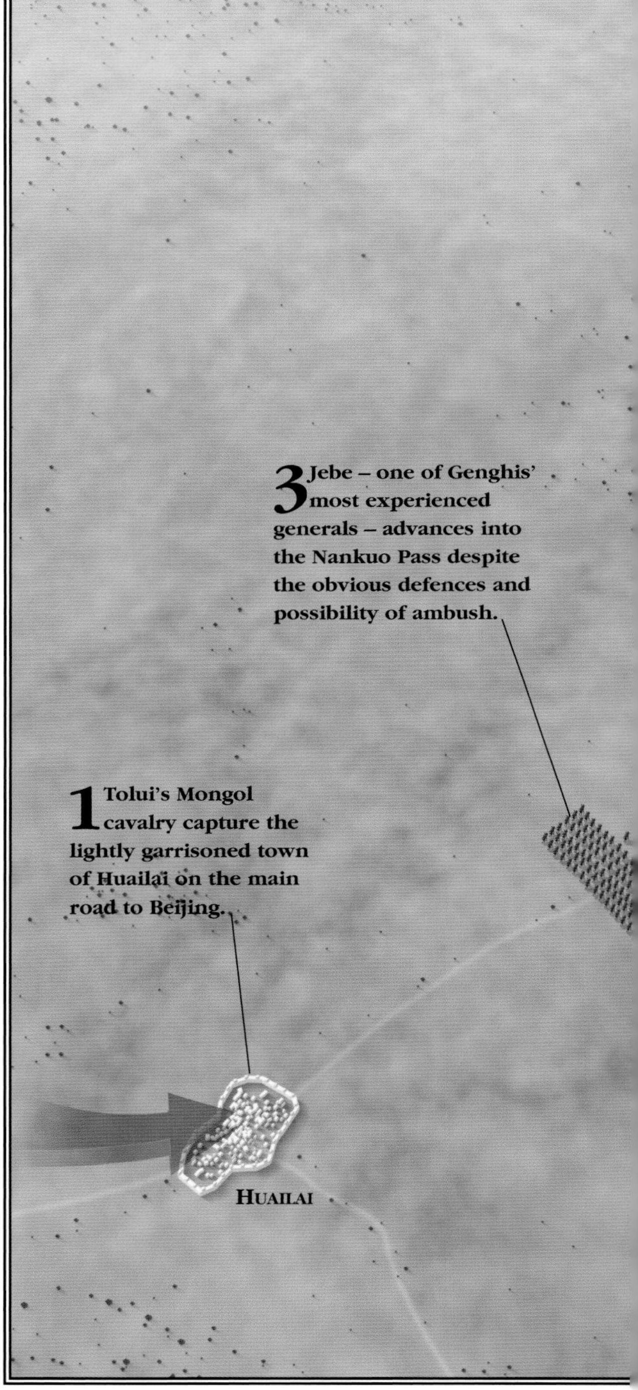
3 Jebe – one of Genghis' most experienced generals – advances into the Nankuo Pass despite the obvious defences and possibility of ambush.
1 Tolui's Mongol cavalry capture the lightly garrisoned town of Huailai on the main road to Beijing.
HUAILAI

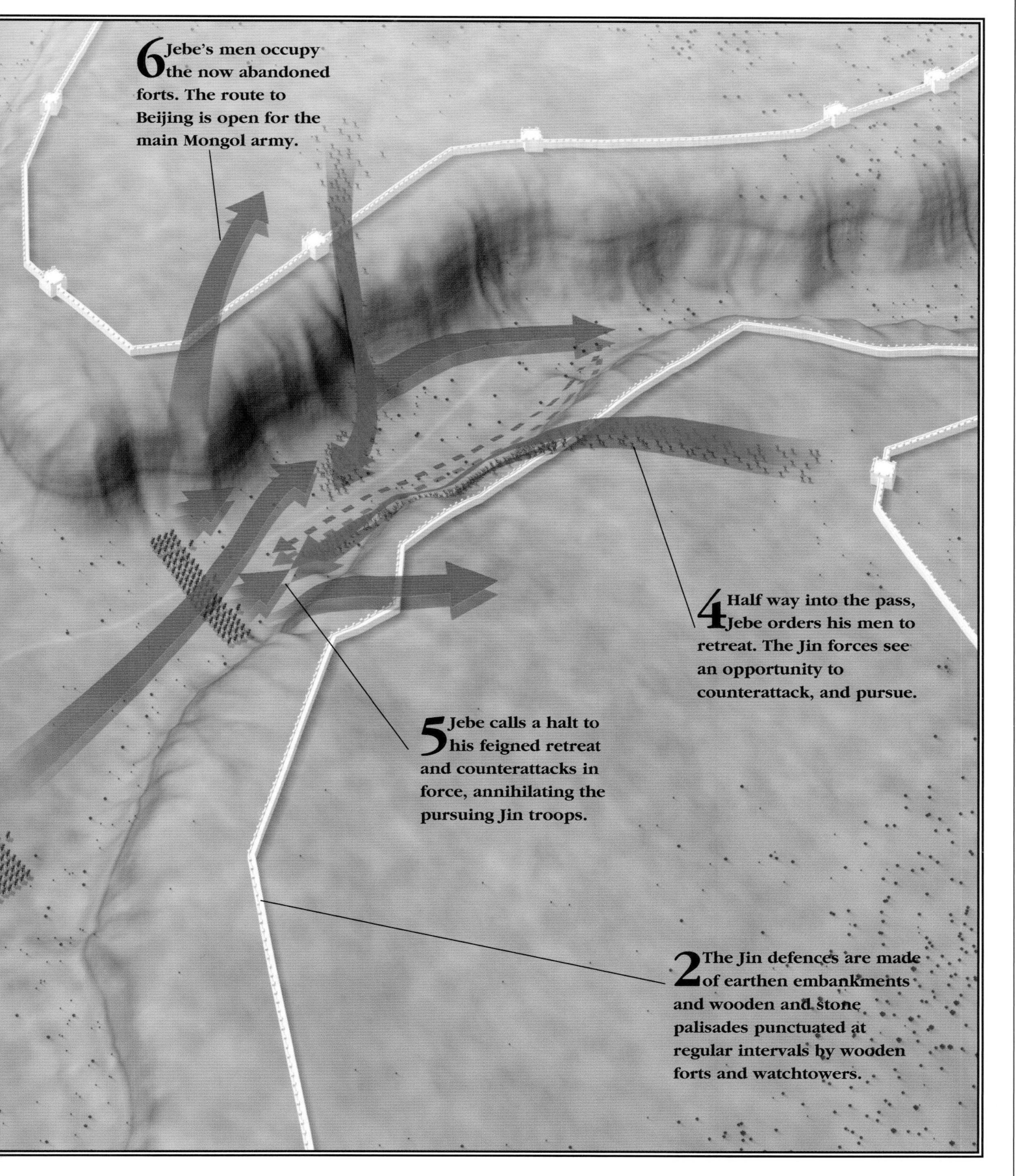
6 Jebe's men occupy the now abandoned forts. The route to Beijing is open for the main Mongol army.
4 Half way into the pass, Jebe orders his men to retreat. The Jin forces see an opportunity to counterattack, and pursue.
5 Jebe calls a halt to his feigned retreat and counterattacks in force, annihilating the pursuing Jin troops.
2 The Jin defences are made of earthen embankments and wooden and stone palisades punctuated at regular intervals by wooden forts and watchtowers.

Mongol Horse Archer (1250)

The swiftly deadly horse archer was the Mongol war machine's most potent weapon and proved devastatingly effective against less mobile or ably led armies. Aiming his composite bow, a skilled rider would fire a series of volleys while galloping - no mean feat. The price the archer paid was that he was lightly protected and vulnerable should he get entangled with the enemy at close quarters. In that case, his head was protected by a metal helmet, he could hide behind his round shield and use his razor-sharp, curved Turkish-style scimitar. See how the archer uses the Mongol saddle (see illustration p.72) to support his back and steady his aim while he fires off an arrow. His mount is protected by hardened leather 'armour' reinforced by metal thread - the familiar lamellar armour.

First the Uighurs and the Ongud Mongols submitted in 1207–1208. Then in 1209, using hitherto little known passes, the Mongols invaded Xi-Xia with devastating effect. Underestimating the Mongols as mere nomad 'barbarians' the Xia had left 70,000 troops to defend the key fortress city of K'ei-men. Seeing the enemy 'retreat' the Xia left their fortifications and pursued them. Only too late did the Tanguts realise they had fallen into a deadly trap – not a single Xia soldier survived and the kingdom surrendered to Genghis.

Thus a dangerous flanking threat to the Mongols had been eliminated while a similarly dangerous western flank for the Jin had been opened up. But the Jin army was vast: 125,000 cavalry and 375,000 infantry. Compared to the Mongols, however, the cavalry were not well mounted, trained or disciplined, and since they were made up of non-Jurchen horsemen their loyalty was doubtful. As for the infantry these were Chinese mercenaries or conscripts who were poorly paid, officered and led. Advised that he should attack the Mongols with his cavalry army as they emerged from the Gobi in a tired state, Jin general Chih Chung chose instead to march out with a combined force. This meant the element of surprise was lost and the advance progressed at a snail's pace so the infantry could keep up. The key to Jin security was now to hold the passes north of Beijing.

Meanwhile, Genghis had taken command of the Mongol centre, his best general Mukhali on the left and Genghis' sons (Ögedei, Chagatai and Jebei) on the right. Border villages fell with alarming speed as the Mongols began their advance in March 1211. Hoping to create an 'iron curtain' to block the Mongol advance down the passes in northern Hebei and Shanxi the Jin high command sent two more armies north. They also urged their Jurchen kinsmen in Manchuria to attack the Mongols in the rear.

However, Genghis was, as usual, one step ahead of the enemy. To forestall this potentially disastrous Jin pincer move Genghis ordered Jebei (d.1225), to invade the sinicised southern Manchurian province of Liaoning (Fengtian) with a detached *tumen* through the Shan hai (Mountain to the Sea Pass). During the winter of 1211–12 Jebei's riders crossed the frozen River Liao and laid 'siege' to the Jin capital of Manchuria: Liaoyang. Then Jebei appeared to abandon the camp, causing the Jin garrison to come out for plunder. During the night Jebei's cavalry came back, stormed through the poorly guarded gate and seized Liaoyang before the garrison had time to realize they had been tricked.

Meanwhile the combined Jin armies had been routed at the Huan pass when Genghis attacked their left flank with overwhelming force and ferocity. A rain of arrows fell on the Jin cavalry followed by an attack from Mongol heavy cavalry. The shocked Jin cavalry were unable to control their horses, fell back in confusion and crushed their own infantry. Soon the Jin armies were in total confusion and falling back. The Mongols set off in hot pursuit all the way to the massive walls of Beijing.

Western Campaign

In the west, the Mongol princes' army had invaded Shanxi to cut off the western Jin armies from Beijing. The Mongols could not take Beijing with its moats and triple set of stone walls but they ravaged the countryside with unparalleled brutality. Having demoralized and worsted the Jin army the Mongols eventually withdrew during the winter of 1211–12.

They returned with a vengeance in the spring of 1213. This time, the Jin were better prepared and had recovered with surprising rapidity. They rebuilt and reinforced their 6000km (3730 miles) of frontier defences made up of forts, barracks, casemates, ditches, moats and deep lines of caltrops and wooden stakes.

The Mongols could not afford a frontal assault, lacked the equipment to force their way through and were hard pushed to find a chink in the Jin armour. But at the village of Huailai, straddling the road to Beijing, one was found. Here was a narrow 22km (14 miles) defile that ended up in a plateau a mere 30km (48 miles) from Beijing's walls.

Genghis sent Jebei down the defile as a decoy to lure the Jin troops out of their forts on the hills overlooking the valley. The Jin abandoned their posts, pursued the Mongols but then the Mongols wheeled their horses around and attacked them.

The Jin were slaughtered where they stood. At the same time Datong had been captured through another ruse. Beijing was now isolated. While Genghis threw a secure cordon around the Jin capital he sent the rest of his army to ravage all lands north of the River Yellow. By April 1214 the Jin had suffered enough and paid off Genghis to quit China. In June Jin Emperor Wei retreated with his court to Kaifeng - well to the south and behind the protective flow of the River Yellow.

The Mongols continued their conquest of Manchuria to prevent a flanking attack from the east and by January 1215 much of the southern part of the Jin homeland was in Mongol hands. A year later most of Manchuria was either directly or indirectly under Mongol control - thus depriving the Jin cavalry of good mounts and reliable horsemen.

In January 1215 the Mongols surprised the Jin with a mid-winter offensive that left Beijing isolated and short of food. The million strong city was soon in the grip of famine and the 43km (27 miles) of vast, thick stone walls were of little use to the defenders. Without a proper field battle or siege having been fought the proud Jin capital of Beijing surrendered meekly in May 1215. The Mongols spent the next two months burning and looting the city while they competed with each other in massacring and raping the populace. Ambassadors from the Central Asian empire of the Khwarezm sent their ruler in Samarkand horrendous reports of the Mongols appalling atrocities. Yet neither defeat nor terror had subdued the Jin empire.

THE MONGOLS' HEAVY CAVALRY *were the shock troops of their army. Their armour was made of metal as were their helmets but the body armour was designed to give maximum protection while allowing the cavalryman the greatest possible flexibility. In this example the horseman is equipped with a composite bow and long sword.*

Crushing the Jin Empire

The Mongol campaign could only be seen, at best, as a partial success since the Jin empire was to survive until 1234. Genghis now turned his back on China, where advances against tough Jin resistance was painfully slow by Mongol standards. As did Napoleon in Spain (1808–14) Genghis left this gruelling and inglorious war of attrition to others while he looked elsewhere, in this case to the conquering of Central Asia and the Khwarezm empire and attacks into Russia (1218–1224). When he did return to fight in the east, before dying on campaign in 1227, it was to crush his personal *bête noire* the Xia and not the stubborn Jin.

The thankless task of fighting the Jin was left to Mukhali with 23,000 Mongols and 77,000 Qidan, Jin and other mercenary troops. Mukhali captured Tai-yuan and had by the time he died in March 1223, taken control of all China north of the River Yellow.

It was up to Genghis' successor, Ögedei (c.1186–1241) to destroy what remained of the Jin empire – Honan and Shenxi provinces. As was their wont the Mongols launched a two-pronged offensive against the rump Jin state. While Ögedei marched south his brother Tolui attacked on the western flank. United during the winter of 1232, their combined cavalry armies crossed the River Yellow on 28 January 1232 and quickly surrounded Kaifeng – the southern capital of the Jin empire. The city surrendered on 29 May 1233 and the last Jin Emperor, Aîzông, committed suicide in early 1234.

*'Suddenly this morning the sky darkened;
wind and rain manifested evil;
catapults and thunder flashed;
arrows descended.'*

– Poem by Song scholar Wen Tianxiang (1275)

Kublai Khan's Yuan Cavalry Army

A new Grand Khan, Mongke, was finally elected in 1251, who had the determination to crush the Mongols' final Chinese enemy, the Song empire in the south. Again there would be no direct assault but an enveloping offensive to the southwest to cut off the Song from trade and contact with India and Burma. This meant invading and conquering the kingdom of Nanzhao in the mountainous, forested and wild region of the southwest where the climate was tropical and detrimental to the health of both Mongol men and mounts.

Mongke sent his younger brother, the Viceroy of North China, Prince Kublai (1215–94) to conquer the Nanzhao with the aid of the brilliant field commander general, Bayan – an Uighur in Mongol service. Invading Nanzhao with three separate columns in October 1253, the 36-year-old Kublai had made meticulous preparations to avoid famine and disease destroying his army while it advanced. A month later the Nanzhao army, led by King Tuan, blocked the River Yangtze but Bayan led a daring night attack across the river. In December the Nanzhao surrendered and Tuan remained on his throne as a Mongol vassal.

In 1259 Mongke set out to conquer the rich Song province of Sichuan but died, as did many of his men, from the hot, unhealthy climate there. In May 1260 Kublai replaced him, proclaiming himself the first Yuan emperor of China and determined to gain legitimacy in Chinese eyes by conquering, at long last, the Song empire.

But first Kublai would have to reform the Imperial Yuan army. Kublai was a good, if cautious, field commander but a first-rate genius at organization. In 1263 he set up the Privy Council to create an entire new Chinese infantry army, a corps of artillery, and, even more impressively, a Yuan navy.

The Imperial army's core, however, remained its cavalry. Both the *Meng-ku* (main Mongol army) and the frontier/provincial forces (*tammachi*) were made up of cavalry troops. A special Imperial department (Court of the Imperial Stud) organized the supply of good mounts by managing the herds, stables, pastures and the feed of the animals. The Court of the Imperial Tack supplied the cavalry with all its equipment such as saddles, bridles, harnesses and armour from their facilities in Shengdu and Beijing. Despite the best efforts of the Imperial Stud to procure horses and riders both remained in short supply, especially good horses. As a result the regime imposed a levy of every hundredth horse held in Chinese hands but these animals were sold at State imposed prices: a measure that was deeply unpopular with the Chinese and undermined support for an alien regime. The Yuan authorities on their part never trusted the Chinese and no Chinese held a supreme military or civilian post in the administration. For example, the new Imperial Guard that Kublai set up included 30,000 Christian Alans but not a single Chinese.

Nevertheless, the logistical power of this Imperial army was formidable. It maintained a postal service across the empire with 20,000 horses and riders, with hundreds of post stations scattered along the main roads. The 1770km (1100 miles) Grand Canal was also kept open thanks to the Imperial army and in 1260 the Quartermaster Office was capable of supplying the army with 10,000 horses, 6,033,000kg (13,300,000lb) of rice and 10,000 outfits of boots, hats and trousers.

That it was able to do so was due to a superlative Sino-Mongol administration, a fusion of Mongol energy and Chinese administrative experience, and the creation of the Armaments Court (*Wu-pei ssu*). This was responsible for supplying the army, including the cavalry, with arms, armour and essential supplies from the arsenals. Furthermore the Yuan army was, thanks to the old Chinese tradition of military colony households, *(aurughs* in Mongol) not such a burden as one might have expected.

The organization was decentralized such that the Khan only kept personal command of the old *Keshik tumens.* The Privy Council had control over the 35 units in and around Beijing itself and the *tumens* stationed in northern China. The rest of the Imperial army was under the control of the local provincial branches of the Privy Council.

Overall Kublai's Imperial army was better organized, supplied, disciplined, trained and equipped than that of his grandfather Genghis. The human cavalry material was better than the wild, ferocious Mongol and Nomad steppe warriors that the first Khan had led into war. Now the awesome power of the modern siege train, a large river flotilla and the huge infantry armies of old China had been combined with the best Mongol cavalry army ever. But this army needed to be good as it faced a range of enemies all around it.

Crushing the Song Empire

Ariq Boke (d.1266) led the traditionalist Mongols resistance to the Yuan regime, forcing Kublai to turn his back on the gains made against the Song and devote his military efforts (1260–64) to an extended cavalry campaign against his own truculent countrymen. Kublai sent 45,000 cavalrymen with 10,000 extra horses into the steppe. Ariq Boke only surrendered in early 1264 and died, probably poisoned, two years later.

This dangerous diversion allowed the Song to take the offensive in the south (1260–62) but in early 1265 Kublai captured 146 Song river ships which were added to the ever-expanding Yuan navy. Now the Mongols could take the offensive by marching a huge combined army on the twin fortified cities of Hanyang and Fancheng on the River Han. These massive fortified cities held the key to the security of the River Yangtze and therefore the entire Song empire. It was only after a supreme effort and endurance that said much about the superb quality of the Yuan army that Hanyang finally fell in late March 1273.

This gave Generalissimo Bayan the opportunity to march on the Song capital of Hangzhou. In January 1275 Bayan crossed the Yangtze, defeated 130,000 Song troops there in March and in January 1276 accepted the surrender of Hangzhou with the Song court. A mission that had begun in 1211 was finally completed. The land war was over in China even though the Song court kept up futile resistance for another three years at sea.

Battle of Vochan, 1277: The Invasion and Subjugation of Burma

It was during Kublai's Nangchao campaign of 1253–55 that the Mongols realised that the porous southwestern border of China posed a special problem to the security of their expanding empire. This was especially true of Burma where the powerful and aggressive military kingdom of Pegu dominated the whole of northern part of the country and had reached an impressive size under its last great king, Narapatisithu (r.1173–1211). When the Mongols took over Yunnan they set up a range of small buffer states between this rich province and the Pagan realm. The Mongols defeated a Pagan army at Ngasaaunggyan leaving the task of revenging this setback to the brutal and boastful King Narathihapate (1254–87). Kublai entrusted the task of dealing with the Pagans to the Viceroy of Yunnan, a capable Muslim commander and official called Nazir ud-Din.

Ud-Din tried diplomacy first with scant hope of success. A clash of civilization took place when the Mongol envoys wore their leather riding boots in the King's presence. Nomads, like the Mongols, wore their riding boots indoors but this was a grave insult in Burmese eyes and especially in the serene presence of a proud, prickly king. They were executed on the spot. Narathihapate then

Yuan Period Chinese Cavalryman (1280)

The peak of Mongol efficiency and professionalism was reached by their cavalry during the Sino-Mongol Yuan empire and especially during the reign of Kublai Khan. Their equipment was a major improvement on the past, with regular, elegant uniforms in the Chinese fashion.

Their professionalism reached new heights as shown in this illustration. Here the fine mount is not covered in lamellar armour, while the rider's uniform and sword is of a decidedly Chinese character. But the rider is wearing the characteristic Mongol-style helmet, the lamellar armour covering his vulnerable neck and sides, while the boots remain exactly the same as in the past - perfectly adapted for riding long and hard.

Battle of Vochan

1277

The flexibility of the Mongol cavalry army was on display yet again at Vochan where a seemingly certain defeat was turned into a great triumph for the northern invaders of the Kingdom of Pegu (Burma). The King of Pegu had brought a massive army into the field, spearheaded by as many as two hundred armed and crewed war elephants. Neither Mongol men or mounts had seen these huge strange beasts before but Ud-Din's steely leadership steadied his shaken troops. Having dismounted, the Mongols, numbering no more than 12,000, trusted their archery firepower as they fired off a series of deadly volleys that saw the war elephants break formation. Remounting, the Mongols gave chase to the disoriented Burmese army and defeated their fleeing enemy with merciless efficiency.

YUAN CHINA
INDIA
VOCHAN
Mandalay
PEGU (BURMA)
SIAM

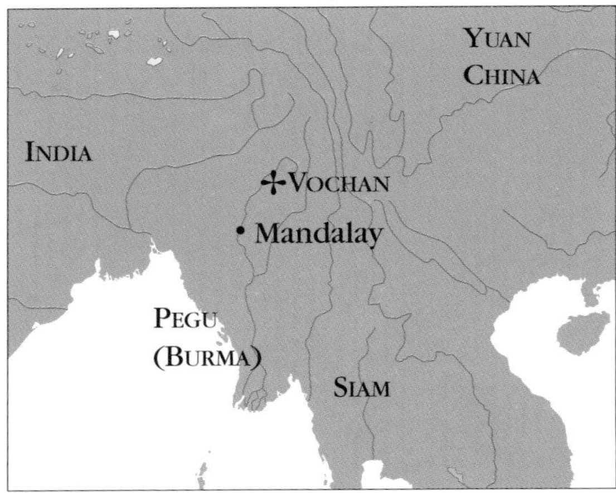

The battle that pitted Burmese war elephants against Mongol cavalry took place near the small village of Baoshan (Vochan) in the northern Burmese province of Bhamo close to the Irrawaddy and north of Mandalay.

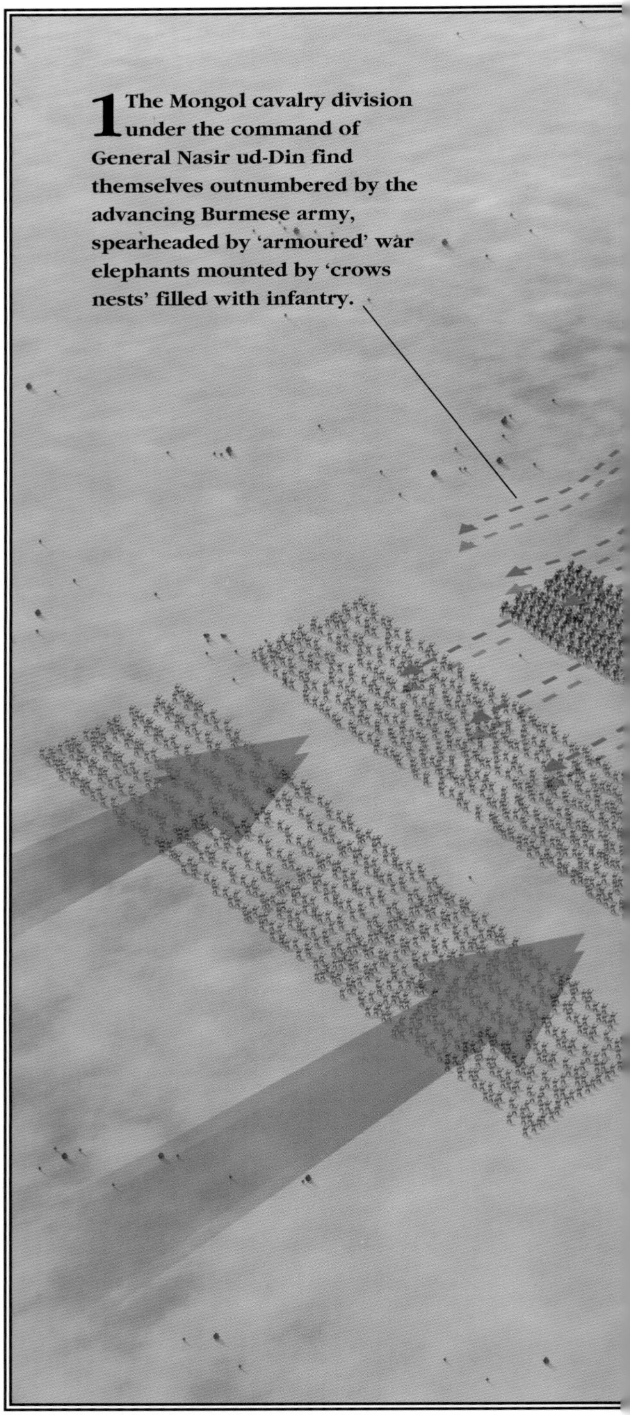

1 The Mongol cavalry division under the command of General Nasir ud-Din find themselves outnumbered by the advancing Burmese army, spearheaded by 'armoured' war elephants mounted by 'crows nests' filled with infantry.

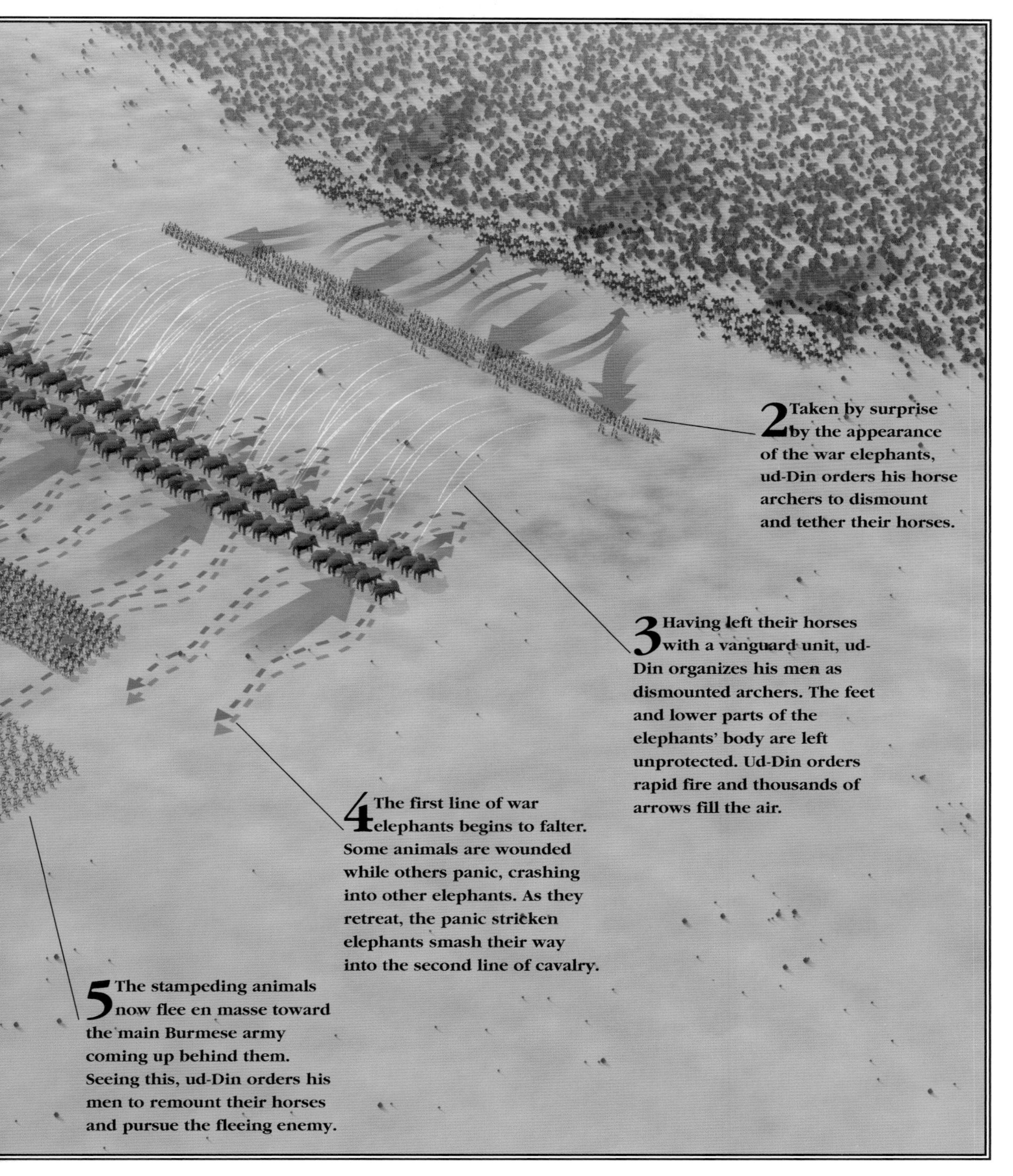
2 Taken by surprise by the appearance of the war elephants, ud-Din orders his horse archers to dismount and tether their horses.
3 Having left their horses with a vanguard unit, ud-Din organizes his men as dismounted archers. The feet and lower parts of the elephants' body are left unprotected. Ud-Din orders rapid fire and thousands of arrows fill the air.
4 The first line of war elephants begins to falter. Some animals are wounded while others panic, crashing into other elephants. As they retreat, the panic stricken elephants smash their way into the second line of cavalry.
5 The stampeding animals now flee en masse toward the main Burmese army coming up behind them. Seeing this, ud-Din orders his men to remount their horses and pursue the fleeing enemy.

invaded the Yuan border state of the Golden Tooth (Kaungai) compounding his earlier insult to Imperial Yuan honour.

Kublai underestimated both the huge problems of campaigning in Burma and the enemy his Mongols faced. The Emperor was sure that ud-Din would be able to crush Pagan with the local *tamma* army of Yunnan province without reinforcements from the north. Ud-Din did not share his ruler's misplaced optimism. He had a single cavalry *tumen* to cross some of the most mountainous, heavily forested and climatically atrocious stretches of Indochina before he faced a Burmese army of 400,000 troops and 2000 heavily armed war elephants! Each beast carried some 12–16 men, mostly archers, but also mounted infantry. The strength of the Burmese cavalry is unknown.

Ud-Din crossed these mountains, passes and rivers to enter the Pagan province of Bhamo where his opponents, Generals Anantpaccaya and Randhapaccagya had built a fortified camp to house their huge army. This army may have been impressive on paper but it was composed of poorly equipped, ill-trained and unmotivated peasant infantry who were up against the finest war machine in the world led by an experienced and ruthless commander. Such was the situation when ud-Din's cavalrymen emerged on the plains along the River Irrawaddy.

The shock of encountering the enemy was probably mutual. The Burmese already knew what a formidable foe the Mongols were while the Mongol riders and their horses were terrified at the appearance of the war elephants. To steady his men in the face of this monstrous apparition ud-Din ordered them to tether their horses behind the battle line and fight dismounted. Their only chance, as the Burmese war elephants advanced on a broad front, was to open relentless fire upon the beasts.

The elephants began to falter in their advance, halted and then turned around and fled into the massed ranks of the Burmese troops behind them. Now ud-Din ordered his men to remount and charge the confused enemy. But the battle proved fierce nevertheless, with the Burmese fighting back with resolution. After a tough fight, eventually the Mongols prevailed. Included in the war booty sent back to Beijing were two hundred war elephants that became the nucleus for the Imperial Corps of War Elephants.

> *'Understand that when the elephants felt the smart of those arrows that pelted them like rain, they turned and fled.... So off they sped with such a noise and uproar that you would have thought the world was coming to an end.'*
>
> – MARCO POLO ON THE BATTLE OF VOCHAN

Aftermath

Vochan was a novel victory but did not lead to the submission of the Pagan empire. Kublai sent his grandson, Yesün Temür, with a large army into Burma with the aim of crushing Pagan once and for all. In December 1283, Yesün's Mongol army defeated the Burmese at Kaungzin and occupied Pegu. This forced Narathihapate to flee to Bassei in the Irrawaddy delta. The unpopular king, now given the name *Tarokpyemin,* 'the king who ran away from the Chinese' was assassinated in 1287. His son, Kyawswa, ruled what remained of Pagan (1287–98) as a Yuan vassal. He was murdered in April 1298, prompting a renewed Yuan invasion that failed to defeat the enemy until early 1301. The Mongol war machine was by this time in steep decline like the rest of the empire.

Japan: the Failed Invasions of 1274 and 1281

In Japan, isolated from the Asian mainland, the cult of the individual horsed warrior – the samurai – flourished in a series of bloody civil wars that racked much of the island nation during these centuries. Only the samurai ('those that serve') had the privilege of fighting on horseback, and looked down on the infantry as lowly commoners.

Like the Mongol horseman, the samurai fought as a mounted archer but he used the native longbow which was less effective a weapon than the composite version. War in Japan was a far more ritualistic and individualistic affair than that being fought by the massed cavalry armies on the mainland. This anachronism proved almost fatal when the brave but less cohesive samurai armies faced the disciplined, massed ranks of the Yuan army in 1274 and 1281.

The Genpei Wars (1180–85) saw the final pre-Mongol triumph of the mounted samurai warrior. In 1182 Yoshinaka Minamoto (1154–84) occupied the imperial capital of Kyoto and proclaimed himself the first Japanese shogun. (The term hails from the Chinese word for general: *jiang jun.*)

On 18 March 1184, a rival Taira (Heike) army was at Ichi-ro Tani on the coast. Trusting his samurai cavalry Yoshinaka was determined to crush the enemy once and for all. He formed a cavalry shock force of some 200 samurai and he would lead the vanguard of the 30 best and toughest warriors himself. Surprise was of the essence so Yoshinaka simply ordered the samurai down a steep cliff just above the enemy positions. A chronicler noted in amazed and shocked admiration: 'The stirrups of the men behind nearly struck the helmets of those in front. The cliff was sandy, so they slid about 120 feet… The sight was so terrifying they shut their eyes.' This example of Japanese derring-do was to become legendary and saw the Minamoto clan become supreme rulers of Japan. They were now able to establish the militarized 'tent government' (*Bakufu*) at Kamakura.

Through their trading contacts with the Song empire and Korea, the Japanese were aware of the rise of the Mongols and the establishment of Kublai's Yuan empire in China. To the isolated and poorly informed islanders the Mongols were nothing but mainland 'barbarians' whose antics had nothing to do with them. That they could pose a deadly threat to Japan was not something that entered their heads.

In 1274, having rejected a series of Yuan diplomatic missions and threats, they faced a massive fleet of 900 ships carrying some 15,000

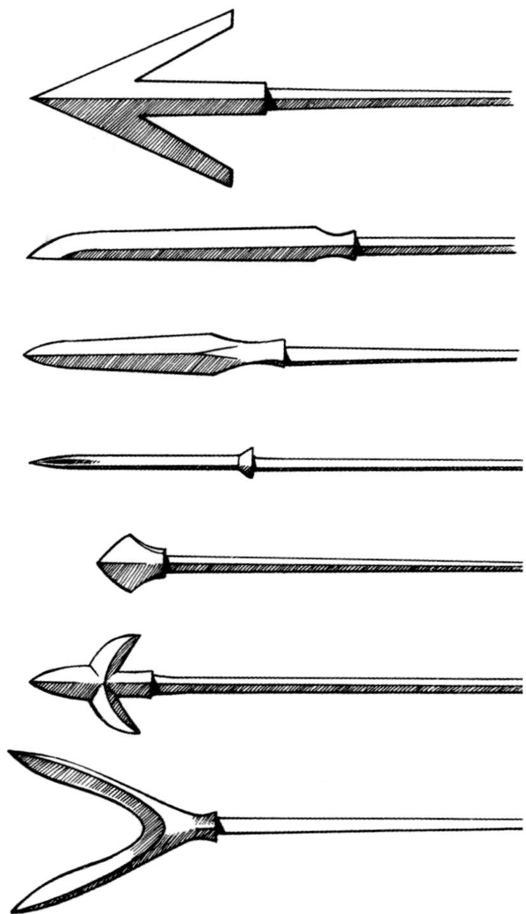

ALTHOUGH THE JAPANESE *were almost completely isolated from continental Asia, their military technology for much of the Medieval period showed healthy signs of keeping pace with developments there, as illustrated here, by seven different types of arrowheads. Like the Mongol ones these, even though designed for longbows rather than composite bows, were given different shapes for very different purposes, from knocking down riders to penetrating armour.*

Korean sailors and 25,000 Sino-Mongol troops onboard. The Japanese were able to field an army of a mere 8000 scraped together from Kuyushu.

The invaders landed on 3 October, on the island of Tsushima. In these first battles the Japanese trusted the mounted samurai archer, but their charges proved a series of costly failures. In

A FINE EXAMPLE *of Japanese military art: the Samurai's armour made of a combination of lace textiles, hardened and lacquered leather combining simultaneously lightness, flexibility and protection. This set of armour dates from 1652 and was made by Japan's master armourer Myochin Munesada.*

one attack only one samurai returned of the hundred who charged. They could make no impression upon the phalanx of shields the Yuan infantry sheltered behind. The enemy used poisoned arrows, javelins, explosives pots and grenades to pelt and cut down the Japanese cavalry. At a single drum signal, thousands upon thousands of arrows – released from deadly composite bows – were unleashed in volleys upon the samurai. After 1000 men had been lost the Mongols moved on to sack Tsushima town and killed 6000 civilians in the process. The Japanese had trusted the samurai warrior but his prowess had been no match against the massed, well-disciplined and well-equipped regulars of the Yuan army.

On 14 October the island of Iki fell as easily as that of Tsushima. Cavalry charges with archers firing arrows with the ineffective longbow fared equally poorly. If the enemy landed in force on Kyushu, as planned, the Japanese would face certain defeat and the country would be conquered by a ruthless and determined invader.

The invasion fleet arrived off Hakata Bay on Kyushu five days later. Governor Shoni-kakye had

NO COUNTRY IN THE WORLD *during the sixteenth and seventeenth centuries, with the possible exception of Imperial Spain, could manufacture swords of the quality, durability, strength and sharpness like Japan. The secret lay in the skill of the country's armourers and the multiple layers and quality of the steel used to create the razor-sharp blade.*

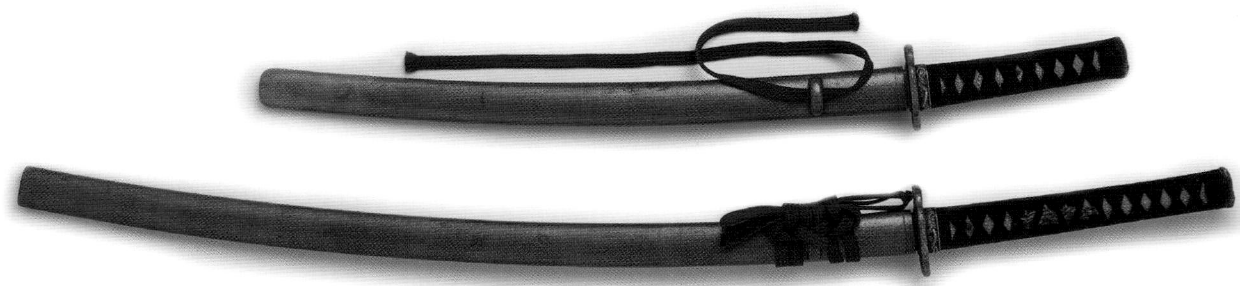

Due to the topography of Japan, like that of densely populated western Europe, cavalry has played a subordinate role in that country's military history. Yet, on occasion, cavalry has in fact played a decisive role in Japanese warfare. Like their European counterparts the Japanese cavalrymen trusted their swords, massed attacks and close-quarters combat.

some 3000 men but only a few mounted cavalry. He faced 40,000 Yuan troops but his small force managed to pin down this vastly superior enemy from the 19 to 22 October. Crucially some 300 samurai cavalry joined in an army of 3000 troops and the battle reached its peak on 20 October. As stated earlier, the flotilla was hit by a typhoon (*kamikaze*) during the last night and lost a third of its ships. Japan was saved, as in 1281, by the forces of nature and the fierce, stubborn resistance of its samurai.

Post-Invasion Japanese Cavalry

Yet the samurai and Japanese warfare had changed and had to change in the face of the Mongol invasions. The mounted samurai wore *yoroi:* a heavy box-like armour made up of small metal plates called *Korzane* laced together with leather strips that had been lacquered and tied together with silk cords. This gave the samurai a robust but light and flexible suit of armour even superior to that of the Mongol heavy cavalryman.

But the samurai no longer remained a mounted archer. He became something like a European knight who trusted his lance and a massed attack to turn the tide of battle. The samurai's *yari* (spear) bore no resemblance whatsoever to the knight's lance as it was lighter and shorter and was not carried in a crouched position. The *yari* was in fact a double-edged weapon with two razor-sharp blades at either end and was used by the samurai when he fought dismounted. When mounted he would use a *naginata* instead – a pole-arm with a long curved blade.

The samurai cavalryman was still the honoured warrior, however, and cavalry warfare, fought between fellow samurais, was still a cherished ideal. In open battle the cavalry could often prevail against infantry as happened at the battles of Minatogawa (1336) and Shijo-Nawate (1348). One chronicler said, quite correctly: 'foot soldiers may be strong, but they are not strong enough to stop arrows; they may be fast, but not fast enough to outrun horses.' These harsh words were about infantry facing a good many mounted archers as late as the Ônin War (1467–76).

This was all set to change during the Sengoku era that dominated the entire sixteenth and early seventeenth centuries. During the Ônin War mounted samurai were more and more supplemented with *ashigaru* infantry who were prone to desertion and looting. Samurai still wore full armour and elaborate helmets but, once firearms were introduced, by Portuguese merchants in 1543, a solid metal breastplate was added for additional protection.

JAPANESE ARMOUR *was not like European armour, which was made of metal, but was made of lacquered leather and tied together with string in sections. It was more solid (shown in top section of the illustration) on the front, facing the enemy, than at the the back (section below). The armour shown here is of the* O-yoroi *type, popular from the twelfth century onwards.*

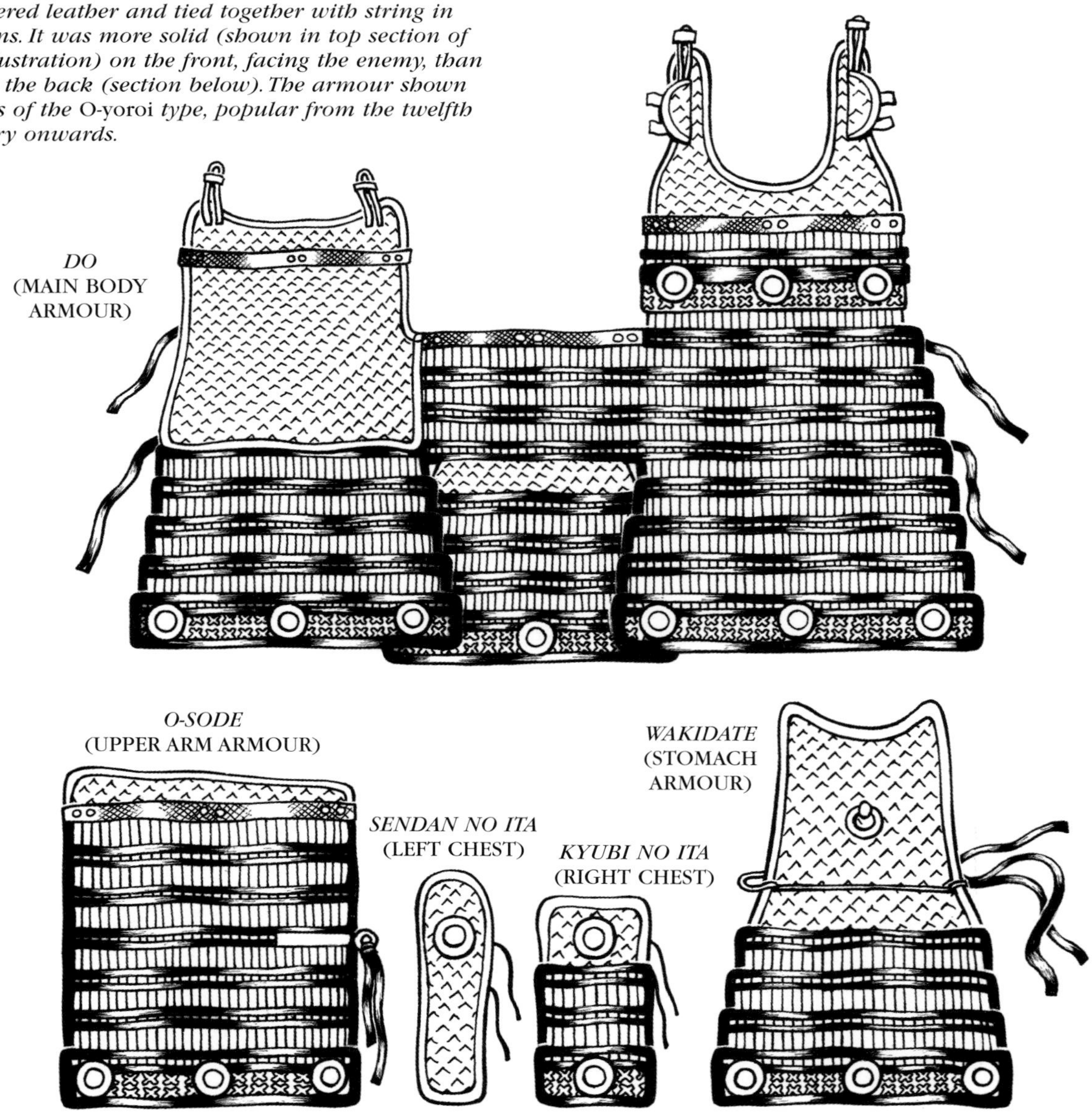

Mounted Samurai (Thirteenth Century)

The Japanese Samurai, armed with his characteristic longbow, in all his glory and showing a false resemblance to the Mongol horsed archer (see earlier in chapter). The Samurai had the same status as a European knight and fought like one. From the eleventh century, he had been equipped with a longbow but this evolved in time to fight with sword or lance-like spears. For a few decades around 1600, Japanese warfare, practised at home, was dominated by cavalry until the rise of disciplined infantry equipped with ever-more sophisticated western-style firearms. Despite these developments, the mounted samurai remained the epitome of Japanese martial valour well into the twentieth century.

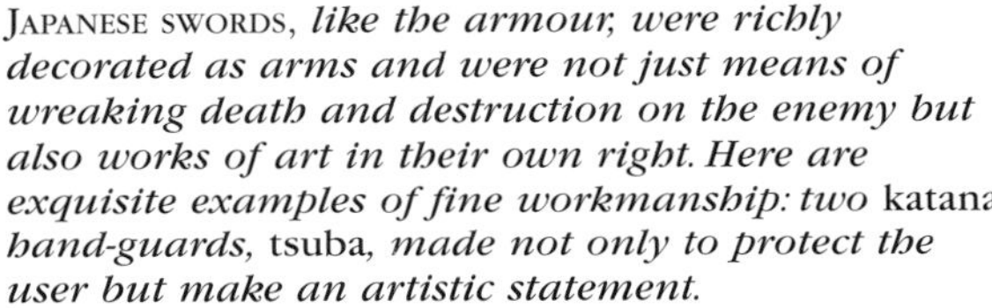

JAPANESE SWORDS, *like the armour, were richly decorated as arms and were not just means of wreaking death and destruction on the enemy but also works of art in their own right. Here are exquisite examples of fine workmanship: two* katana *hand-guards,* tsuba, *made not only to protect the user but make an artistic statement.*

It would be wrong to see Japanese cavalry as in terminal decline from that point onward. Their role simply changed from the mounted archer/knight in armour to a lance-wielding professional. Some mounted archers were maintained as mobile sharpshooters but most cavalrymen were by this time equipped with *naginata* or lances. This was an inevitable change as from 1560 the bow was being gradually replaced by the arquebus.

The most effective cavalry response to this challenge came from the brilliant and innovative cavalry commander Lord Takeda Shingen whose cavalry army became as feared as it was legendary. At Uedahara in 1548 Lord Shingen crushed an army equipped with primitive handguns. Better still, at Mikatagahara, a famous double battle in 1572, the Takeda massed cavalry charge broke infantry supported by arquebusiers. For a few decades during the middle part of the sixteenth century the cavalryman dominated the Japanese battlefield. Yet this triumph was far from total. Many battles won required Takeda samurai to dismount their horses to dislodge the enemy remaining on the battlefield. The shock tactics were less formidable than we might have imagined. Since the samurai were accompanied by retainers who had to run alongside them, the pace of the advance could not be very high. Furthermore the Japanese horse of the time was only half the height and weight of a modern one.

End of an Era

Finally the cavalry era was over before it had even begun. Oda Nobunaga had drawn the right lessons from Mikatagahara (1572). To neutralize the Takeda advantage in horse and shift the initiative back to the infantry, arquebuses had to be used on a mass scale, with specialized *ashigaru* trained to provide disciplined, close range volley fire. Lord Nobunaga formed a special core of pikemen *ashigaru* to protect the arqubusiers while they reloaded their firearms and had archers to fire off volleys while this was taking place.

At Nagashino (1573), Oda Nobunaga put his theories into practice. Three thousand arquebusiers were shielded by *ashigaru* spearmen and wooden shields. They managed to blunt and break the Takeda cavalry charge but not the individual samurai. The outcome of the battle was still hours of bloody hand-to-hand combat. But hereafter Japanese armies would be a mix of *ashigaru* spearmen, arquebusiers and cavalry. The golden era of the Japanese cavalry had come to an abrupt, untimely end.

Kublai's Mongolian Campaign

The failed invasions of Japan were not symptoms of failure or decline but rather supreme efforts that were wasted due to weather and bad luck. Kublai did score one last victory, ironically against his own Mongol nation. When the Grand Khan's sworn enemy, Chaghatai Khan Kaidu (1230–1301), supported the rebel Nayan, Kublai had to muster one last effort to save his empire from this dire threat. He mobilized three massive armies to deal with the combined forces of Kaidu – who had occupied the old Mongol capital of Karakorum – and Nayan, who had invaded Manchuria. It was vital to expel Kaidu from this vital region before he became a threat to the exposed Yuan capital of Beijing. One army was sent to cut communications between the two allies, another to retake Karakorum and the third, under Kublai's personal command, rode into Manchuria. They rode, because three-quarters of its numbers were cavalry, accompanied by only a small force of Chinese infantry. It took Kublai's army 25 days to cover the huge distance to the River Liao, where the Imperial army caught Nayan's rebels unprepared for battle in their camp. Ever the cautious commander, Kublai stayed in his hillside camp for two days, despite having a vast numerical and qualitative superiority over the enemy.

THIS ILLUSTRATION FROM *a block book published in 1855 by the Japanese artist Geijutsu Hideu Zue shows how old and new blended so well in late Shogun and early Meiji Japan in the nineteenth century. The mounted samurai is firing at his assailants with two matchblock pistols with amazing agility thanks to his light and flexible armour.*

Mongolian Heavy Cavalryman (1300)

The Mongols' heavy cavalry were often decisive in routing and destroying a defeated enemy army. In contrast to the horsed archer, the heavy cavalryman was well protected and equipped for close quarter combat. He could use either, as here, his curved sword or a steel-tipped mace or lance for fighting the enemy at close quarters, having first unnerved him with a heavy, shock attack. The lamellar armour was linked by numerous laces, allowing the cavalryman to move easily in his saddle when in close combat.

Seventy-two and suffering from arthritis Kublai was taking no chances.

Then on the third day his army broke camp, marched forward in three divisions and yet again caught the rebels off guard. Nayan had probably thought that Kublai would avoid a battle or make further preparations. He was in for a surprise as the Imperial army extended its flanks across the steppe to trap the rebel army and its camp in one huge lunge – as if they were taking part in the old *nerge* war games. What was new was that the advance was spearheaded by infantry armed with swords and short stabbing spears. Their task was to attack the rebels' horses if they counter attacked, and if need be they could be withdrawn behind the protective wall of the cavalry. At a signal the cymbals and drums were sounded as the right and left flanks advanced to spring the trap. Both sides poured arrows into each other's ranks. The cavalry then advanced until they could engage in ferocious hand-to--hand combat with spears, swords and maces. In this Mongol civil war no quarter was given or asked. As a consequence this battle of the River Liao was one of the most savage ever fought by the Mongols. Here, for once, more men than mounts were killed. Despite being outnumbered Nayan's men fought tenaciously from early morning to noon. But in the end the resources and numbers of the Imperial side prevailed. Kublai stood as victor one last time and had the satisfaction of seeing Nayan smothered. According to ancient Mongol custom no royal blood could be split so royal personages were strangled or smothered to death.

'My bannerman was first. His horse was shot and he was thrown down. I and my three retainers were wounded. Just after my horse was shot and I was thrown off, Michiyasu attacked with a formidable squad of horsemen and the Mongols retreated...'

– Samurai Takezaki Suenaga, who fought the Mongols in 1274

The Rise of the Ming

Kublai died in 1294 and with him ended the golden age of the Mongol empire. He was the last great Mongol commander and Yuan emperor. His grandson, Yesün Temür, the last emperor of any great ability, died on 15 August 1328, leaving a weakened empire to be torn apart by civil war. Here cavalry would decide which side won. El Temur, who led the rebel side, gained control of the *Khesik* cavalry and thereby tipped the scales in the rebels' favour. The loyalists, who supported Yesün Temür's heir, mobilized their infantry to hold the frontier. Having gained the support of the Eastern Mongols (Khalka) and the Jurchen the rebels were vastly superior in cavalry. A simultaneous cavalry attack from Manchuria and eastern Mongolia managed to repeat Genghis' feat of 1211–15. On 14 November 1328, the rebel Mongol cavalry surrounded Beijing where the legitimate Yuan court surrendered the following day, only to be massacred by El Temur's men. Loyalist Yunnan was the last province to surrender to the rebels in March 1332.

During the 1340s China was hit a by series of plagues, famines and floods that provoked popular rebellion across the country. The Yuan administration in Beijing had never had anything more than indirect control over the army garrisons and *tammachi* scattered across China. Under the stress of the crises Beijing lost control completely over the provincial governors and their *tamma* armies. By the early 1350s these commanders and governors had become independent warlords who might suppress rebels, or just as well march on Beijing to topple the ever more enfeebled Yuan dynasty.

The Imperial Chancellor, Toghto, tried to stop the rot and mobilized troops, including both Chinese and Mongol cavalry. He managed to restore some control over central and southeastern China down to the River Yangtze before he was dismissed from his post in 1354. Now the empire's disintegration accelerated with alarming speed. There were several reasons for this but the most important was the lack of military power as the empire entered the 1350s. The key here was a lack of good cavalry: Mongolia was marginalized while the Chinese and sinified Mongols who had settled in China were no longer willing or able to serve as cavalry. The old military colonies had declined in importance while the supplies of good mounts from the steppes were also reduced: suppliers were no longer willing to sell them to the authorities who paid them in worthless paper money. This crucial lack of cavalry ensured that the south Chinese rebel armies, although weak in cavalry, would be able to defeat the Yuan armies sent against them.

During 1351 the secret White Lotus society spread rumours that the Song would be restored, setting off the rebellion of the Red Turbans whose leader Kuo led armed attacks against the Imperial authorities. When he died in 1355 his place was taken by a former vagabond, Zhu Yuanzhang (1328–98), who combined enormous military leadership skills and political acumen. Zhu led his increasingly professionalized army into the valley of the River Yellow (1352–59) and managed to take Kaifeng, mainly because the Mongols had failed to repair its walls. By 1360, the last Yuan garrison in the Yangtze valley had been expelled.

Like all the southern Chinese rebel armies, the Red Turbans were short of cavalry. The south Chinese rode into battle like European cavalry – with straight legs, backs and riding with leather stirrups. They wore buffalo hide hardened or glazed to give them the pretence of 'armour' and were armed with spears, crossbows and shields. Both in appearance and armament these horsemen were more like mounted infantry than real cavalry. In combat, the south Chinese horsemen would simply dismount and fight the enemy on foot. Zhu compensated this grave weakness by having a strong navy, artillery corps and well-trained infantry in large numbers.

Zhu was too busy fighting the other Chinese rebel states in southern China to bother with the Yuan to the north. In combined operations, where the cavalry only played a minor role, Zhu crushed the Han state on the middle Yangtze River between 1360 and 1363. Chu declared his state to be the Ming ('Brilliant') dynasty and announced his ambition to unite the whole of China under his new dynasty's rule.

Next he invaded the state of Wu at the mouth of the Yangtze, whose ruler, Prince Chang, was

ONCE PAINTED IN VIVID, *florid colours, this statue from the Ming Dynasty (1368–1644) shows a horseman raising his missing lance or sword in a dramatic attack pose. It was nomad horsemen of this kind that conquered an empire obsessed with civilian pursuits, including the arts, to the detriment of military and defensive skills.*

THE MONGOLS WERE SUPERB *craftsmen when it came to leather work. On the left, a thirteenth-century Mongolian quiver made of hardened but supple leather that could contain up to 60 arrows. On the right, a Mongolian leather bowcase from the fourteenth century.*

defeated at the battle of Hsin-ch'eng. Here, for once, the outcome of the battle was decided by the Ming cavalry which charged and defeated the Wu infantry. In October 1367, the Wu capital of Suzhou fell after a ten month siege and its 250,000 strong army was absorbed into the expanding Imperial Ming army.

Zhu sent General Hsu Ta with 250,000 men north. His 'campaign' was more of military promenade than a real campaign. The last Yuan Emperor, Toghon Temür (r.1333–68) fled to Karakorum, leaving Beijing to Hsu Ta's Chinese troops. Beijing, called Shang-tu by the Mongols was renamed Pei-ping ('The North is Pacified') showing a seismic shift in Chinese history: for the first time a south Chinese ruler had conquered the mighty north. Zhu changed his name to Emperor Hongwu (r.1368–98) but did not relent in his campaign against the Mongols and the remaining rebels.

The Rise and Fall of a Mobile Army

Hongwu never had any illusions that his dynasty's power was built on military strength and during this reign he never allowed his army to decline. Yet he committed a major mistake by allowing the Ming princes to set up their own armies and states within the empire. Thus he allowed a state of affairs that would encourage a civil war once Hongwu died.

The unwarlike Zhu Yunwen (1377–1402) who succeeded Hongwu, set up his court in the old capital of Nanjing but left his uncle, Prince of Yen, in charge in Beijing. Yen made good use of this north China base to topple his nephew. Yen controlled a third of the regular Ming army, some 100,000 troops, and had a wide margin of superiority when it came to cavalry – the north Chinese were better cavalrymen and they were supplemented by old Yuan cavalry formations. Steppe Mongols and Jurchen were also recruited by Yen to beef up his cavalry arm.

Nevertheless, Yunwen had a greater number of troops, controlled the richest part of the empire and was far stronger than Yen when it came to artillery and firearms. As a consequence, the Imperial army won over Yen's rebels at the battle of Baoding but the rebels simply retired northwards to lick their wounds. During 1400–01, there was inconclusive fighting along the Grand Canal. Using the frozen state of the canals and rivers to good effect, Yen suddenly went on the offensive in January 1402 with his superior cavalry spearheading the attack. Although the loyalist forces were able to beat his army in several battles, Yen was able to retreat, counter-attack and outflank the Imperial troops utilising his fast-moving cavalry. Yen was acting more like Genghis Khan or Kublai than a Ming commander and this, together with his many Mongol mercenaries, now ensured that he won the civil war.

Yen's army crossed the Yangtze thanks to ships that had deserted from the Ming flotilla and his supporters opened the gates to Nanjing. Yunwen was killed during the street fighting that ensued and Yen now seized power, declaring himself the Yongle Emperor (r.1402–24). He killed thousands of his real and perceived enemies among the Yunwen loyalists, purged the Imperial administration and restored the militaristic policies of the Hongwu era. With that he moved the court, administration and the army high command to his own preferred

CHINESE SOLDIERS PRACTICE THEIR ARCHERY *skills in this eighteenth-century illustration. Long after the advent of gunpowder weapons, the bow remained an integral part of the armoury of most oriental armies.*

capital – Beijing. This move committed Ming China to almost three centuries of frontier wars against the growing menace of the Mongols and Manchurians. Yongle and Hongwu borrowed their military organization and policies from their Yuan predecessors with a most un-Chinese-like emphasis upon military strength built around a professional core of cavalry in combination with other arms.

Until 1364 the Ming Army was in fact nothing more than a collection of veteran Red Turban troops, hired mercenaries, volunteers, armed peasants and a host of former enemy troops (Yuan, Wu and Han). Building on the Yuan model Hongwu broke up the existing units replacing them with a system of military districts each having one or two brigades (*wei*) of about 5,000 men each – each brigade being made up of five battalions. Quite sensibly Hongwu wanted no undue burden placed on the civilian populace to pay for his troops so he created – on vacant or abandoned farm land – military colonies.

During the fourteenth and fifteenth centuries, these proved to be a roaring success, since they occupied large tracts of fertile land in central and northern China, supplying all the needs of the troops and officers wages while even providing a surplus for the State grain stores. There were five hundred *weis* (brigades), giving the Ming Army an overall strength of some 2,800,000 men. This was not only the largest but also the best equipped, led, officered and most professional army in the world.

Fighting the Mongols deep inside their steppe lands, Yongle came to appreciate the need for a strong, independent cavalry arm. After all as Prince of Yen he had won in 1402 because he had 100,000 Mongol and north Chinese cavalry and his nephew did not. Neither Beijing nor the rest of China would be secure without a strong cavalry army that could match the Mongols, even though Yongle had no illusions that his cavalry would ever be as good as the enemy's. That he had a long way to go was shown in 1372.

Commanded by the otherwise brilliant Ming general Hsu Ta the Imperial army penetrated deep inside Mongolia and fought the Mongols at Karakorum – but suffered a catastrophic defeat. With one blow all Ming hopes of controlling Mongolia in the same way as the Yuan dynasty was gone, condemning the Ming to two and a half centuries of cavalry wars against the perennial threat of a new nomad invasion.

'...when they reached Hsuan-fu there came a great storm...some of the officials tried to persuade the monarch to return. Wang Chen [the Chief Imperial Enuch blamed for this military debacle] was exceedingly angry. Chu Yung, Duke of Ch'eng-kuo, and some others happened to arrive to report on certain affairs and they were made to enter on their knees...'

– Ming chronicle, prior to the Ming defeat at Tumu 1449

'New' Cavalry Army

With customary energy and determination, Yongle set about creating a powerful cavalry army. The core of his new force – the *San-ch'ien* – was built around some 3000 captured Mongols he had taken prisoner during one of his many expeditions into the steppe. These prisoners were used in three training (*Ying*) camps around Beijing to teach north Chinese – they were deemed to be better with horses than their southern countrymen – elementary handling of horses, grooming, riding, signalling, horse archery and the other Mongol 'tricks of the trade'. A similar number of horse camps to train Chinese horses in the Mongol fashion were also set up using the Mongol prisoners. The Ming cavalry arm was, therefore, a straight copy of the Mongol model and could hardly be told apart. The Ming had also learnt from the Yuan about convergent offensives from several separate directions deep inside enemy territory as shown when the Ming captured Kaifeng in 1368. The Ming would use the old Mongol tricks of burning grass, feigned retreat, dummies on horses and so forth to good effect.

A Ming field army would never have less than three or four of these independent cavalry divisions on campaign to outflank, envelop and annihilate Mongol cavalry. The Ming's greatest headache was, of course, to pin down the swift moving Mongol cavalry and they were only able to do so when they either surprised the enemy in a camp or were able to surround the Mongols with their back against a lake.

Wang Yueh (1426–99), a scholar by training who had turned himself into an outstanding Ming general, copied the Mongols and their style of warfare down to the last detail. Until now the Ming would usually lead a large combined army into the steppes where they either ran out of supplies and water or lost contact with the enemy. General Yueh spent a long time training his Chinese cavalry to imitate the Mongols and when he was ready only picked the best to accompany him on a campaign deep inside enemy territory. Here the force would be small, mobile and use the swift 'hit and run' tactics in which the nomads excelled.

In early 1473 before the spring grass had come out properly, preventing a nomad army from active operations, Yueh led a cavalry army of 4600 horse archers in the hope that he would find his Mongol enemy – Prince Bag Arslan – holed up inside a camp waiting for warmer campaigning weather. They found Arslan's camp at a point called Hung-yen-ch'ih.

The Mongols were away hunting and foraging when the Ming cavalry attacked, capturing a huge booty of captives, horses and livestock. Arslan was soon aware of the Ming raid, rushed back only to

TIME SEEMS TO STAND STILL *on the steppes of Outer Mongolia. A group of herdsmen with their characteristic pole lassos pass a group of yurts. Had it not been for the modern round hats on some of the men, this scene with the starkly beautiful landscape could date from the thirteenth century and not the twenty-first century.*

fall in a carefully laid trap where he and his men were slaughtered.

Decline of the Cavalry Arm

General Yueh's campaign was a brilliant exception to an otherwise increasingly defensive posture by the Ming army. After the death of the warrior Emperor Yongle in 1424 there was a gradual decline in the military competence of the Ming army as exemplified by the growing influence of eunuchs and courtiers over the military. The Tumu disaster two decades later exemplified this. But this decline was only gradual. Wang Yueh was leading campaigns even when aged 71 but after his death, in 1499, the forward aggressive policy of taking the war to the Mongols was in steep decline.

Yueh had trusted in an offensive strategy, based on elite cavalry troops with light equipment, to keep the Mongols on their toes. Already in 1504 the military courtiers dissuaded Emperor Hung Chih from emulating Yongle and Wang Yueh's bold strategy by leading a campaign in person. They pointed out, quite rightly, that the Ming cavalry had gone to seed as the empire had failed to keep their cavalry forces in good order and well trained, disciplined and paid. This despite the glaringly obvious fact that the security of China and their own imperial regime depended upon having a large and fit force of cavalry available to meet and crush the growing menace from the nomads of the north.

The fate of the Imperial Guards cavalry, called the 'Brave Men', was an obvious example of this dangerous neglect and wilful lack of military competence. Yongle had placed four divisions of this Special Imperial Guards cavalry (*siwei jun*) under the supervision of the Directorate of the Imperial Horses (*Yuma jian*) and the Eunuch Directors (*taijian*). As late as Emperor Hongzhi (r.1488–1502) the Brave Men numbered 11,780 horsemen in good order with another 30,000 regular banner (infantry) troops to support them. In other words a good mobile strike force with adequate backup. By 1537 there were only, and then only on paper, some 5000 'Brave Men' left. By the reign of Emperor Tainqi (r.1621–27) the Inspector of the Armed Forces recommended that the Brave Men be further reduced in numbers and broken into smaller units.

As the cavalry went into decline so did the rest of the Ming army. In 1528 most of the real generals had been replaced by courtiers, favourites and eunuchs – all of whom had two things in common: inexperience and incompetence. Originally there had been twelve cavalry divisions in the all-important Beijing military district, with 107,000 troops and 150,200 horses. By 1528 there were only 54,000 troops and a mere 19,300 horses left. From the 1540s the Ming were reduced to scraping to together troops on an *ad hoc* basis and putting their trust in fortified positions. Out of this defensive attitude would grow the conviction that only a continuous stone wall – the Great Wall – could protect China from the nomad invaders. This was a dangerous delusion since a wall garrisoned by infantry but without adequate cavalry reserves, or better still a Wang Yueh-style strike force, did not stand a chance against the nomad horse archer. When in 1550 the Mongols threatened a direct attack upon Beijing the President of the Board of War was only able to scrape together a reluctant force of 50,000 infantry. A disaster was only just averted. But if the Mongols made a more determined effort would the Ming army be able to protect the dangerously exposed capital situated on the very frontier of the empire? More and more Ming officials were not sure and this doubt began to gnaw at the very foundations of the empire's survival.

> *'…neighbouring states will take advantage of your crisis to act. In that case, no man, however wise, will be able to avert the disastrous consequences that ensue…'*
>
> – *Sun Tsu,* The Art of War

Traditionally, the Chinese, following the Song example, would simply bribe the nomad tribes but the Ming chose, like the Jin and Yuan dynasties, not

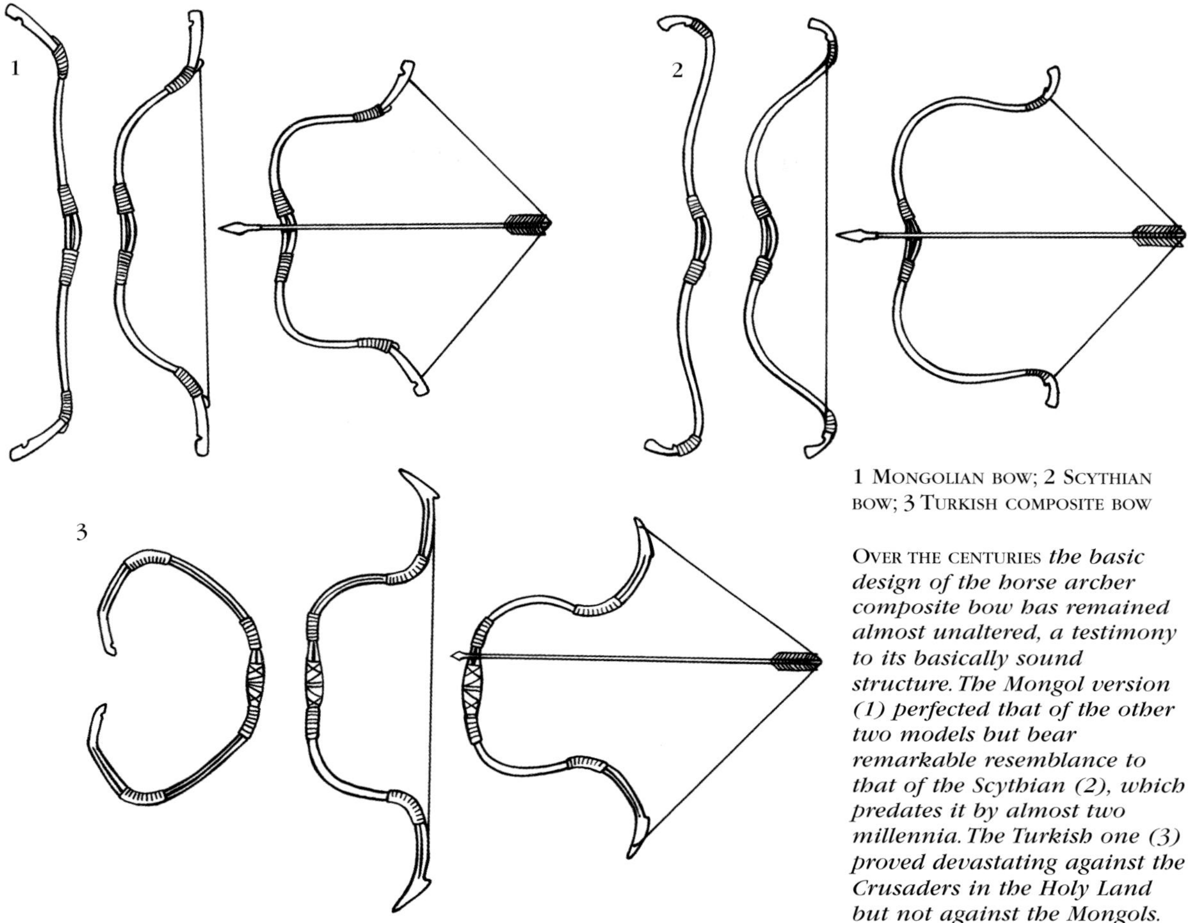

1 Mongolian bow; 2 Scythian bow; 3 Turkish composite bow

Over the centuries *the basic design of the horse archer composite bow has remained almost unaltered, a testimony to its basically sound structure. The Mongol version (1) perfected that of the other two models but bear remarkable resemblance to that of the Scythian (2), which predates it by almost two millennia. The Turkish one (3) proved devastating against the Crusaders in the Holy Land but not against the Mongols.*

to placate their northern foes. They were convinced, wrongly as it turned out, that the Mongols were not merely tribal blackmailers but were would-be conquerors of China. The continued existence of Yuan pretenders at Karakorum only served to reinforce their assumption that bribing the Mongols would only whet their appetite for plunder and bring on the conquest of China.

Bribes and Blackmail

On the other side, the Mongol leaders knew that the Golden Age of the Mongols had gone with Kublai Khan and their expulsion from China in 1368. They reverted to the simple nomad 'outer' strategy – practised since Attila the Hun blackmailed the Roman empire – of posing a potential threat in order to extract vast tribute and trade from their more civilized neighbours. This the Mongol leaders would do by raiding and plundering the border areas of Ming China. Uncompromising to a fault Emperor Yongle epitomised the military solution to the Mongol problem. By 1400, the steppes were divided between the western (Oirat) Mongols led by Mahmud Khan and the eastern (Khalka) Mongols led by Arughtai. Yongle played these off against each other and when one seemed to be gaining the upper hand he would intervene to restore the balance between the two Mongol 'kingdoms'. By 1412, Mahmud had defeated his rival and was threatening to establish a unified Mongol state. Yongle invaded the steppes in 1416 and defeated the Oirat cavalry with the use of artillery. While

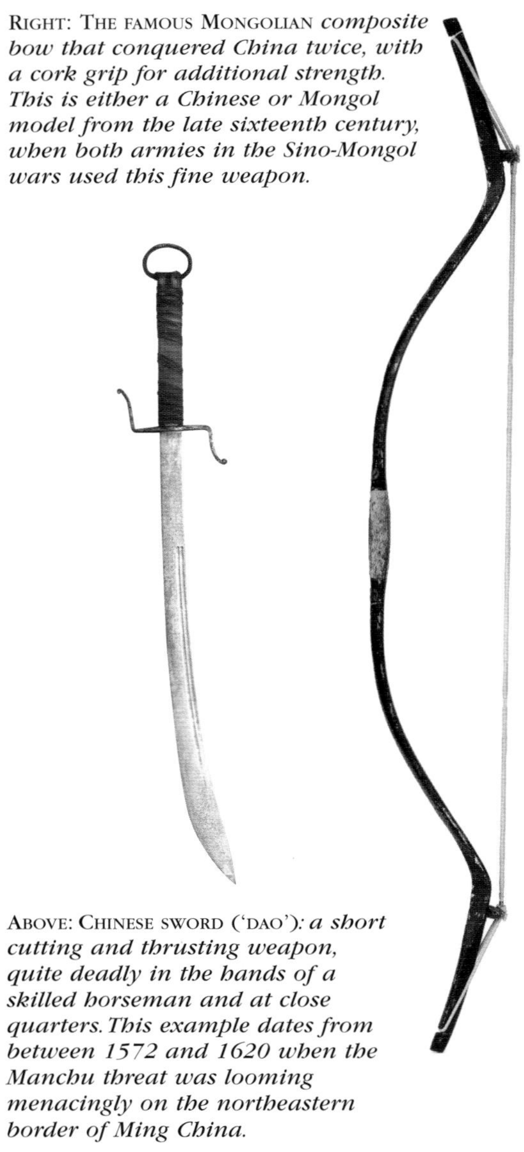

RIGHT: THE FAMOUS MONGOLIAN *composite bow that conquered China twice, with a cork grip for additional strength. This is either a Chinese or Mongol model from the late sixteenth century, when both armies in the Sino-Mongol wars used this fine weapon.*

ABOVE: CHINESE SWORD ('DAO')*: a short cutting and thrusting weapon, quite deadly in the hands of a skilled horseman and at close quarters. This example dates from between 1572 and 1620 when the Manchu threat was looming menacingly on the northeastern border of Ming China.*

walls had always been able to halt the Mongols, now artillery, when used on a large scale with enough precision and range, could defeat cavalry in open battle. As cannon quality improved this superiority would increase until even the elite Mongol and Manchurian cavalry, as shown in 1860, could no longer prevail on the open battlefield.

Yet nothing was permanent on the steppes. Arughtai returned, forcing Yongle to invade with a new army that required some 21.8 million kg (48 million lb) of grain and rice. This huge store of food was transported for the invasion army in 117,000 carts, pulled by 340,000 donkeys and 117,000 human cart pullers. Arughtai simply moved with his swift moving mobile army out of range and thus avoided having his cavalry worsted by the Ming artillery.

The resurgence of the Mongols

Only members of Genghis' family, the Genghisid dynasty, could hold the exalted title of Khan or even still more that of Khagan (Great Khan). Others had to make do with that of *tayishi* (lieutenant) as did Esen who seized power in 1439 extending his realm westward to Hami (1443) and Manchuria in the east by 1444. He not only forced out the Ming garrisons from Sinkiang's oases but also added Kansu to his equestrian empire. He declared himself Khan in 1453, having already removed the Genghisid incumbent, Tokhto Bukha, through assassination. Appropriately enough Esen was himself assassinated two years later.

For three decades, the Ming army declined in combat readiness and prowess, and it was fortunate it did not have to face a serious Mongol threat. But in 1488 Beijing's nightmares were back. Dayan Khan was not only a superb cavalry commander, a brilliant organizer and a great political leader – he was also a legitimate Genghisid Khan. He was the worst threat the Ming had encountered so far. Dayan created an elite *Khesik* cavalry strike force of 15,000 men for his raids deep inside Ming territory and to facilitate these operations he created some 30 supply depots in the steppe at a safe distance from the frontier. Thus his cavalry would not have to return or retreat through lack of supplies but could instead keep up a relentless series of attacks, feints and plundering raids.

A dangerous lack of cavalry prevented the Ming from burning the depots and retaliating against the Mongols. Worse still the combination of elite cavalry and depots enabled Dayan, who was a first-rate cavalry commander and strategist, to isolate Beijing and drive the Ming defenders to distraction.

In 1516 Dayan and 70,000 cavalry launched a massive offensive against the border defences east of Beijing. The Mongols could dominate the field and ravage the countryside at will since there was hardly any Ming cavalry in the field. Dayan's ambitious offensive was hardly a mere 'raid' but seemed to confirm Chinese fears that the Mongols, especially a Genghisid Khan like Dayan, sought to revive the ancient glory of the past by conquering China. The following year, as if to reinforce this impression, Dayan was back, albeit with less manpower – some 50,000 horsemen. It was only with the greatest difficulty that Dayan's ferocious assault on Beijing was repulsed. A lack of cavalry prevented the Ming from launching a counterattack, diversionary or flanking attacks or even staging a proper pursuit of the Mongols once they began their inevitable retreat. In 1522 Dayan launched a two-pronged invasion of northern China with a diversionary attack against Kansu while his main army marched on Beijing. This time he occupied and burnt down the capital's suburbs. Thereafter he settled for annual raids that yielded a rich booty of slaves, silks, silver and other Chinese treasure.

'They lined up the women, children and the aged of the villages and towns in groups and, in order to save trouble for the executioners and avoid spoiling their clothes with their spilled blood, they forced them to undress. Then the executioners plunged their knives into the miserable people...'

– Archdeacon Thomas of Spalato on Mongol atrocities, Hungary 1241

This seemed to confirm that even the ambitious and capable Dayan was nothing more than a Mongol warlord. His realm did not extend across the steppes like Esen's; his true base remained the eastern half of Mongolia and when he died in 1533 his 'empire' simply fell apart like a house of cards.

Eventually his grandson, Altan Khan (1507–82), was to replace him as the Ming's Mongol bogeyman. Altan's position was even weaker than his grandfather's as the Khagan was one Tümen Khan who ruled the Chahar Mongols in the easternmost part of Mongolia, leaving Altan as the subsidiary Khan of the Tümed Mongols. Fortunately for Altan his tribe controlled the central sector of the frontier including the northern part of Shanxi. Altan's ambition was far more modest than his grandfather's, showing clearly how the Mongols were in decline. Altan wanted, through raids, to force the Mings to the negotiating rug and get the proud Ming Emperors to engage in trade.

The Mongols would exchange their sturdy horses for a range of Chinese luxuries that made their harsh life a bit more pleasant, such as tea, silks, porcelain and textiles. The cost of such markets, bribes, gifts and tribute were only a tenth of the military cost. It was only after two decades of war the Ming finally, in 1570, agreed to such markets being set up.

Ming strength continued to decline, however, and as well as facing Mongol raids they were being assailed along the coast by pirates from Japan who combined with unscrupulous Chinese renegades. Warlord Daimyo Hideyoshi's invasion of Korea (1592–98), a loyal Ming vassal state, could not be tolerated but proved an expensive diversionary effort when Ming military resources were already stretched.

Defensive Posture

By 1570 the Ming were wholly defensive in their posture along the frontier and had failed to keep their cavalry forces up to strength. This was a fatal weakness that sapped the empire's ability to defend itself against a growing number of nomad enemies. A host of cultural prejudices including a traditional Confucian contempt and distrust of the military that permeated both Ming intellectuals and bureaucrats also served to fatally weaken the Ming army. There was also the practical fear that a capable general was also an ambitious politician who might use his army to topple the dynasty.

Grand Secretary (Chancellor) Chang Chü-cheng had recognized the threat this dangerous anti-militaristic delusion posed to an empire that was being threatened from several directions. He had found a capable general, Qi Jiguang (1527–87), who might be able to stop the rot and put some new life into the Ming army. General Qi had created a fighting army of professionals out of conscripted peasants and had, surprisingly enough, crushed the pirates that infested the coasts of the south, especially on the Fukien coast. Facing the existing Mongol threat and a growing Manchurian threat in the northeast, Chancellor Chang called Qi to take over the command of the Chi-chou military district – one of nine such districts along the northern border. These combined to form the frontier defence army that, on paper at least, numbered 80,000 men and 22,000 horses. Granted the title 'Commander in Chief', Qi was being asked to fight the Mongol cavalry menace with a single infantry brigade of 3,000 men. These men were tough south Chinese troops but they were infantrymen and had no idea how to fight cavalry. Eventually Qi created an army of 20,000 men in the Chi-chou district but his plans for creating an entire new frontier army of 100,000 fell on deaf ears in Beijing. Qi was a serious military intellectual who created his new frontier army around the concept of the 'battle wagon': a huge two-wheeled mule cart with a wooden screen made of eight separate sections which could be folded flat when not in use.

During battle the mules would be unhitched and the battle wagon turned sideways with one wheel facing the enemy and the screen opened and raised behind the wheel to cover an area of 4.5m (15ft). The wagons in battle position were lined up next to each other to form a continuous wall, while the end sections of each screen could be swung aside to allow the infantry to enter or exit the wall.

Each wagon carried two pieces of light artillery called *fo-lang-chi* and made of bronze or iron, resembling large-calibre rifles more than a cannon. These, together with four muskets, were fired from the battle wagon through holes in the screen. Each wagon also had a troop of some 20 men. Half were used to man the *fo-lang-chi* and manoeuvre the wagon, both on the march and on the battlefield. The other half consisted of an assault team including four musketeers. When the Mongol cavalry attacked the entire crew would put aside their firearms to grab rattan shields, fork spears and long-handled swords to fight off the horsemen. Lancers were trained to stab at the riders while the swordsmen slashed at the horses' knees and hooves. The assault teams were admonished not to advance too far from the wagons if it was necessary to advance.

> *'A state that has to fight on four fronts should concern itself with defensive warfare. In defending walled cities, the best way is, with the strength of the worn out men, to fight fresh strength of the invaders. It is assaults upon walled cities that wear out the strength of men.*
>
> – *Sun Tzu,* The Art of War

Qi suggested creating a mixed brigade of 3000 cavalry, 4000 infantry, 128 heavy and 216 light battle wagons. Being an infantryman the general did not see the full potential of the cavalry who ended up playing a subsidiary role protecting the wagon train during the march to battle, screening the wagons as they set up the battle line, then, once the Mongol horsemen appeared nearby, retire inside the wagon line. To allow the cavalry through there would be gaps in the line filled by moveable objects. Once the Mongol cavalry reached some 76m (250ft) from the wagon line they would be fired on. Now would have been the time to unleash the Ming cavalry in a shock attack and scatter the Mongols to the four winds. But Qi used

Ming Era Standard Bearer (1550)

The weakness of the Ming army's cavalry doomed their efforts to conquer and colonize the steppes to the north. The shortage of horses, skilled trainers, reliable Mongol or Manchu mercenaries, pacifism and an increasing reliance on fixed defences only made matters worse. Denied good pasturage and skilled riders, Ming cavalry was vastly inferior in quality, if not in numbers, to their nomad enemies. This Ming standard bearer is entirely equipped according to Chinese, rather than Mongol, custom and is as a consequence quite inferior to the nomads that he faced in battle. He wears brigandine armour - iron plates attached to a fabric garment - and would have carried a sword similar to that featured on p. 106. On his head is an iron helmet with leather neck guard.

instead the infantry for that. When the momentum of the Mongol cavalry charge had been broken, then and only then, did Qi allow the cavalry to charge and give chase.

The Manchurian Menace

In their obsession with the Mongols the Ming lost sight of a far worse menace brewing on their northeastern frontier in Manchuria.

Ming Emperor Hongwu chose not to emulate the Yuan by occupying Manchuria. He settled for occupying Liadong mainly as a supply base for horses. The Court of the Imperial Stud and the Pasturage Office had 20 pasturage areas in the province to rear, breed and train horses. The Jurchen in Manchuria proper were more than willing to supply the Ming with horses and serve as cavalry troops for them against their ancient Mongol foes.

In the face of growing unrest in Manchuria from the mid-1400s it was increasingly difficult for the Ming to defend the exposed Liaotung province, so it was decided to build the Willow Palisade (*Liutiaobian*) to protect the frontier with Manchuria. The western section was built 1441–47 and the eastern was completed by 1468. It consisted of a linear earth ridge with willow trees planted along it with fortified gates and stone forts at regular intervals. Regular troops manned the Palisade, its forts and 21 gates (*bianmen*). It proved no more capable of stopping a determined nomad invader than did the Great Wall proper.

In 1466 a Ming army of 50,000 troops supported by their Korean vassal with 10,000 men invaded Manchuria, swiftly defeated their leader Prince Li Manchu and burnt down Jurchen settlements. During the following century the northeastern frontier remained generally quiet especially after General Li Ch'eng-liang managed to kill the growing menace of Jurchen warlord Athai, at the fort of Gure. But a far worse menace was brewing. The Ming appointed Nihan Wailan as their puppet Khan of the Jurchen and agreed to his main rival, Taksi, being murdered. Unfortunately Taksi's son, Nurhaci (1559–1626), thereby became the sworn enemy of the Ming and created a new, formidable nomadic empire – emulating the Jin Jurchen – that would eventually destroy the Ming empire. In 1586 Nurhaci had Nihan murdered in revenge and set out to unify all the tribes of southern Manchuria. First, he brought his father's tribe under his personal control, then the Hada (1599–1601) and then the Hoifa. That left only the Ula tribe, led by his brother-in-law Yehe.

Unity and strength was Nurhaci's motto. He began by changing the name of the Jurchen to the name that they would be known in history – the Manchu (Brave). Out of these disparate tribes Nurhaci, like Genghis, created a nomad cavalry army organized along traditional lines. A hundred and fifty families supported the creation of a company ('arrow'), 50 arrows made up a regiment and five regiments, in turn, made up a banner (division or *tumen*).

Like his Mongol counterpart, Nurhaci had a strong dislike and distrust of the tribes and their independent chiefs. By 1601 there were four banners (Yellow, Red, White and Blue) each

THIS 'ZHOU' (HELMET) *dating from the sixteenth century show clearly the strong, if not overwhelming, nomadic and Mongol influence upon Chinese weapon and armour design. It is made of metal.*

commanded by his sons including his eventual heir Abahai. Fearing a plot by his sons Nurhaci created another four banners for his remaining sons and each of these eight banner princes (*belies*) were placed under the supervision of a trusted commander (*amban*).

In 1613 Nurhaci finally crushed the Ula tribe, had its ruler Prince Chu-yeng executed and declared himself, as a consequence, Khan of the Manchus in 1615.

In 1619 an army of 100,000 Ming troops (mainly infantry) invaded the Manchu state at four separate points. This dispersal of forces and the army being composed of slow moving infantry allowed Nurhaci to defeat each column in detail – the main army being defeated at Sar-Hu. Ming miscalculations allowed Nurhaci to take the whole of the rich Liaotung province including the peninsula of Liaotung under his control. By 1623 the Ming were reduced to only two forts inside the province where Mukden (Shenyang) became the new capital.

Nurhaci died, however, from a Ming artillery shot when besieging one of these forts in 1626. His place was taken by Abahai (r.1626–43) who had extensive imperial ambitions including the conquest of China. Abahai had himself declared the first Qing emperor Hung Taiji, marginalized all Manchu leaders who did not support him and expanded his new Imperial Qing army. First he created eight wholly Chinese banners and by taking over Inner Mongolia in 1633 added another series of banners there.

THE MACE WOULD *have a frightening metal head with spikes or nails for crushing the enemy's helmets or break his arms, legs and hands. This one dates from the fourteenth century and is of Mongolian or Chinese origins and has a finely decorated, leather casing for the head.*

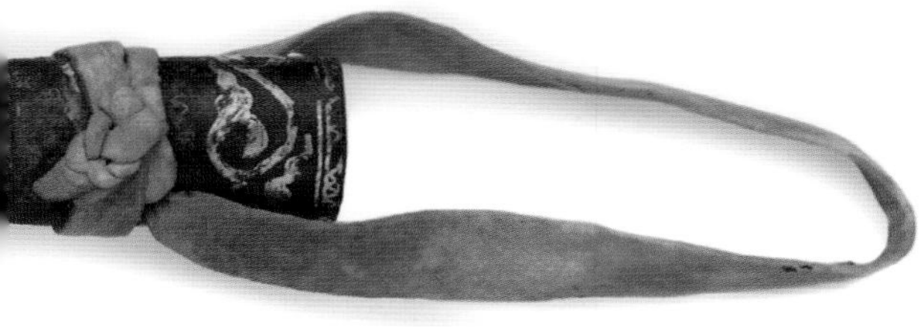

New Alliance

To remove the threat of an alliance between the Mongols and the Ming, which would isolate Manchuria or expose her to a flanking attack once the Manchu armies invaded China, Abakai made an alliance in 1629 with the Kharachin Mongols. Then he destroyed Lighdan Khan's (r.1604–35) attempt to recreate a Mongol empire.

By incorporating the southern (inner) part of Mongolia Abakai not only added valuable cavalry banners to his rapidly expanding army but also gave the Manchu the option of invading China from along the entire frontier. But already in 1635 this army was too large and costly for the slender resources of the Manchu state. Only the rich could afford the better-fed and kept mounts that would be able to stand up to the rigours of a prolonged campaign. Most Manchu warriors were reduced to buying inferior horses that reduced the mobility, speed, range and power of the Qing army – which at this point was entirely made up of cavalry.

One way to complement this was to increase the number of Chinese troops in the Army but this was a far from popular politically and militarily not particularly useful. The tough Manchurians, who were even better and more disciplined cavalrymen than the more famous Mongols, were quite sure that the Chinese were effete, cowardly, unreliable and quite useless when it came to fighting – especially on horseback.

The Manchu Conquest of China: 1644–50

Beijing, the 'City of the Yellow Dragon', had never been a good site for the Ming, whose origins were south Chinese and whose economic base lay in the Yangtze basin. Now they were to pay a high price for their capital's exposed position near the nomad frontier.

On 8 February 1622, Beijing was struck by an exceptionally strong desert wind that covered the city, ominously for the superstitious Chinese, with red dust that made the city look like it had been drenched in blood. That same day the Ming were bankrupted and one of their generals, Li Zicheng (1605–44), declared himself the first Shun emperor in the old Tang capital of Xian.

On 24 April 1644, the Shun marched, unopposed, into the suburbs of Beijing and the last Ming emperor, Chongzhen (r.1628–44) committed suicide by hanging himself in a tree outside the city's walls. The populace willingly submitted to their new Shun emperor and was relieved that his infantry army, in green and white uniforms, maintained such good order in the capital.

A loyalist Ming general, Wu San-kuei, now made a deal with the Manchus, prompting 'Emperor' Li to quit Beijing on 18 May. Two days later Manchu cavalry rode through the gates of the Great Wall. It was supremely ironic that this hugely expensive line of fortifications had not only helped to bankrupt the Ming but had also failed to protect China from the nomad 'barbarians'. On 27 May the Manchu army, supported by Wu's ex-Ming troops, defeated the Shun army at Shan-hai-kuan or the battle of the passes. Again the side with the strongest cavalry force, the Manchus, had beaten a native Chinese army with weak or non-existent cavalry. The Shun army limped back to Beijing but was forced to evacuate the city as the Manchus, led by Prince Regent Dorgon (1612–50), approached. Dorgon's cavalrymen marched into the stricken capital on 5 June, greeted by startled Ming loyalists who had expected General Wu's Imperial troops instead.

The Manchus, like the Mongols, were to find out that taking Beijing was only the beginning of the conquest of China. But what is surprising is how quickly and painlessly the Manchus managed to take mainland China. A new Ming emperor, Longwu, had fled to Nanjing where he had appointed his energetic relative, the Prince of Fu as Generalissimo. Fu concentrated 60,000 elite Guards troops around Nanjing and along the southern bank of the Yangtze. Further north he organized five military districts between Kaifeng and Yangchou with a massive army of half a million troops – almost all infantry – to keep out the Manchu invaders.

Dorgon had eight banners (80,000 men) under his direct command. An equal number were under the command of Prince Bolo further east. This army was small with slender resources to back it up, and lacking in infantry. But what it lacked in numbers it more than made up in discipline, *esprit de corps*, leadership, mobility and manoeuvrability. Before they could deal with the Ming the Manchus had to crush the Shun army of 200,000 men under Li's command at Xian. Dorgon moved his banners through Honan and he concentrated his cavalry east of the Tongguan pass: the gateway to Xian from the east. To the north stood Prince Ajige's banners ready to strike. The Manchus would defeat the Shun army in a pincer offensive against Xian. Li had little experience fighting nomad cavalry and committed the cardinal error of leading his infantry out of their prepared positions at the Tongguan pass. On 8 February 1645 the Shun army was badly defeated at the pass by elite Manchu guard (*keshik*) cavalry troops led by General Tulai. The Shun army fled back to Xian where they found to their horror they were surrounded. They

Manchu Cavalryman (Seventeenth Century)

A Manchu (Qing) horseman is attacking the enemy with raised sword. Judging by his arms and equipment, he's a heavy horseman leading the vanguard into the stricken Ming empire. Although dating from the seventeenth century, his soft leather boots, covered metal helmet, curved sword and light armour could just as well have been taken from five centuries earlier – a clear sign that in eastern Asia, military technology had not followed the revolutionary developments in Europe.

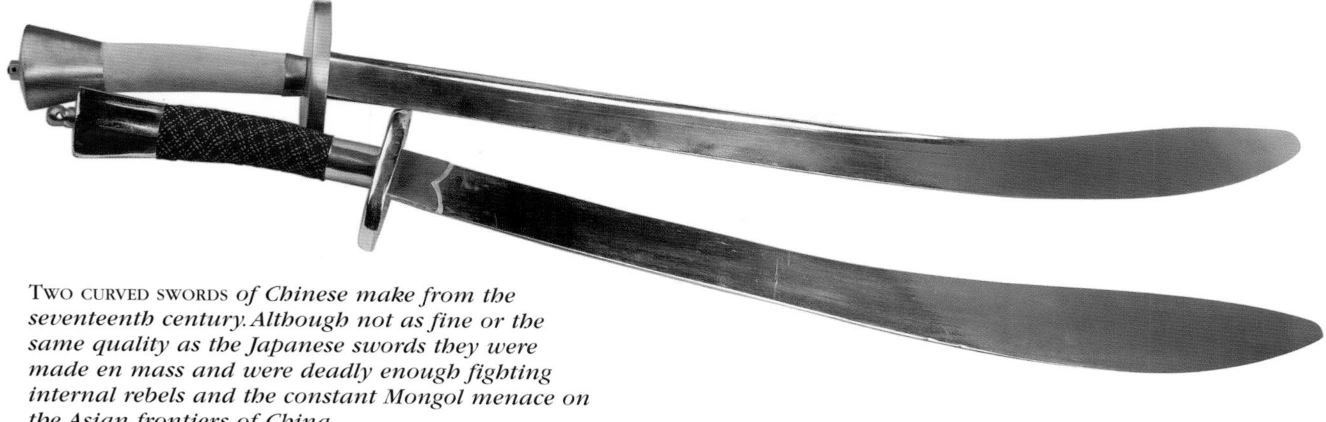

TWO CURVED SWORDS *of Chinese make from the seventeenth century. Although not as fine or the same quality as the Japanese swords they were made en mass and were deadly enough fighting internal rebels and the constant Mongol menace on the Asian frontiers of China.*

could not turn back for there was Dorgon's banners, they could not escape west since there stood Tulai's guards, and in the north was Ajige's horsemen.

Dorgon could now turn his attentions to the Ming. Fortunately for him the Ming chancellor Hung-Kuang and Prince Fu had been bitter enemies since the 1620s. Their petty quarrels paralysed the new Ming government in Nanjing and made a shambles of Hung-Kuang's solid defence cordon plan drawn up earlier. Dordon formed a special banner (5000–6000 strong) of his best horsemen to spearhead the attack southward. In total his army would number 40,000 and began advancing southward in three columns on 1 April. They encountered little resistance except at Yangchou which fell on 20 May. For once the Manchus emulated Genghis' use of terror by having the city sacked. On 30 May the Manchus, using a ruse, managed to cross the River Yellow and attack the shocked Ming infantry in the back. When the key fortress city of Chen-ching fell it set off a panic in Nanjing. The Emperor Longwu fled from the city on 3 June and by 16 June Dorgon arrived in triumph. A hundred thousand Ming troops surrendered. Prince Nihan's cavalry force went after the slowly retreating Ming infantry and managed to capture Prince Fu alive.

OPPOSITE: A HEAVY CAVALRYMAN *wearing the characteristic Mongol lamellar armour ('zub can') dated to the seventeenth century that gave both rider and mount good protection without impairing their mobility or speed too much. Although supposedly 'heavy' this cavalry was in fact lighter and nimbler than their occidental counterparts.*

At the same time, Prince Bolo with 80,000 troops, both infantry and cavalry, had been marching down the Grand Canal and had crossed the Yangtze further east, and already on 6 July had captured Hangchow – the old Song capital – without much resistance. By January 1647 Bolo had captured Canton, then Sichuan. By the winter of 1648 the whole of China, except for the extreme southwest, was in Manchu hands.

Ulan Butung, 1690: The Camel Wall

Cavalry had been the key to the success of the Qing army's conquest of China but it became, as had its nomadic predecessors, a hybrid force including artillery, naval forces and infantry armies. Yet here in the east, cavalry was to play a role much more significant than it did in the west. Due to their paucity of horses and men the Ming cavalry had been very weak, and this had, in military terms, sealed their fate. With the Qing the Chinese would have the first combined army since the Yuan era that was confident in fighting on the steppes. It also helped that the Manchu already had control of the southern (inner) part of Mongolia where, by 1670, there were some 49 banners of cavalry. These were to be an invaluable complement to the Manchu's forces.

By now, new threats had appeared in the east, including a most unexpected player on the scene:

A LARGE CHINESE SHIELD ('PO') *dating from the nineteenth century. Richly decorated, in this case with a tiger or lion's face with snarling fangs, these lovely pieces of armour were no match for western artillery, rifles and explosives.*

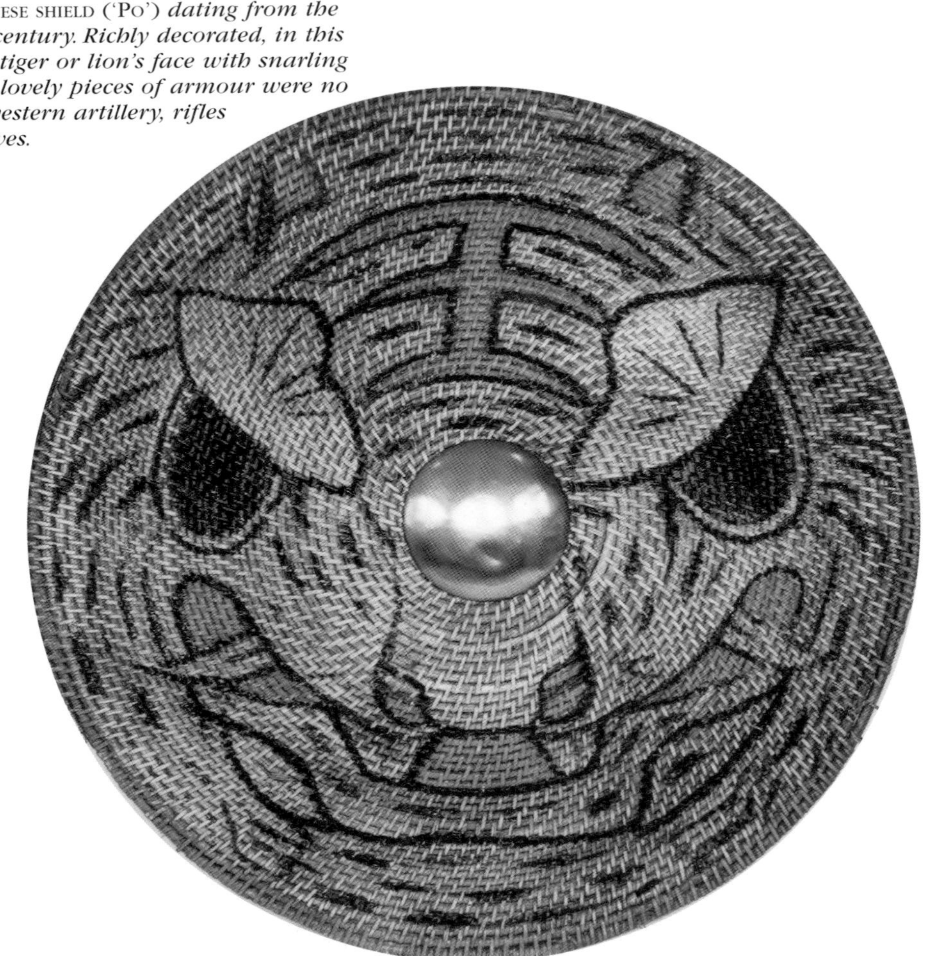

Imperial Russia. Russia had been the victim itself of Batu Khan's invasion during the winter of 1237–38 with 150,000 men. It was only at the 'Field of Snipes' (Kulikovo) in 1380 that 30,000 Russian troops (mainly Mongol-style cavalry) managed to defeat the Golden Horde Mongols.

In 1552 Tsar IV the 'Terrible' conquered Kazan and in 1582, using artillery and muskets, the Cossacks defeated a Mongol cavalry charge outside Isker (Sibir). This deadly combination of firepower would herald the demise of the mounted nomadic warrior. Yet these same Cossacks were the illegitimate occidental heirs of the Mongols themselves and became the world's most famous western horsemen.

Establishing a string of forts (*ostrogs*) across Siberia, the Cossack horsemen had founded Irkutsk and Yakutsk in 1632 prior to reaching the Pacific in 1646. Russia had, therefore, become a neighbour of Manchu China and in 1650 the ruthless Cossack chief, Khabarov, raided Dauria in northern Manchuria. He founded a fort at Anchansk that the Qing army seized on 24 March 1652. The Russians were stronger in muskets but the Qing artillery proved more potent. In July 1686, the Manchu stormed the last Russian fort in the Amur valley, Albazin, forcing the Russians to sign in September 1689 the humiliating treaty of Nerchinsk. For the next 150 years, until the Russians established Nikolayevsk in the River

Amur delta, the Yablonovoi range constituted the Sino–Russian border.

It was just as well that the Russian threat had been neutralised. In 1634, Batur Khuntaidzhi took control of the western (Oirat) Mongols who had formed the powerful Zungharian (Dzungarian) confederation under his father. The following year Batur defeated the Kazakhs in the west and established cordial relations with the Russians in Siberia.

The Attack from Zungharia

In 1671 Galdan Boshughtu became ruler of Zungharia and during 1679–80 Galdan's superb Mongol cavalry occupied Hami, Turfan and Kashgar in short order. This made Zungharia a threat not only to the trade and communications but also to the security of China's westernmost province of Kansu and of Manchu-occupied eastern Mongolia. The latter was a vital buffer zone for the Manchu empire against both Zungharian and Russian expansionism. Should these two empires combine, given their fine cavalry armies, the Qing empire would be seriously threatened like never before.

'From a total of 2647 officers...and brave-men only half are present. There should have been 1043 horses but not even one can be obtained. From a total of 7240 commanding officers and regular banner-soldiers only 4600 are left...'

— *Report on corruption in the Imperial Ming army in 1614*

When Galdan invaded eastern Mongolia in 1688 with 30,000 cavalry Emperor Kangxi (1654–1722) decided to meet the challenge head on. In 1690 Galdan marched down the strategic Kerulen river with his army in a bid to seize control, or so Beijing believed, of inner Mongolia. Perhaps native Chinese dynasties like the complacent Song or the isolationist Ming would not have cared about such a move but to the nomadic Manchu it was a direct challenge and one that could not be tolerated without retaliation. In this long drawn out conflict the Zungharians would have, like all nomad cavalry armies, speed, mobility and nimbleness on their side while the Qing would have numbers, arms and firepower on theirs.

The Manchu Prince of Yu marched his combined Qing infantry-cavalry army with its strong artillery contingent into the steppes hoping to catch, as did the Ming in the past, the Zungharian either in a camp or with a physical barrier at their backs. The slender hope was that the enemy would be cornered, make a stand and allow themselves be destroyed. In a set piece battle the lightly armed and armoured nomads would be at a disadvantage.

Even for the cavalry minded and steppe-wise Qing a campaign deep in the wilds of Mongolia was an exhausting, expensive and dangerous venture calling for meticulous planning and preparation. The expeditionary army would have to bring with it an enormous amount of supplies carried in carts or on camels' backs. It would also need every cavalryman and his horse they could find to screen the army's movements, and to scout, harass and then, in a battle, charge the enemy's own cavalry. The strength of the cavalry force could be the factor that would decide the fate of an expeditionary army.

As with previous nomad regimes that had moved their base to China, the Qing had problems with their cavalry. Horses were expensive to purchase and feed and the state only offered a paltry sum, 12–20 *taels* for each animal. Qing officials were reduced to simply seizing the horses they needed from the Chinese – who answered, in time-honoured fashion, with violent riots. Despite every captain in the army being responsible for feeding and fattening ten horses and then turning them over to the Ministry of War's officials there were never enough. Inspectors reported before

Battle of Ulan Butung

1690

Having finally caught up with the Zungharians amidst the wooden hills of Ulan Butung, the Qing Chinese forces employed a combined force of artillery, infantry and cavalry in a traditional combined-arms attack. Caught by surprise, the Zungharian commander, Galdan, decided to stand and fight. After suffering an artillery bombardment, the Zungharians protected their camp in a highly unconventional fashion, forming a defensive barrier by trussing up their camels into a camel wall, or *tuo cheng*. Although the Chinese commander Yu claimed a victory when his troops made a breakthrough on the left flank, the main attack proved a failure, allowing the bulk of the Zungharian cavalry army to survive and withdraw intact after a negotiated ceasefire. The camel wall proved decisive, in that the Zungharians were able to hold their positions while at their most vulnerable, allowing them to survive to fight another day.

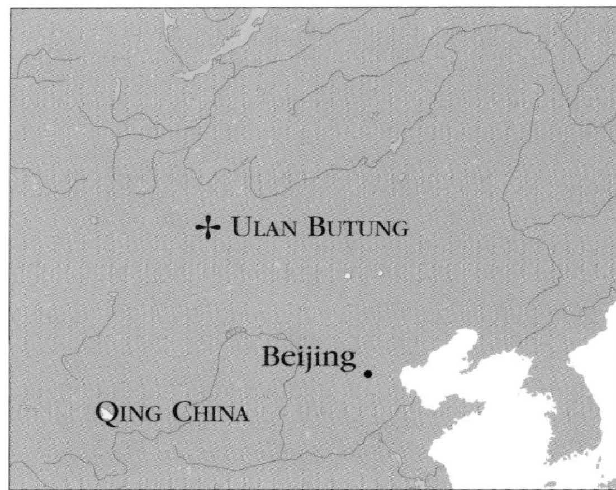

Ulan Butung is situated in Outer Mongolia, and proved to be a difficult place for the conventional Qing Chinese army to access, with supply lines stretched to a maximum and food and water supplies proving difficult to manage from early in the campaign.

2 **The Zungharians are pushed back on the left flank and surrounded in the wooden hills behind. They suffer some casualties but escape using their superior mobility.**

1 **The Chinese open the engagement with an artillery bombardment and then massed attack.**

CHINESE CAMP

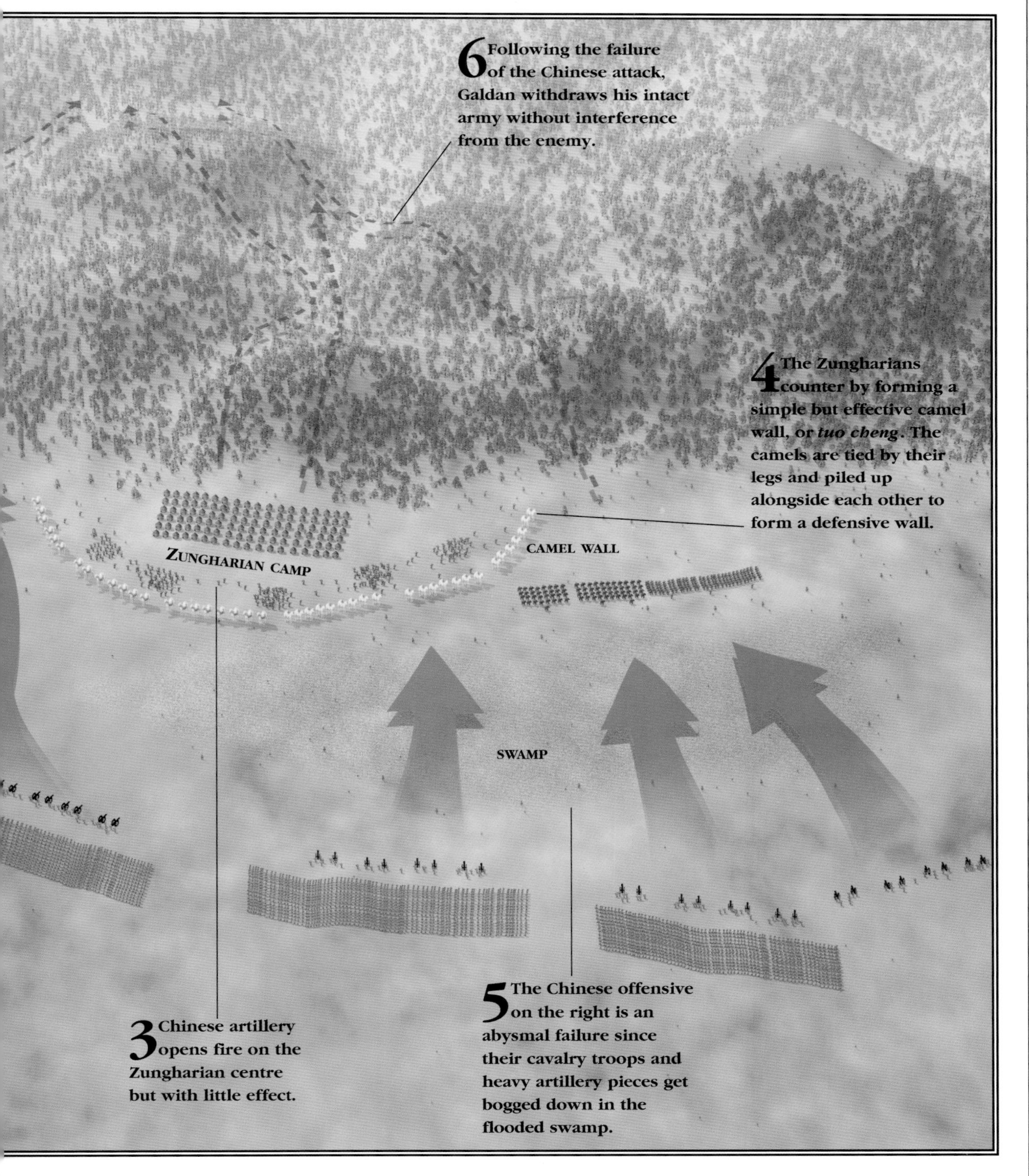
6 Following the failure of the Chinese attack, Galdan withdraws his intact army without interference from the enemy.
4 The Zungharians counter by forming a simple but effective camel wall, or *tuo cheng*. The camels are tied by their legs and piled up alongside each other to form a defensive wall.
ZUNGHARIAN CAMP
CAMEL WALL
SWAMP
3 Chinese artillery opens fire on the Zungharian centre but with little effect.
5 The Chinese offensive on the right is an abysmal failure since their cavalry troops and heavy artillery pieces get bogged down in the flooded swamp.

the expedition set out that the frontier depots were filled with thin, sick horses.

A Manchu officer in the Qing army, Maska, described how the expeditionary army was in trouble as it crossed the bone-dry Gobi desert where the troops were forced to dig wells to find any water. This water turned out, after hours of back-breaking labour, to be undrinkable. While the men began to die from thirst the horses found neither water nor fodder. As if this was not bad enough the grain ran out when they had crossed the Gobi. Not surprisingly the army's advance was reduced to a snail's pace – 300km (186 miles) in twelve days.

Early in the morning of 3 September Qing scouts reported that they had found Galdan's camp nearby on a slope of wooded hills at Ulan Butung. Yu could not believe his luck and ordered a swift advance to prevent the Zungharians escaping. Galan seemed to be willing to stand and fight as the Qing artillery swung into position and, at noon, opened fire.

Another eyewitness to the unfolding battle was the Russian Ambassador, Kiribev, still smarting from the humiliation of signing the Nerchinsk treaty. Galan, noted Kiribev, had come up with an ingenious way of protecting his camp and army from the Qing artillery. He had ordered his men to tie the legs of their camels, some ten thousand beasts, place them alongside each other in a lying position in an unbroken line and then cover them with felt. There were gaps between some of the animals so the Zungharians could fire arrows at the enemy. Maska was most impressed with the *tuo cheng* (camel wall). Yu tried to claim a victory when the Qing left flank attacked and created a breakthrough. They were supposed to have pushed part of the Zungharian army into the mountain range and surrounded them. The claim seems to be an exaggeration on the part of Yu in his reports since the camel wall held and on the right the advance of the Qing right was halted. They were in fact forced back into their own hastily-improvised camp and by evening Yu ordered his whole army back to camp. It seemed to have been anything but a 'great victory' as Yu claimed.

Kiribev noted that Galdan's army had suffered some casualties but was intact thanks to the camel wall and the protection of the trees on the hill slopes. Galdan would be ready, the ambassador claimed, for a new battle the following day and Kiribev wasn't sure that the Qing could hold their line should the Mongol cavalry charge with customary ferocity.

On 7 September, after three days of negotiations, Yu allowed Galdan to march off unharmed into the steppes. Yu's generals urged him to pursue Galdan while in the open but by the time the Kharchin Mongol reinforcements had joined with Yu's army at Shengjiang Ula, the Zunghars were long gone.

> *'All warfare is based on deception. Hence, when able to attack, we must seem unable... when we are near, we must make the enemy believe we are far away; when far away, we must make him believe we are near. Hold out baits to entice the enemy. Feign disorder, and crush him.*
>
> – *Sun Tzu,* The Art of War

Jaghun Modo, 1696

Yu and his generals were given a violent dressing-down by an infuriated Emperor less than impressed with their 'victory' at Ulan Butung. A few more of those and the empire would be finished. Kangxi thundered at them that a campaign that had cost his hard pressed subjects 16 million *taels* had been wasted thanks to their incompetence in letting Galan off the hook.

Previous native Chinese dynasties had no idea about what campaigning in the steppes entailed as Mongolia was *terra incognito* for them. Not so Kangxi who not only knew the steppe but appreciated hunting there. He wrote:

'The condition of the northern areas can be understood only if you travel through them in person; and as you move you must pay attention closely to details of transport and supply. You can't just make guesses about them, as they did in the Ming dynasty – and even now, the Han Chinese officials don't know much about the region.'

Kangxi's words stung hard and deep since they were simple truth. The Qing were now determined, in the fashion of the nomad wars, to finish off the Zunghars as their nomad rivals once and for all.

After putting down internal rebellions, Galan was back in central Mongolia by August 1695 but this time with a mere 5000–6000 horsemen. The Qing intelligence service reported from their Mongol spies that Galdan was camped at Bayan Ula. Kangxi wanted to strike immediately but was persuaded by the tough and experienced general Fiyanggû to wait until the following spring when there would be adequate pasturage for the Qing horses.

Kangxi planned a massive, sweeping invasion of Mongolia to trap Galdan and destroy his small army. He would take command of the Capital (Central) Army, setting out from Beijing with 33,000 troops due north along the River Kerulen. Fianggyû was to block Galdan's western escape route with his Western Route army (30,000 troops) holding the River Tula but if he could cover the 1160km (722 miles) to Bayan Ula he was to do so and battle it out with Galdan. Finally, in the east, General Sabsu with 10–12,000 troops was to block the River Khalka. In August 1695 their Mongol allies supplied another 8000 horses for the coming campaign.

The Western Route army left their headquarters at Guihua on 22 March 1696. It was only by abandoning those with lame or sick horses and by leaving behind his infantry that Fianggyû

IT WAS EXPERIENCED, *tough and resilient horsemen like this Manchu warrior that allowed the Qing armies, successfully combining Chinese infantry and nomad cavalry, to defeat the 'menace' of the Zungharian Mongols and conquer much of eastern Central Asia in the process.*

finally reached the River Tula on 6 June. During this campaign the Qing troops came to terms with how horses worked. Horses had to be tended and cared for with the utmost consideration. They could die from over-exertion, a sudden halt could kill them and if they sweated heavily they had be galloped to increase their body temperature. Experienced Mongols pointed out that the horses were not to be watered unless they were completely dry. Hundreds of horses were lost through inexperience or neglect – the army was yet again reduced to eating horsemeat.

Finally on 12 June the Qing found their quarry at a place called Jaghun Modo ('a hundred trees' in Mongolian). It was a small valley with a river at the bottom surrounded on all sides by hills. Finally Fianggyû had Galdan in a trap. Galdan had at most 6000 cavalry but he faced 14,000 Qing troops, mainly Mongol and Manchu cavalry, supported by a strong corps of artillery.

It was imperative to capture the high ground before the Zungharians did so and thus gain the upper hand. If the artillery could be placed on the high ground then the Zungharians below would be destroyed. Both sides realised this and this made the fighting a savage life-and-death battle in the nomadic tradition: no quarter expected or granted.

Fianggyû ordered his men to use the wooden screens that they had constructed and stuff their armour with cotton – all to reduce casualties from the hail of arrows the Zungharians unleashed upon them. He then ordered that most of his men to dismount and advance on the Zungharians behind the wooden screens, as his artillery fired round upon round against the enemy. The Qing had only solid shot and no explosive rounds but the effect at close range was still devastating. The Zungharians, showing that the Mongol mettle was still there, stood their ground. But the Qing screens showed themselves to be effective protection while the artillery fire wreaked havoc on the Zungharian horsemen and camp.

Then a rumour, which may have been planted by Qing agents, spread in the Zungharian ranks: Emperor Kangxi was approaching with his own army. Their position was hopeless as they would be trapped between two strong cavalry armies with artillery.

Whether Fianggyû knew of this rumour or not ,he signalled that the Manchu heavy cavalry were to move up. They prepared themselves and then as a flag was waved charged as only Manchurians and Mongols knew how to: with loud, guttural shouts in dense ranks. The Zungharian cavalry rode up to meet the charge but their spirits, already low by the false rumour that Kangxi was on his way, broke completely when their best commander, Arubdan, died fighting. They wilted, fled and were killed where they stood. Galdan, seeing his army buckling and cracking, fled with only some 40 loyal retainers. The Qing attacked with relentless fury and spared none of the enemy. For once a

DECKED OUT *in all his ceremonial glory, Qing (Manchu) Emperor Kangxi (1654–1722) radiates calm resolve and all the self-confidence of the most populous, ancient and possibly civilized empire in the world. But the Zungharian Mongols were as yet undefeated by his vast armies and resources.*

IN THIS FINE COLOUR *painting by the eighteenth century artist Chaio Ping Chen, the Emperor Kangxi is shown entering the town of Kiang Han in the year 1699. The town's population turns out to gaze as the vast entourage enters the town.*

battle on the steppes had been decisive. The Zungharian army had been well and truly destroyed in a field battle.

Emperor Kangxi had marched northwards in October but he was nowhere near Ulan Butung at the time of the battle, so upon his return to Beijing, on 12 January 1697, he ordered two Manchu-Mongol cavalry columns to set after Galdan. Kangxi was taking no chances. As long as Galdan was alive he and his men posed a threat to the empire's security and he wanted the warlord brought to Beijing dead or alive - preferably dead! On 4 April Galdan died of poisoning, probably at the hands of Zungharians in Qing pay. Exactly four months later the Emperor staged an enormous military parade in Beijing to celebrate his phenomenal victory over the dreaded Zungharians.

'Chiao' – Extermination

Emperor Kangxi may have celebrated the fact that his arch-rival Galdan was finally gone and that he now occupied Hami and had secured the empire's northern and northwestern flanks. But the victory was dangerously incomplete since the Zunghar empire was still intact and the Qing victory at Jaghun Modo had only been secured with the assistance of Galdan's sworn enemy and nephew, Prince Tsewang Rabdan. As ruler of Zungharia (r.1697–1727) Tsewang was to prove an even greater menace than Galdan had ever posed.

In 1710, Zungharian cavalry raided into Russian Siberia thus alienating a potential Zungharian ally in the coming life and death struggle against the Qing. However, Kangxi, now 60, was no longer able, as in 1696, to take the field himself, while the malaise of political corruption was growing throughout the empire. The Emperor laid bold plans for three cavalry armies, each of 10,000 men, to invade Zungharia from different directions. The first objective was to capture Hami, 500km (310 miles) west of the nearest Qing depot at Sizhou. The ultimate aim was to take Ürümchi, another 500km (310 miles) further west from Hami which was the key to the eastern defences of the

Zungharian Khanate. By this time the Chinese officers in the Qing army had learnt more about cavalry warfare while the Manchurians had learnt about the need for proper supply lines, so the Qing command set up a supply depot at Barköl (Balikun). The Manchurians had learned a thing or two from fighting the Zungharians about not taking unnecessary risks.

A planned offensive for 1715 was cancelled and it was only in April 1717 that 8500 Qing troops (predominantly cavalry) set out from Barköl for the advance on the oasis city of Turfan, while a second force marched directly on Ürümchi. The two armies defeated smaller Zungharian armies covering their eastern frontier, at the oases of Mulei and Pizhan. Boldly the first army commander, Furdan, wanted to advance on the Ili valley – the core of the Zungharian realm – but a swift attack by 300 Zungharian cavalry put that thought out of his mind.

Furdan became more cautious since he established forts and posts deep inside central and western Mongolia and a series of fortified military agricultural colonies before he began planning a second-stage move against Xinjiang. As it was Turfan only fell in 1719 to General Fiyanggû but the victory did mean that the Chinese were back in Xinjiang after an absence of 1000 years.

Meanwhile, during the summer of 1717, Tsewang sent his best and most ruthless general, Tsering Dondub, with 3000 cavalry, into Tibet. The Qing could not believe that these exhausted and famished men, who had been forced to eat dogs en route to Lhasa, could capture the Tibetan capital. But they did, sacking the monasteries and installing a Zungharian puppet as Dalai Lama.

On the March

In response, Manchu General Erentei marched with 7000 men from Xining through the desert while his colleague, Namujar, with 10,000 Manchu cavalry marched due west from Sichuan.

A MANCHU (QING) *army uniform (*ding jia*) from the middle of the nineteenth century, showing how military technology, like everything else in the Middle Kingdom, had remained rooted in the past while the West was at the same time increasing its technological superiority over China.*

Unfortunately for the Qing, Erentei's army was trapped and destroyed by Tsereng's army in September 1718. Kangxi ordered his son, Prince Yinti, to invade Tibet with 300,000 troops divided into three armies. Yinti's forces occupied Lhasa on 24 September 1720.

Fianggyû had created two massive depots at Barköl and Turfan where he positioned a combined garrison of 33,000 troops, while he placed a further 25,000 in garrison forts across the Altai. The Zungharians were now faced with Qing troops on two sides. Given their superior artillery the Qing should, believed the general, invade Zungharia in a vast pincer move spearheaded by the best Manchu-Mongol cavalry they had. The problem was that Prince Yinti, who should have became Emperor and supported Fiyanggû's bold plans, was recalled on 21 December 1722 and placed under house arrest by his rival, Yinzhen, who ascended the Dragon throne as Emperor Yongzhen (1678–1735).

Five years later his Zungharian enemy, Tsewang, died leaving his place to Galdan Tsereng (reigned 1727–45) who proved every bit as formidable as his father. Yongzhen concentrated all the troops at Barköl, recalled those in Tibet, placing them instead in Sichuan and Xining. When the rough and outspoken Fiyanggû protested he was removed from office. Yongzhen, who wanted peace with Zungharia and reduced military costs, trusted a line of static defences in Sinkiang. In 1729 Yongzhen finally changed his mind and having gone from one extreme, pacifism and appeasement, veered to the other pronouncing the dreaded word: *chiao* (exterminate).

In April 1729 Yongzhen appointed Furdan commander of the Northern Route Army (18,000) while General Yue Zhongqi became commander of the Western Route Army of 324 officers and 26,500 troops. During 1730–31 the Zungharians raided the area around Turfan but it was not until March 1731 that the Emperor agreed to Yue's demand that the two armies should march against Ürümchi, 170km (105 miles) northwest of Turfan. Ominously he told Yue: 'I am beginning to lose faith in you.'

Furdan advanced in the north then occupied and fortified Khobdo. This forced Tsewan to divide his army. Some 10,000 cavalry were in the west keeping an eye on the restless Kazakhs while he placed the lion's share of his force, 30,000 men, under Tsering Dondub. Leaving 7500 men at Khobdo to block the northeast route for the Zhungarians and prevent them linking up with Eastern Mongol rebels, Furdan advanced against the Zhungarian heartland with three divisions from the rest of his army.

Mongol Trap

Tsering's son attacked 2000 Manchu troops on 20 July with his 3000 and three days later Furdan fell into a classical Mongol trap. He had advanced to Hoton Nor, a small lake 210km (130 miles) west of Khobdo. Furdan had been harassed all the way while Tsereng with 20,000 men hid between a nearby mountain range. Furdan had only half that number of cavalry, mainly allied Khalka and Khoshot Mongols. Furdan managed to escape the Mongol trap with only 2000 of his men while the rest were massacred. General Yue also abandoned Ürümchi for Barköl.

The defeat drove Emperor Yongzhen to despair and he made peace (1731–41) with his over-mighty nomad neighbour. It was his son and heir, Qianlong (1711–99), who was to complete the quest to exterminate the Zhungarians: something he was only able to do when Galden Tseren died in 1745. A savage war of succession gave him the opportunity to invade Zungharia in March 1755 with 50,000 men. The northern army (25,000) was commanded by General Bandi while an equal number, under Yong, marched from the west at Barköl. The two armies met up, after little nomad resistance, at Bartala in June 1755.

Amursana, who had refused to rule the Zhungarians as a Qing puppet, fled northwards. Of his subjects, 600,000 in total, some 30 per cent, were massacred by the Qing army, 40 per cent died of smallpox and the remainder (120,000) only survived by fleeing into Kazakh- or Russian-controlled territory.

In early 1757 Amursana hoped to join up with Prince Chingünjav's Khalka Mongols, who had risen in rebellion in the summer of 1756. Qianglong ordered that all grasslands in rebel hands be burnt, and by January 1758 the rebels –

Jaghun Modo

1696

Six years after the 'victory' of Ulan Butung the Qing army finally got to grips with Galdan's Zungharians. Finding himself trapped with only 6000 cavalry in a narrow wooded valley surrounded by hills, Galdan could only, like his men, stand his ground and fight. Qing general Fianggyu caught the enemy off guard, seized the high ground and pummelled Galdan's army with intense fire before ordering his dismounted men to advance behind wooden screens. This proved effective as a diversion for the coup de main: a full and ferocious charge by his heavy Manchu and Mongol cavalry. The charge proved decisive, breaking the already weakened Zungharian morale, and forcing Galdan to flee with the remnants of his shattered army. Although the Zungharian war would continue for another half a century, their strength was permanently weakened by the defeat at Jaghun Modo.

JAGHUN MODO
Beijing
QING CHINA

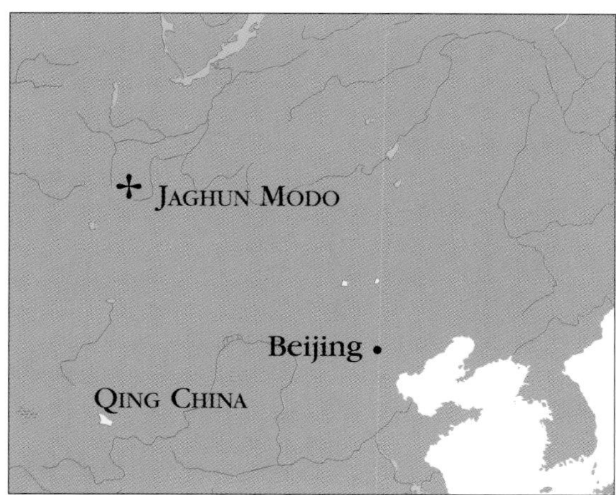

As in the case of Ulan Butung, the exact geographic location for Jaghun Modo is not known. Experts estimate that it must have stood somewhere near Bayan Ula in Mongolia.

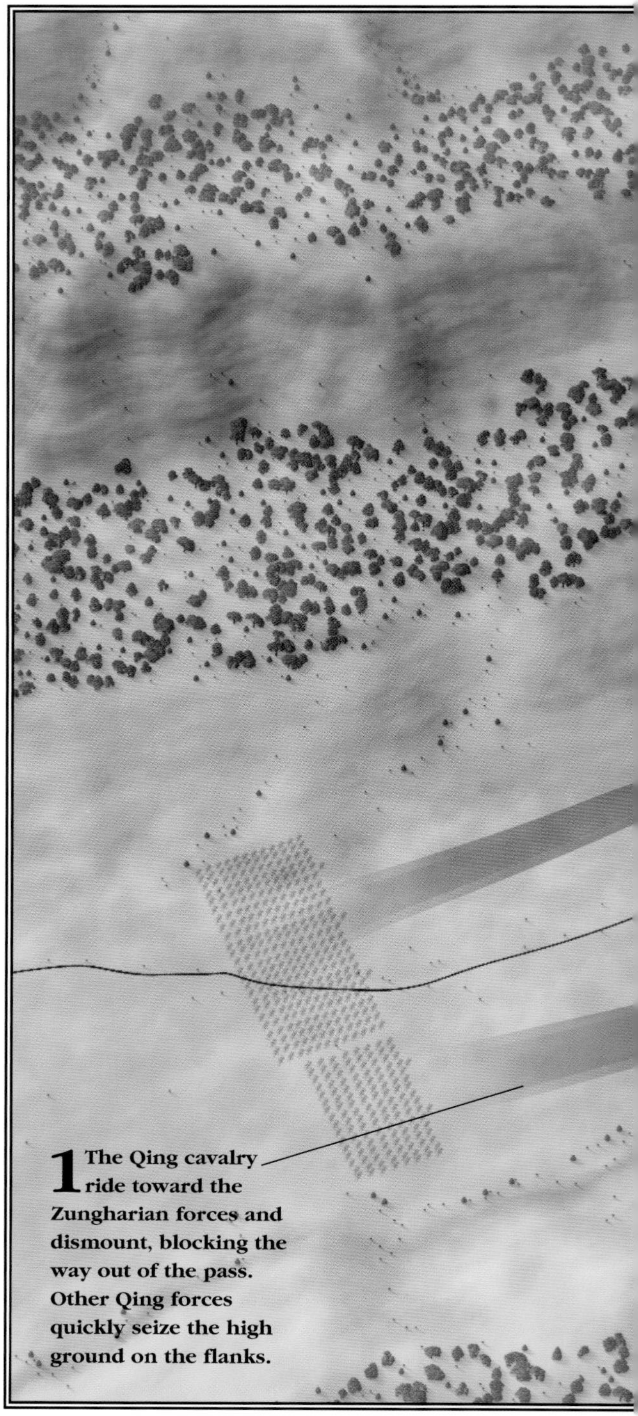

1 The Qing cavalry ride toward the Zungharian forces and dismount, blocking the way out of the pass. Other Qing forces quickly seize the high ground on the flanks.

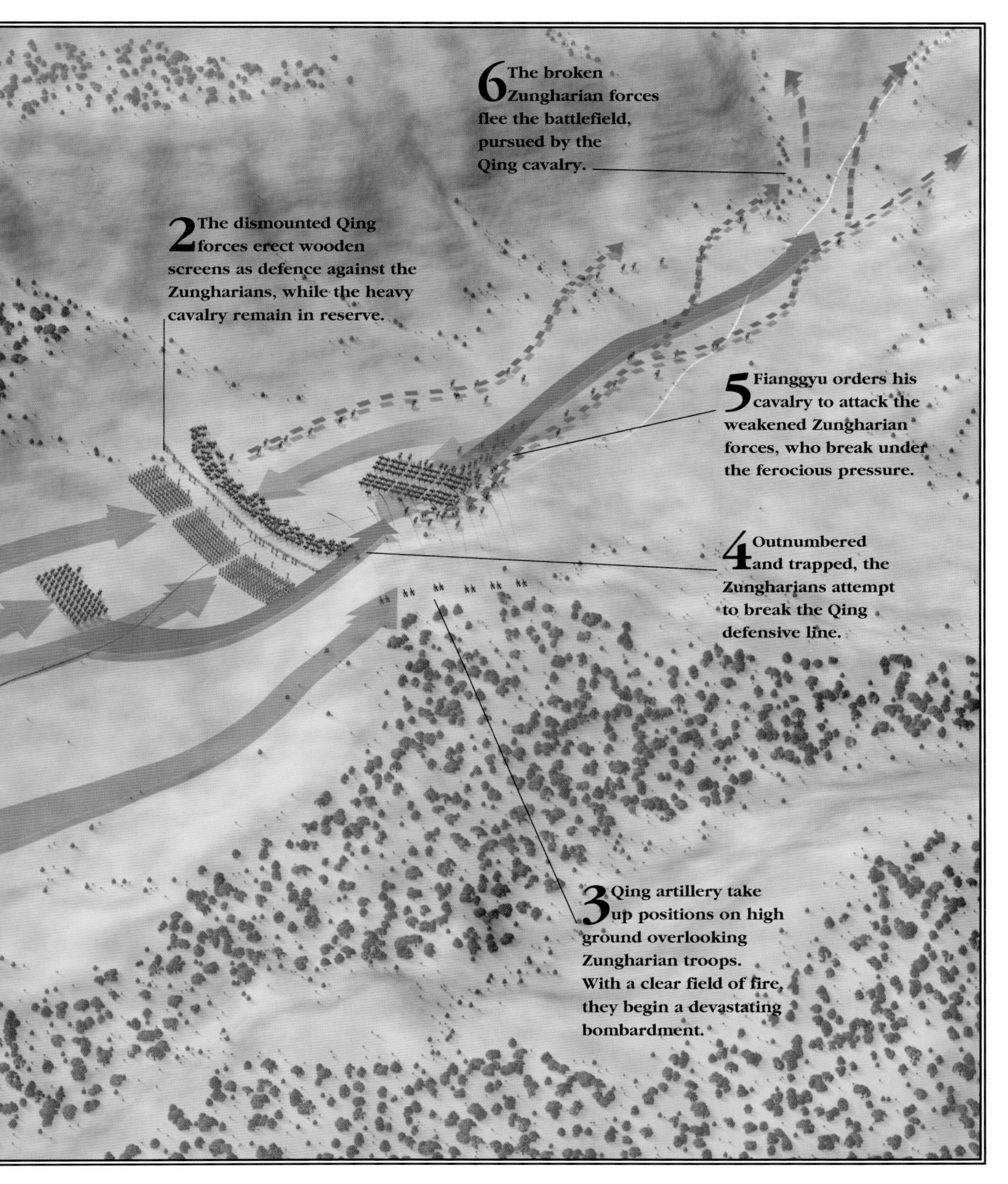
6 The broken Zungharian forces flee the battlefield, pursued by the Qing cavalry.
2 The dismounted Qing forces erect wooden screens as defence against the Zungharians, while the heavy cavalry remain in reserve.
5 Fianggyu orders his cavalry to attack the weakened Zungharian forces, who break under the ferocious pressure.
4 Outnumbered and trapped, the Zungharians attempt to break the Qing defensive line.
3 Qing artillery take up positions on high ground overlooking Zungharian troops. With a clear field of fire, they begin a devastating bombardment.

who had been reduced to 2000 cavalry - were crushed, with Chingünjav and other leaders being executed. But before then Amursana, after being hunted like a wild animal, had managed to reach safety in Semipalatinsk where, on 12 September 1757, he died of smallpox aged only 35. Qianlong demanded the Russians turn over the body. To their credit the Russians refused.

March on Beijing, 1860: A Gallant Last Stand

Qianlong, the Emperor that finally crushed the Mongol menace once-and-for-all, extended the empire to include, by 1760, the whole of Central Asia all the way west to Sinkiang. He was also the first emperor to receive, both reluctantly and rudely, a British ambassador in 1793.The people he dismissed with such arrogance returned in 1840s, when his descendants were defeated by the iron warships and modern gunnery of the British Royal Navy. In 1842, China was opened to western trade and missionaries - an unbearable affront to the proud, xenophobic Chinese rulers.

After the Qing had refused to honour the Nanjing treaty, and after earlier repulses, in August 1860 France and Britain sent an expeditionary army of 17,000 men to China. Commanded by Lieutenant General Sir Hope Grant, the British expeditionary army (10,000) had a strong contingent of British and Anglo-Indian cavalry, the latter made up of tough Sikh, Punjabi, Bengali and Pathan *sowars* (cavalry troopers). This Anglo-Chinese war would see the traditional

THIS PRINT FROM A WESTERN *journal dated September 1860 shows how Manchu horsemen clash with Anglo-Indian cavalry. In this clash between East and West the proud Manchus were defeated and forced to accept humiliating peace terms.*

nomadic mounted archer pitted against the finest western and south-Asian cavalry of the era. This was a classical horse warrior's war, and both sides proved disciplined and brave. The allies landed and took possession of Beidang fort in August and almost immediately made contact with 2300 Qing cavalry. These extended their lines to envelope and outflank the European infantry forcing their commander, de Montuban, to order a hasty retreat back to Beidang. The Qing commander reported prematurely that his brave cavalry had driven back the barbarians.

During 5–6, August the British 1st King's Dragoon Guards (KDG) numbering 326 men and 339 horses landed at Beidang together with the Indian cavalry of Probyn's Horse (456 men, 495 horses) and Fane's Horse (363 men and 383 horses).

Previously, on 12 August, the Anglo-Indian cavalry had met a 3000 strong Qing cavalry force that sought to envelop their infantry. The Anglo-French infantry formed squares, as if fighting under Wellington, while the cavalry charged with drawn swords and lowered lances. One Bengali skewered a Manchu rider with one blow but the nimble Manchu horses were more agile and outpaced the heavier British and Indian mounts when pursued.

Having taken Tientsin on 24 August the allies began the advance on Beijing on 8 September on semi-paved roads. Off the roads were fields of cut millet with short stalks, which were dangerous for both horses and riders. On 18 September, Probyn's Horse (down to 106 *sowars*) and the KDG confronted and defeated 2000 Qing cavalry with a violent charge. One old Sikh *sowar* insultingly compared the proud Manchu cavalry to chickens: difficult to catch, but once caught, quite harmless. Genghis would have spun in his grave at the Sikh's insult of troops that had conquered much of India.

The final battle of the campaign took place at Palichao bridge across the strategically crucial Yang Liang canal. Qing general Sankolinsin, with 20,000 men, placed his infantry in the centre and his strong cavalry on the right. The Chinese cavalry attacked into the gap between the French and British formations. Again, the allies were saved by their cavalry, where the ferocious Pathans fought alongside the elite KDG.

End of an Era

Once the allied cavalry charge had broken that of the enemy and driven him back, Sankolinsin signalled, using the traditional flags and banners, for his army to retire. It was the last, gallant stand of the ancient nomad cavalry army that now had been defeated by alien cavalry and modern technology. On 24 October, the imperial Manchu-Chinese capital Beijing fell to yet one more barbarian army.

CHAPTER 3

COMMAND AND CONTROL

Without doubt, the military thinkers, strategists and battlefield leaders of the Orient have been profoundly influenced by a collection of ancient Chinese texts known historically as the Seven Military Classics. Although many of the tenets put forth in these manuscripts may be considered quite basic and even oversimplified, their enduring application and tremendous influence on command decisions during two millennia cannot be discounted. In fact, their impact on military command both east and west is acknowledged to this day.

The success of a military commander, both on the strategic and tactical levels, must be attributed to a combination of native intelligence, temperament, decisiveness and discernment. These attributes and others, practical and theoretical, are discussed in the Seven Military Classics, which include *The Art of War* by Sun Tzu (c.544 BC–496 BC), Jiang Ziya's *Six Secret*

DURING THIS 2003 REENACTMENT *of a battle from the* Sengoku, *or Warring States period in Japanese history (1467–1573), at Onizawa, Honshu, costumed samurai prepare the night before a battle.*

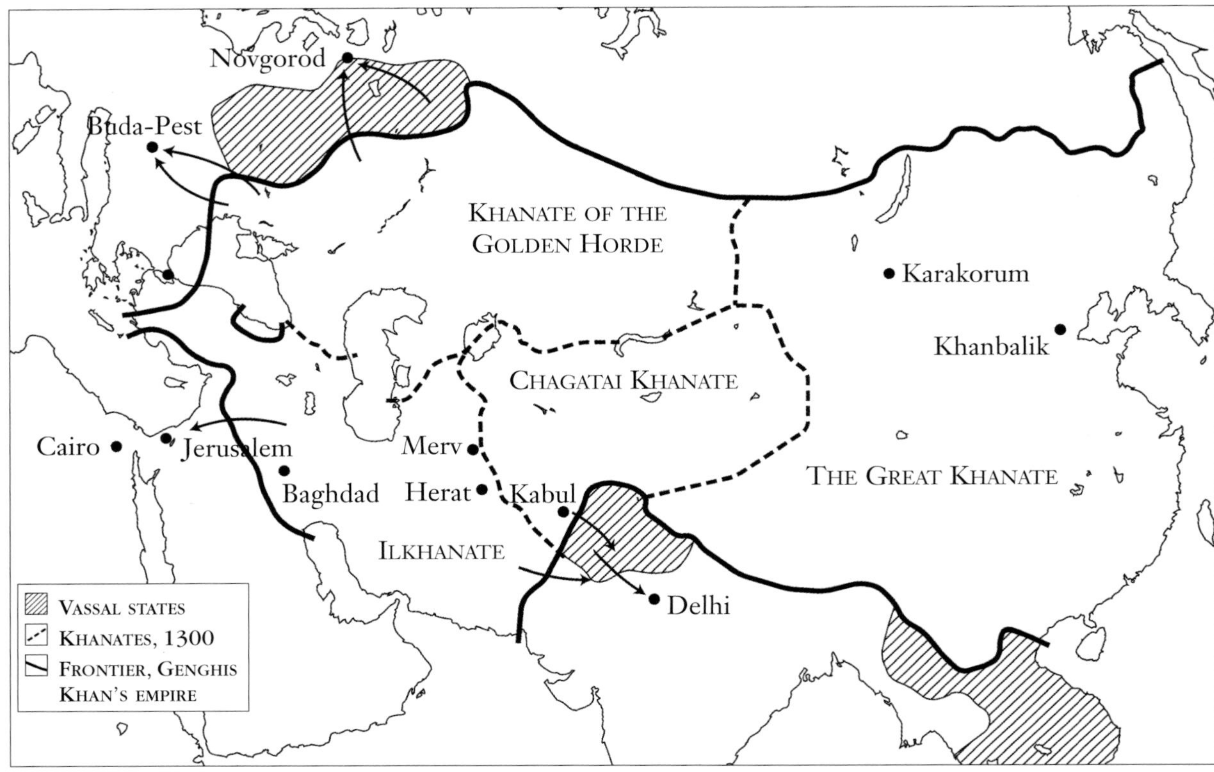

THE MONGOL EMPIRE *in 1300 was the greatest land empire the world has ever seen. At its height, it stretched from Southeast Asia in the east to Poland in the west. Under the leadership of Genghis Khan's grandson, Batu, the Mongols first entered Europe in 1237, capturing the plains of western Russia in just a single winter.*

Teachings, The Methods of the Ssu-ma, Wu-tzu by Wu Qi (d.381 BC), *Wei Liao-tzu, Questions and Replies between T'ang T'ai-tsung and Li Wei-kung* and *Three Strategies of Huang Shi-kung.*

The appointment of a military commander, whether the ruler of an empire or principality or the lieutenant of such a ruler, requires an understanding of those qualities which will lead to victory.

In *The Art of War,* the best known of the Seven Military Classics, Sun Tzu advocates nine principles of war, encompassing those of manoeuvre, objective, offensive, surprise, economy of force, mass, unity of command, simplicity, and secrecy. In discussing manoeuvre, Sun Tzu advises:

'Let your rapidity be that of the wind, your compactness that of the forest. In raiding and plundering be like fire, in immovability like a mountain. Let your plans be dark and impenetrable as night, and when you move fall like a thunderbolt.' He admonishes the commander to:

'Refrain from intercepting an enemy whose banners are in perfect order, to refrain from attacking an army drawn up in calm and confident array – this is the art of studying circumstances.'

Conquest and Unification

During more than 600 years of warfare from 1200 to 1854, the great campaigns and battles waged by rulers and generals of the Orient were to achieve varied purposes. The scope of these confrontations and the numbers of people involved often dwarfed contemporary conflicts in the west. The great Genghis Khan (1167–1227) forged an empire which encompassed millions of square miles and sired a dynasty which reigned for generations following his death. Tokugawa Ieyasu

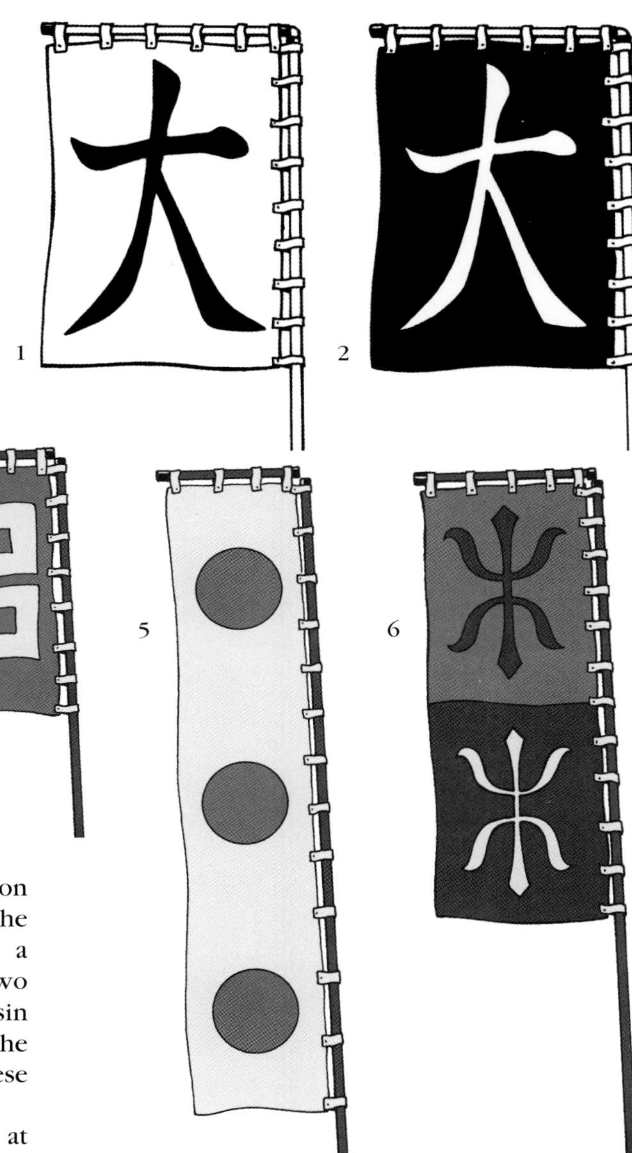

BANNERS OF RECOGNITION *were inspiring for soldiers and served as points of reference during often confused fighting. These banners are representative of the type which were carried in battle by feudal Japanese armies.*

1 & 2 BANNERS BEARING THE CHARACTER TAI, MEANING 'GREAT', USED BY TAKEDA KATSUYORI (1546–82); 3 *SASHIMONO* OF MAKINO TADANARI DEPICTING A THREE-LEAVED FLOWER (1615); 4 *SASHIMONO* OF TOKUGAWA'S RETAINER KYOGOKU TADATSUGU; 5 *NOBORI* OF SAKAI TADATSUGU (1527–96); 6 *VAJRA* DESIGN, YAMAMOTO KANSUKE (16TH CENTURY).

(1542–1616) prosecuted a campaign of accession to the rule of a unified Japan and established the Tokugawa shogunate in the island nation, a dynasty which endured for more than two centuries. The genius of Admiral Yi Sun-sin (1545–98) resulted in astounding victories for the Korean navy and forced the invading Japanese to retreat.

The vast expanse of Eurasia offered little at times to provide sustenance to an army of thousands, except that which could be plundered and taken by force from conquered peoples and territories. The sheer logistics of warfare were daunting. The Mongols, for example, campaigned from Southeast Asia to the interior of China, the Middle East and the steppes of Russia. Their nomadic existence, while an advantage in that they were able to travel relatively lightly, nevertheless required that thousands of people and even larger numbers of animals were continually fed, clothed and provided with rudimentary shelter. The successes of the Mongol campaigns under Genghis Khan are testimony to the recognition of the necessities of lengthy campaigns and the ability of the Mongols to adapt to a variety of extremes.

Advancing Technology

Inherent in the success of a general's battle plan is the effective and efficient employment of technology. The Mongols became conquerors of much of the known world owing to the organization of their forces, particularly light cavalry mounted on their diminutive but hardy horses, and their sturdy bow, with its range of 320m (1050ft) easily outdistancing the English longbow by up to 90m (295ft). The magnificent performance of Admiral Yi's famed turtle ships annihilated the Japanese and resulted in the celebrated naval victory at Hansando. A single volley of musket fire at Sekigahara set in motion the undoing of the Western Alliance and propelled Ieyasu to victory.

In order to exert command and control on the battlefield, great generals of the Orient were masters of organization and deployment of their forces, as well as in communicating before, during and after a major battle. Communication by signal flags, couriers, drums or horns was increasingly difficult at times due to distance, noise, and the proverbial fog of war and rattle of battle.

'Generals have five critical talents... What we refer to as the five talents are courage, wisdom, benevolence, trustworthiness, and loyalty. If he is courageous he cannot be overwhelmed. If he is wise he cannot be forced into turmoil. If he is benevolent he will love his men. If he is trustworthy he will not be deceitful. If he is loyal he will not be of two minds.'

– SIX SECRET TEACHINGS, *THE TAIGONG*

Those who consistently communicated clearly and entrusted orders to capable subordinates were most often victorious. Subutai Bahadur (1176–1248) and Jebei Noyan (d.1225), two of Genghis Khan's famed 'Dogs of War', carried Mongol imperialism far and wide, employing an innovative military doctrine which was clearly superior as their forces moved inexorably westward, beating all oppositions.

Across the centuries, the military maxims tested during the great battles of the Orient have stood the rigours of time. Great generals have proven their worth both tactically and strategically through meticulous planning and execution. Sun Tzu summed up the key elements of a victorious general in *The Art of War:*

'He will win who knows when to fight and when not to fight; he will win who knows how to handle both superior and inferior forces; he will win whose army is animated by the same spirit throughout all its ranks; he will win who, prepared himself, waits to take the enemy unprepared; he will win who has military capacity and is not interfered with by the sovereign... But when the army is restless and distrustful, trouble is sure to come from the other feudal princes. This is simply bringing anarchy into the army and flinging victory away.'

The simple yet timeless truths of the great Seven Military Classics resonate today as clearly as in centuries past. Generals who aspire to greatness must continue to heed their advice and their warnings.

The Mongol Invasion

To many of the Europeans they encountered, the Mongols were known as Tartars, perhaps an allusion to the fear that these nomadic Asian warriors sweeping toward them had emanated from the gates of Hell itself (the Latin word for Hell being 'Tartarus'). Indeed, the Mongol juggernaut was like nothing the world had ever seen, possibly eclipsing even the military conquests of Alexander the Great.

For two centuries, the Mongol Empire was pre-eminent across most of the known world. At its height, the empire stretched for an incredible 12 million square miles (32 million sq km) and

included more than 100 million people. The chief architect of the Mongol Empire was the legendary Genghis Khan, who emerged as leader of a unified confederation of peoples following years of ethnic infighting. Legend has it that the Great Kahn traced his lineage to his father, a grey wolf, and his mother, a white doe. It was said that the child came into the world clutching a large clot of blood, an omen that he would one day become a great conqueror.

While much of the Great Khan's early life is shrouded in myth and folklore, it is known that his ascendancy to the leadership of the Mongols and the beginning of the empire occurred about 1206. Subsequently, he led a three-year campaign into northwest China and Tibet. By 1216, his gaze had turned toward Central Asia, the Middle East and Russia. The neighbouring kingdom of Khwarezm, which encompassed the modern states of Tajikistan, Uzbekistan, Turkmenistan and Afghanistan, lay to the west. Relations between the two empires were contentious and open warfare became inevitable when Mongol ambassadors to Khwarezm were assassinated.

The 800th anniverary of Genghis Khan's unification of the Mongol tribes was celebrated in Sergelen County, Mongolia in 2006. Five hundred re-enacters dressed in thirteenth-century period costume for the occasion.

Although the war in northern China was not yet concluded, Genghis Khan led a powerful army westward in the spring of 1220. This campaign was marked by a determined Mongol advance from several directions, which was punctuated with successful sieges of major cities. The ruler of Khwarezm fled to an island in the Caspian Sea but was hunted down and killed. Genghis Khan returned to his Mongolian homeland and later waged yet another campaign against China; however, he fell ill and died in the summer of 1227 at the height of his power and prestige.

Effective Subordinates

Throughout his campaigns, Genghis Khan was well served by a cadre of trusted subordinate generals. Foremost among these were Subutai and Jebei. Subutai had risen rapidly through the ranks of the Mongol army and distinguished himself during the fighting in China. At the age of only 25, he was in command of a force of Mongol cavalry. He pledged his loyalty to Genghis Khan by swearing, 'As felt protects one from the wind, so I will ward off your enemies.' Although he had at one time been a man of athletic build, he is said to have gained weight to the extent that he was hauled by wagon during the campaigning of his army.

Jebei, on the other hand, was believed to have had a somewhat less than auspicious beginning in the service of the Great Khan. *The Secret History of the Mongols,* a primary source on the life of Genghis Khan and the campaigns of his Mongol horde, recounts that when the Khan attacked the Taichud tribe, to which Jebei belonged, he faced his future master in battle. Jebei (who was at that time known as Zurgadai) is said to have either killed Genghis Khan's horse or wounded Genghis himself in the face, or both. Captured during the fighting, Jebei was brought before the Khan and admitted to committing these deeds.

'It is the business of a general to be quiet and thus ensure secrecy; upright and just, and thus maintain order. He must be able to mystify his officers and men by false reports and appearances, and thus keep them in total ignorance. By altering his arrangements and changing his plans, he keeps the enemy without definite knowledge.'

– *Sun Tzu,* The Art of War

Faced with the prospect of execution, Jebei declared that the choice of killing him in retribution rested with Genghis Khan. However, should the Khan spare his life, Jebei would serve him loyally. Impressed with the display of bravado, Genghis bestowed upon his new lieutenant the name Jebei, which means 'The Arrow' in Mongolian.

Following the great victory in Khwarezm, Subutai and Jebei retained joint command of a Mongol force in Azerbaijan. In 1221, a reconnaissance mission took their army into Russia for the first time. There the Mongol forces performed brilliantly. After plundering and destroying numerous settlements, the Mongols were paid a substantial tribute to spare the city of Tabriz. At Tbilisi, an army of Georgians under King Giorgi the Brilliant was defeated, and the winter of 1221 was spent in Armenia. After defeating a second Georgian army, the Mongols captured Astrakhan in southern Russia. The winter of 1222 was spent along the banks of the Black Sea and in the spring the Mongols crossed the River Don and entered the Crimea and the Ukraine.

March Westward

The advance of the Mongols brought them into contact with the Kipchaks, a nomadic people who lived in the region of the River Dniepr. Khan Kotyan, a Kipchak ruler, raised the initial warning that a grave threat from the east was surging toward the principalities of Russia and the Ukraine. Kotyan reached the court of his son-in-law, Mstislav of Galich (d. 1228), also known as Mstislav the Bold, and warned: 'Today they will slaughter us. Tomorrow they will come for you.'

Although these Russian princes were wary of one another and sometimes cast an envious eye upon one another's territory, it was apparent that an alliance was necessary for mutual self-preservation. Mstislav Romanovich (d.1223) hosted a summit at Kiev, which Mstislav the Bold and Mstislav Svyatoslavich of Chernigov both attended. Fatefully, they concluded that a combined force should march eastwards and confront the approaching Mongols.

Historians theorize that while the Mongols had won numerous victories their ranks were depleted somewhat by attrition. Therefore, Subutai and Jebei decided to send envoys in order to parlay with the Russians. These diplomats offered to spare the Kievan Rus' from the ravages of war and withdraw from their land in exchange for a free hand to exploit their gains against the Kipchaks. Msistlav the Bold rejected the peace overture out of hand and, to his everlasting regret, murdered the Mongol envoys.

The Strategic False Retreat

One of the favoured tactics of the Mongols was the strategic false retreat, which often played on the overconfidence of an adversary who might logically assume that the Mongol army was in full flight. An opposing commander all too often found his force overextended and himself the victim of a clever ruse when the Mongols stood and fought on ground of their choosing.

One European observer cautioned his fellow commanders against being taken in by the Mongol practice. 'Even if the Tartars retreat, our men ought not to separate from each other or be split up, for the Tartars pretend to withdraw in order to divide the army, so that afterwards they can come [at us] without any let or hindrance and destroy the whole land.'

When news of the fate of their envoys reached Subutai and Jebei, the enraged Mongol leaders did indeed order such a retreat. Confident of victory, the Kievan Rus' and their allies gathered along the Dnieper at Khortytsva and initiated a pursuit. At first, they encountered only small, lightly armed bands of Mongols whom they swept up and quickly defeated. Those Mongols along the eastern bank of the mighty river fell back quickly, but their commander, Gembayek, was taken prisoner and summarily put to death. A second Mongol force was also attacked by a contingent from Volhynia. Bolstered by these easy but inconclusive victories, the Kievan Rus' pressed ahead while Subutai and Jebei lured them toward the banks of the River Kalka (now renamed the Kalmyus River) near what is now Donetsk in the Ukraine. In his work *All the Khan's Horses*, author Morris Rossabi described the Mongol strategy of retreat and the important role played by their ubiquitous mounts:

> 'Mobility and surprise characterized the military expeditions led by Chinggis [Genghis] Khan and his commanders, and the horse was crucial for such tactics and strategy. Horses could, without exaggeration, be referred to as the intercontinental ballistic missiles of the thirteenth century. The Battle of the River Kalka in southern

A MONGOL HEAVY CAVALRYMAN *of the thirteenth century is depicted astride his sturdy mount. Protected by leather helmets and thick armour, these troops often swept into a disorganized enemy and inflicted severe casualties with spear and sword.*

Russia is a good example of the kind of campaign Chinggis Khan waged to gain territory and of the key role of the horse. After his relatively easy conquest of Central Asia from 1219 to 1220, Chinggis Khan had dispatched about 30,000 troops led by Jebe [Jebei] and Subedei [Subutai], two of his ablest commanders, to conduct an exploratory foray to the west. In an initial engagement, the Mongols, appearing to retreat, lured a much larger detachment of Georgian cavalry on a chase. When the Mongols sensed that the Georgian horses were exhausted, they headed to where they kept reserve horses, quickly switched to them, and charged at the bedraggled, spread-out Georgians. Archers, who had been hiding with the reserve horses, backed up the cavalry with a barrage of arrows as they routed the Georgians.'

RECOGNIZED AS ONE *of the great military conquerors of history, Genghis Khan eventually ruled an empire larger than that of Alexander The Great. Genghis Khan is also remembered as a brutal and ruthless ruler.*

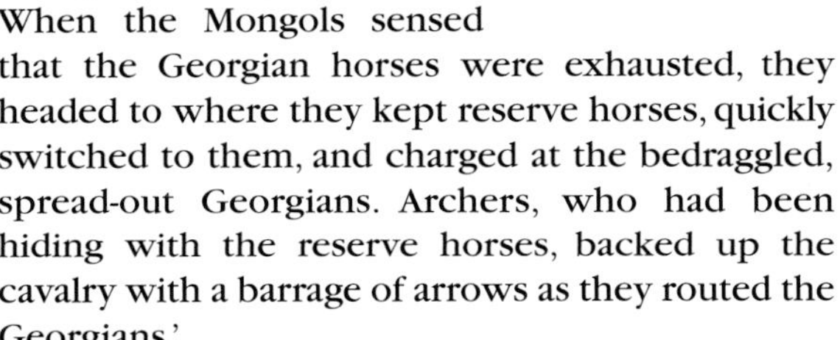

During extended campaigns, Mongol warriors preferred to ride mares, usually stringing together three or even four of the animals to be interchanged when one became fatigued. The horses grazed on the short grass of the steppes, so transporting fodder was not necessary. The Mongols, in turn, could subsist at least partially on the mare's milk and even on its blood if necessary. Mongol lore relates that no horse which had braved the sting of battle with its rider was ever slaughtered for food. At the end of its usefulness, the battle horse was retired to pasture, and when a warrior died, his horse was killed and the two were interred together, riding as spirits.

Battle of the River Kalka, 1223

The intent of the Mongol incursion into Russia had not originally been to conquer and hold territory. It was, according to most historians, a reconnaissance in force. Numbering at peak strength no more than 40,000, the Mongols under Subutai and Jebei were heavily outnumbered by a gigantic Russian army nearly twice their size. The Kievan Rus' were, however, prone to internal strife and discord, which played into the hands of the Mongols. They failed to achieve unity of command, with each general advocating his own strategy for the annihilation of the invaders. To compound the problems faced by the Russians, only about a quarter of their troops had received any proper military training. Those who had been trained or had actually experienced combat had faced only Western style opposition. The tactics of the Mongols would prove very different from anything previously encountered. For nine days, the Mongols gave ground. The divided Russians began to fragment even further, weakening their decided advantage in numbers. Finally, on 31 May 1223, the advance elements of the Russian army, the Volhnyians and Polovtians, drew up along the River Kalka.

With a predetermined signal, the light cavalry of the Mongols assumed the offensive. The initial contact, however, seemed to be favouring the Russians, and the Mongols executed a coordinated retreat across a bridge spanning the Kalka. The Russians, though, were unable to press their temporary advantage and their pursuit faltered.

Subutai sensed that the time was right for a decisive blow and committed the Mongol heavy cavalry to the fray. These more heavily-armed and armoured horsemen bowled into the Russians and put their main body to flight, while the light cavalrymen fired withering barrages of arrows as they repeatedly crossed the enemy's route of advance. Turning in confusion, the Polovtians and Volhynians panicked and attempted to cross the bridge to the western bank of the Kalka. In doing

THIS MODERN DEPICTION *of a Mongol campsite conveys the immensity of the following which accompanied the warriors who ventured far from their homeland in East Asia and conquered much of the known world. The mobility of the Mongols proved a great asset during long campaigns.*

so, they became hopelessly tangled with an allied contingent from Galicia which had just arrived. Milling about at the crossing point, the Russians were easy targets for the sledgehammer blows of the Mongol heavy cavalry. Soon the Galicians had either been trampled or panicked and had joined the retreat.

With the enemy pressed in the centre, Subutai ordered his subordinates Tsugyr and Teshi Khan to launch attacks from the left and right flanks. The Chernigov army and the Kievan contingent had arrived, but came upon the scene too late to alter its outcome. The Chernigov soldiers were cut down in great numbers and fell back, while the Kievans attempted to stem the Mongol tide by building makeshift fortifications with their wagons and equipment. The effort was futile.

One contemporary account of the battle reveals that the Kipchak warriors: '...failed and retreated in such haste that they galloped over the Russian camp and trampled it underfoot. And there was not time for the Russian forces to form ranks. And so it came to complete confusion, and a terrible slaughter resulted.'

Come Retribution

The swift light cavalry of the Mongols was said to have pursued the fleeing Russians for more than 96km (60 miles), inflicting horrific casualties and showing no mercy. The Kievan soldiers who had made a defensive stand managed to hold out for two days but were forced to surrender.

Although Msistlav the Bold escaped with his life, he is remembered as abandoning the other princes to what can only be described as a ghastly fate. The Mongols had promised that they would not spill the blood of any prince who was captured, but six Russian princes were taken prisoner, bound hand and foot, and stretched beneath large wooden planks. The Mongols then enjoyed a festive victory banquet, eating and drinking atop the planks while their captives slowly suffocated beneath them. Although the execution of the princes was horrific, the Mongol

Dismounted Mongol Archer (1220)

Although they were quite adept at archery while on horseback, the versatile Mongol horse archers often took advantage of tactical circumstances during battle. Under favourable conditions, the cavalrymen would dismount and loose barrages of arrows against a confused and disorganized foe. The bow itself was of composite construction, made from horn, sinew and wood.

James Chambers in The Devil's Horsemen *describes it thus:*

'The bow was easily the Mongols' most important weapon. The mediaeval English longbow had a pull of 34kg (75lbs) and a range of up to 229 metres (250 yards), but the smaller bows used by the Mongols had a pull of between 45-68kg (100-150lbs) and a range of over 320m (350 yards). The velocity was further increased by the difficult technique known as the Mongolian thumb lock: the string was drawn back by a stone ring worn on the right thumb which released it more suddenly than the fingers. A soldier could bend and string his bow in the saddle by placing one end between his foot and the stirrup and he could shoot in any direction at full gallop, carefully timing his release to come between the paces of his horse, so that his aim would not be deflected as the hooves pounded the ground.'

pledge that no royal blood would be spilled if the Russians surrendered had been kept.

While Mongol casualties at the battle of the River Kalka had been light, the Russians suffered heavily, with estimates of their casualties running as high as 50 per cent. The defeat of their combined forces hastened the irreparable fracture of the Kievan Rus' and one of the byproducts of their disintegration was the rise of a Russian Empire centred on Moscow far to the north.

The path to the conquest of vast territories in Eastern Europe and Russia was laid open by the Mongol victory. Even so, after covering more than 6400km (4000 miles) during a three-year campaign the Mongol vanguard turned back to the east to rejoin the main army of Genghis Khan. Jebei died, apparently of disease, during the return trek. With the death of Genghis Khan himself in 1227, it would be nearly a decade before a Mongol army under Batu Khan (1205–55) returned to the west to complete the conquest of Central Asia and Europe.

'...failed and retreated in such haste that they galloped over the Russian camp and trampled it underfoot. And there was not time for the Russian forces to form ranks. And so it came to complete confusion, and a terrible slaughter resulted.'

– Contemporary chronicler at the Battle of River Kalka

An Irresistible War Machine

In less than two decades, the Mongols had achieved significant successes, conquering vast amounts of territory and ruthlessly destroying or subjugating those who opposed them. Mongol military doctrine was a spectacular combination of a nomadic people, superbly organized and capably led. The very nature of Mongol existence contributed to their success in battle. A Persian author wrote of these fierce warriors from the East: 'Their eyes were so narrow and piercing that they might have bored a hole in a brazen vessel, and their stench was more horrible than their colour... Their chests, in colour half black and half white, were covered with lice, which looked like sesame growing on a bad soil. Their bodies, indeed, were covered with these insects...'

Regardless of their appearance, the fighting prowess of the Mongols was undeniable. An Arab writer observed: 'They had the courage of lions, the patience of hounds, the prudence of cranes, the cunning of foxes, the long-sightedness of ravens, the wildness of wolves, the passion of fighting cocks, the protectiveness of hens, the keenness of cats and the fury of wild boars.'

Perhaps most important in the wave of Mongol imperialism which swept over much of the known world during the thirteenth and fourteenth centuries was the fact that Mongol commanders utilized every advantage, strategic and tactical, which was available to them. The elements of surprise, mobility, and concentrated firepower were very effective during westward campaigns, as their adversaries had never seen such a combination and were limited in their experience to the set-piece battles fought in Europe.

The Mongol people themselves were highly mobile, living in felt-covered wooden tent structures called *yurts,* which could be assembled and torn down quickly. They lived in the high altitudes of Mongolia and quite probably possessed a larger number of red blood cells in their systems, which resulted in greater stamina.

According to *The Secret History of the Mongols,* in the clear morning air they could see easily for six to eight kilometres (four to five miles) and even distinguish a man from an animal at a distance of 29km (18 miles). Mongol warriors were said to have slept in the saddle and sometimes covered up to 97km (60 miles) per day on the march. They grew up on horseback, literally learning to ride before they could walk. The men were subject to service with the army from the age of 14 to 60.

Battle of the River Kalka

1223

Numbering only 40,000 at its peak strength, a Mongol force under Subutai Bahadur and Jebei Noyan, trusted lieutenants of the great Genghis Khan, ventured into Russia in the spring of 1221. Although their incursion may have been intended only as a reconnaissance in force, the Mongols were confronted by a much larger but fragmented army from Kievan Rus'. Employing classic Mongol strategy, Subutai and Jebei gave ground for nine days before turning on their pursuers along the banks of the River Kalka, near Donetsk in the Ukraine. Swift Mongol cavalry and deadly archers decimated the enemy ranks. The resounding Mongol victory at the Battle of the River Kalka opened the door to full-scale invasion and the conquest of vast territories in Eastern Europe.

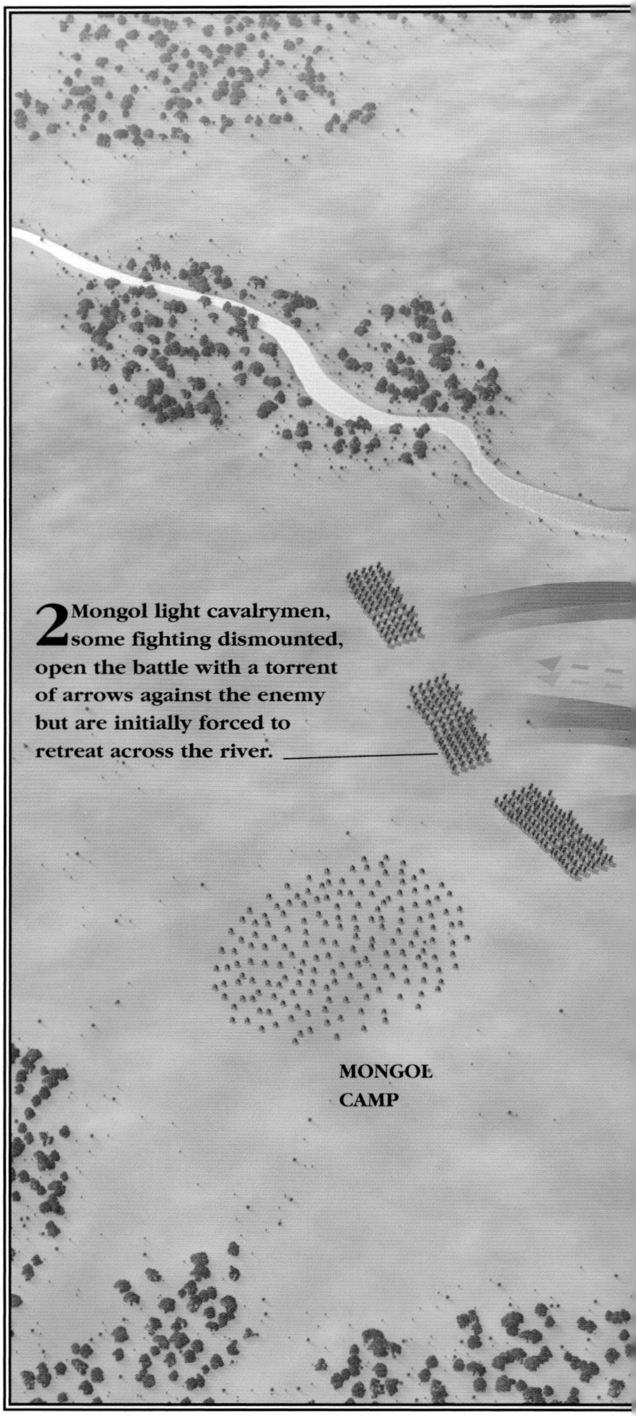
2 Mongol light cavalrymen, some fighting dismounted, open the battle with a torrent of arrows against the enemy but are initially forced to retreat across the river.
MONGOL CAMP

NOVGOROD
POLOTSK
POLOVTSIAN KHANATE
HUNGARY
RIVER KALKA

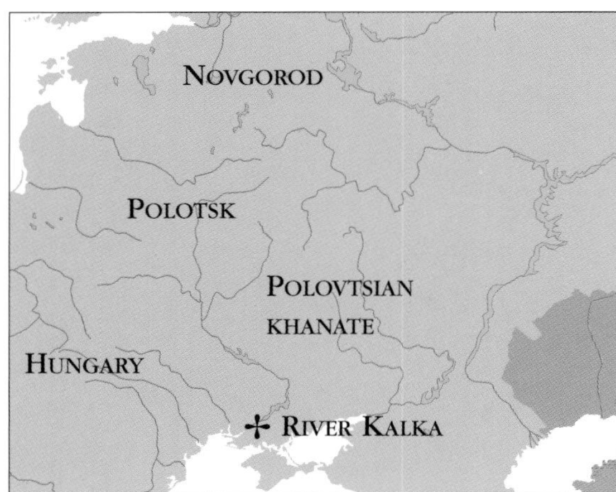

Following a series of victories in Central Asia, the Mongols under Subutai and Jebei crossed the River Don, advancing into the Crimea and the Ukraine. For the first time, an army of Genghis Khan had ventured into Europe.

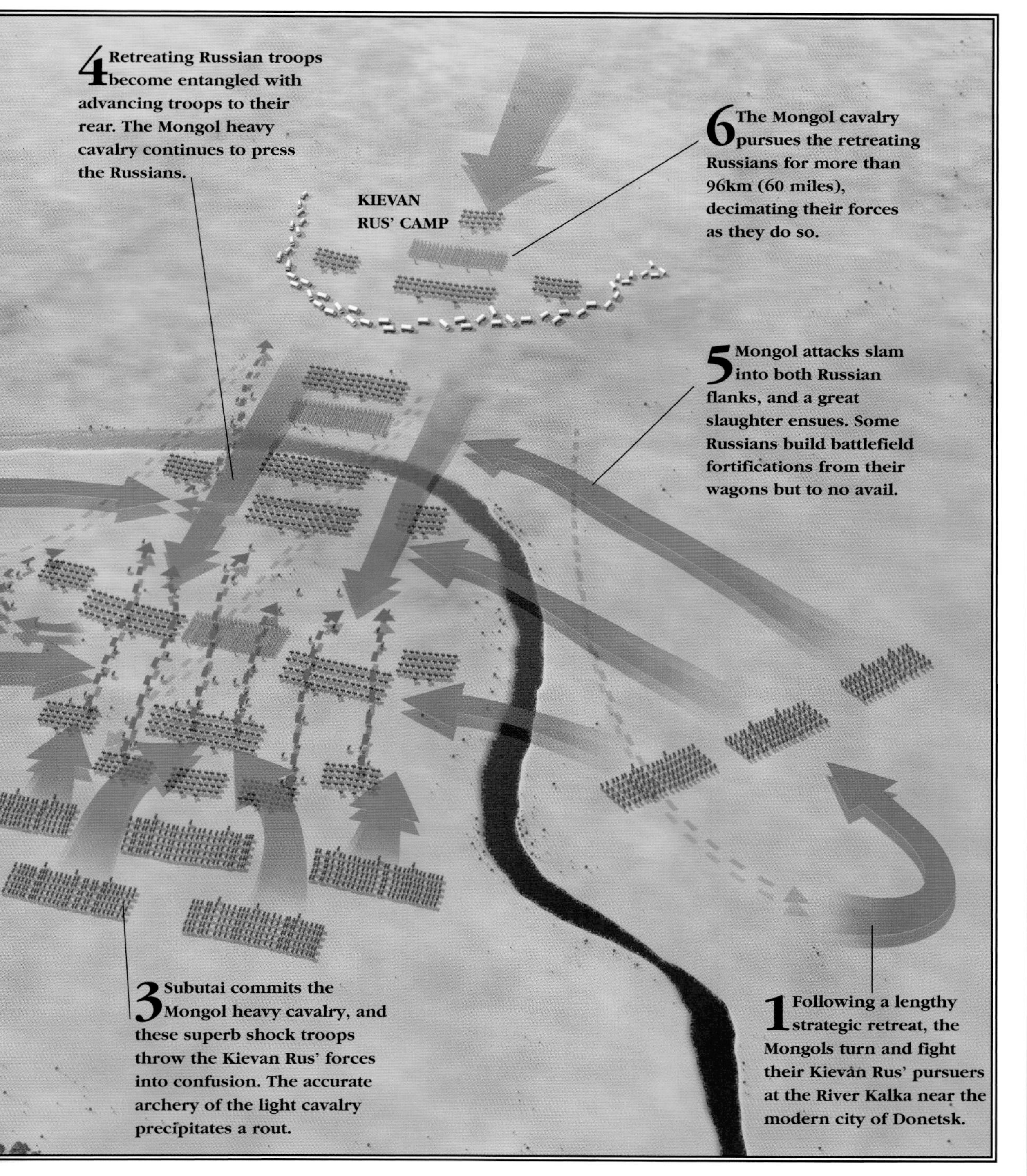
4 Retreating Russian troops become entangled with advancing troops to their rear. The Mongol heavy cavalry continues to press the Russians.
KIEVAN RUS' CAMP
6 The Mongol cavalry pursues the retreating Russians for more than 96km (60 miles), decimating their forces as they do so.
5 Mongol attacks slam into both Russian flanks, and a great slaughter ensues. Some Russians build battlefield fortifications from their wagons but to no avail.
3 Subutai commits the Mongol heavy cavalry, and these superb shock troops throw the Kievan Rus' forces into confusion. The accurate archery of the light cavalry precipitates a rout.
1 Following a lengthy strategic retreat, the Mongols turn and fight their Kievan Rus' pursuers at the River Kalka near the modern city of Donetsk.

The Sixteen Military Tactics

The Secret History of the Mongols relates that Genghis Khan employed the elements of speed, suddenness, ferocity, tactical variation, and discipline to conquer far and wide. Ruthless and relentless in pursuit of empire, the Mongols effectively used terror as a weapon as well, attaching shrill whistles to their arrows to frighten the opposing army and cutting of the ears of the enemy dead to count the tally of the slain.

With his 16 military tactics, several of which are evidenced at the battle of the River Kalka, Genghis Khan became one of the most powerful rulers in history. These tactics, according to *The Secret History of the Mongols,* include the following:

1 Crow soldiers and scattered stars:
Splitting the Mongol force into small units prevented the entire force from being surrounded, while the Mongols appeared quickly on the battlefield and evaporated just as rapidly following a fight. A force of but 100 Mongol cavalrymen could surround an enemy force 10 times its size, and 1000 mounted Mongols were capable of controlling a front 53km (33 miles) long.

2 Cavalrymen charge:
A direct charge against an enemy line was preferred and followed up by successive charges if necessary. At a predetermined signal, Mongol cavalry would charge simultaneously from all sides.

3 Archer's tactic:
While some fired their arrows on horseback, others, protected by small shields, dismounted to shoot. Once the withering fire had taken its toll, the cavalry would charge.

4 Throw into disorder:
The Mongols sometimes herded horses or oxen toward the enemy's lines or fortifications to create confusion.

5 Wearing down tactic:
Against a strongly held enemy position, the Mongols continued harassing fire with arrows but waited until the enemy was forced to relocate due to exhausted provisions before launching a decisive attack.

6 Confusion and intimidation:
Lighting numerous fires in their camps, sending horses to raise clouds of dust and mounting children or women on horseback to create the illusion of greater numbers were favourite tactics.

7 The feigned retreat:
Extending the enemy's force for miles and giving the appearance of fleeing, the Mongols were often successful in luring their adversaries, fatigued and disunited, to battle on favourable ground.

8 Arc formation tactics:
Sending two units of cavalry in sweeping arcs while a concealed force remained astride the enemy centre, the Mongols eventually assailed their enemy from at least three sides.

9 Outflanking:
Sometimes leaving a token force in front of an enemy formation, Mongol generals sent cavalry ranging far to the rear, across mountain treks and paths worn by wild animals, to take the foe from behind.

10 Encircling:
Taking advantage of exposed enemy flanks or undefended rear areas, the Mongols surrounded their enemy, in the field or holding a city.

11 Open end tactic:
Leaving the enemy an escape route often

contributed to the rout of a fleeing army and enabled the Mongols to fall upon the panicked survivors and slaughter them in great numbers.

12 Combining swords and arrows:
The Mongols avoided hand-to-hand combat wherever possible, preferring to employ the bow from a distance. Once the archers had inflicted serious casualties, the heavier units subdued the enemy in total.

13 Hot pursuit and dispersing tactic:
During the course of a victory, the Mongols were relentless and barbaric in their pursuit of the enemy, intent on annihilating them. If they experienced a reversal, Mongol forces dispersed in all directions to avoid being encircled and suffering a major defeat.

14 Bush clump tactics:
Keeping in contact, small groups of Mongol soldiers advanced under cover, often completely unobserved. These tactics were particularly effective at night or during low light.

15 Strategic outflanking:
The Mongols were undeterred in their efforts to outflank an enemy, sometimes holding a battlefront for an extended period while additional forces marched hundreds of miles across the steppes or even desert to accomplish a flanking manoeuvre.

16 Lure into ambush:
Used particularly during a difficult fight against fortifications or in narrow or rugged terrain, the Mongols sometimes chose to lure their enemies out of strong defensive positions with a slow, deliberate retreat. At the appropriate moment, they turned on their pursuers and attacked.

The Mongol army of Genghis Khan proved to be the scourge of Eurasia, terrorizing and subjugating diverse populations and conquering vast territories. Through perseverance, superb organizational skills and the superb strategic and tactical ability of their generals, the Mongols shaped the course of world history.

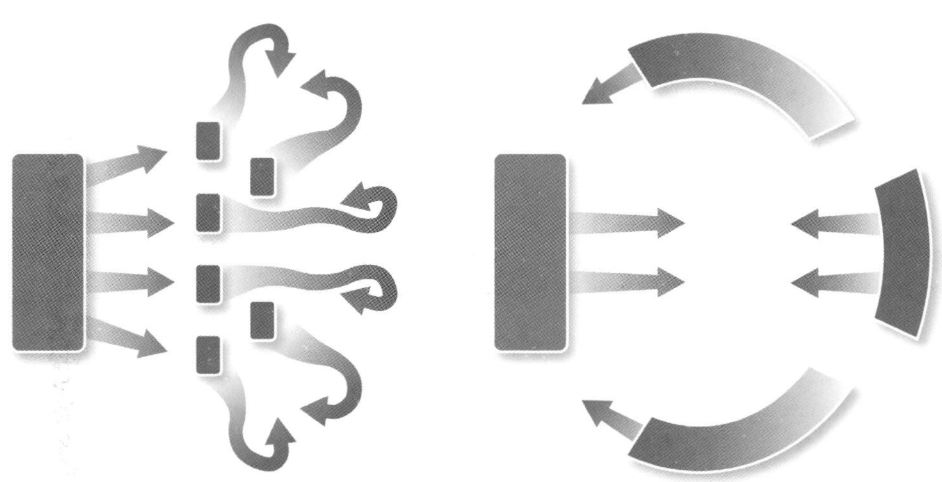

THE MONGOLS *employed a number of effective tactics to beat their enemies. The feigned retreat (left) was particularly effective in drawing the enemy out of a strong position, while the employment of an arc formation (right) allowed the Mongols to outflank and encircle their enemy.*

After they had gained the initial advantage at the River Kalka, the light cavalry and archers ushered in the Mongol heavy cavalry, which wreaked havoc on the disorganized Russians. The heavy cavalrymen wore chain mail over a shirt woven from raw silk, a cuirass, and a helmet of beaten iron. Armed with two bows, a lasso, a lance which was easily 3.7m (12ft) in length, a battle axe, and a dagger, these troops were deadly in close combat. Their horses were also frequently protected by a form of armour.

Rossabi notes that once the rout of the Russians was on the archers and light cavalry combined with the heavy cavalry to decimate the Russian ranks. The battle, he says, was generally a fine example of the standard Mongol battle plan authored by Genghis Khan. Although the army at the time of the Great Khan's death numbered no more than 130,000 men, it was superbly organized into 10 divisions called *tumen* and then further divided to the smallest operational group of just 10 soldiers, called an *arban*. Typically, the army was divided into three grand wings and an imperial guard for Genghis Khan which grew over time from about 150 men to more than 10,000, including 400 archers and an inner guard of 1000 hand-picked warriors.

'The greatest pleasure is to vanquish your enemies and chase them before you, to rob them of their wealth, and see those dear to them bathed in tears, to ride their horses and clasp to your bosom their wives and daughters.'

— *Genghis Khan*,

The Secret History of the Mongols

Expanding Eastward

Half a century after the landmark victory at the River Kalka, Kublai Khan (1215–94), a grandson of Genghis Khan, undertook an expedition to conquer the neighbouring islands of Japan. Following the death of Genghis Khan, the Mongol Empire began a decades-long process of unravelling. The most famous of his many sons were Jochi, Jagatai, Ogodei, and Tolui. Batu Khan, the son of Jochi, ruled the Golden Horde, a Mongol realm in Eastern Europe, while Ogodei, the designated successor to Genghis Khan, held court in Mongolia and northern China. Jagatai ruled in Central Asia. Mangu Khan, the son of Tolui, briefly reigned over a united Mongol Empire in the mid-thirteenth century, and another son, Hulagu, was the founder of the il-Khanid dynasty in Persia. The second son of Tolui, Kublai Khan founded the Yuan dynasty in China and located his seat of government at Beijing. It was Kublai's struggle with a younger brother, Ariq Boke, to succeed Mangu Khan, which precipitated the eventual fragmentation of the Mongol Empire.

In 1260, both Kublai and Arik Boke claimed the title of Great Khan. A three-year war ensued, and by the time Kublai Khan prevailed he had been confronted with and had ruthlessly put down a revolt by the ethnic Hans. The other kingdoms, or khanates, however, were being established. As for Kublai, the nominal Great Khan extended Mongol rule to its greatest distance with the subjugation of the Song dynasty in 1279. During this time, however, the forces of Kublai Khan would fail in attempts to conquer lands encompassing the modern nations of Vietnam, Burma, Indonesia and Japan.

Turning Towards Japan

Early in his reign, Kublai Khan attempted to strong-arm Japan into submitting to the dominion of the Mongols. On two occasions, diplomatic missions ended in failure, and the Japanese responded to the threat by bolstering their defences on the island of Kyushu, which was closest to Mongol land on the Korean peninsula and thus most likely to be invaded. The ruling Kamakura shogun decreed that landholders on Kyushu should return there while forces were marshalled in the west. An impatient Kublai Khan had wanted to begin the

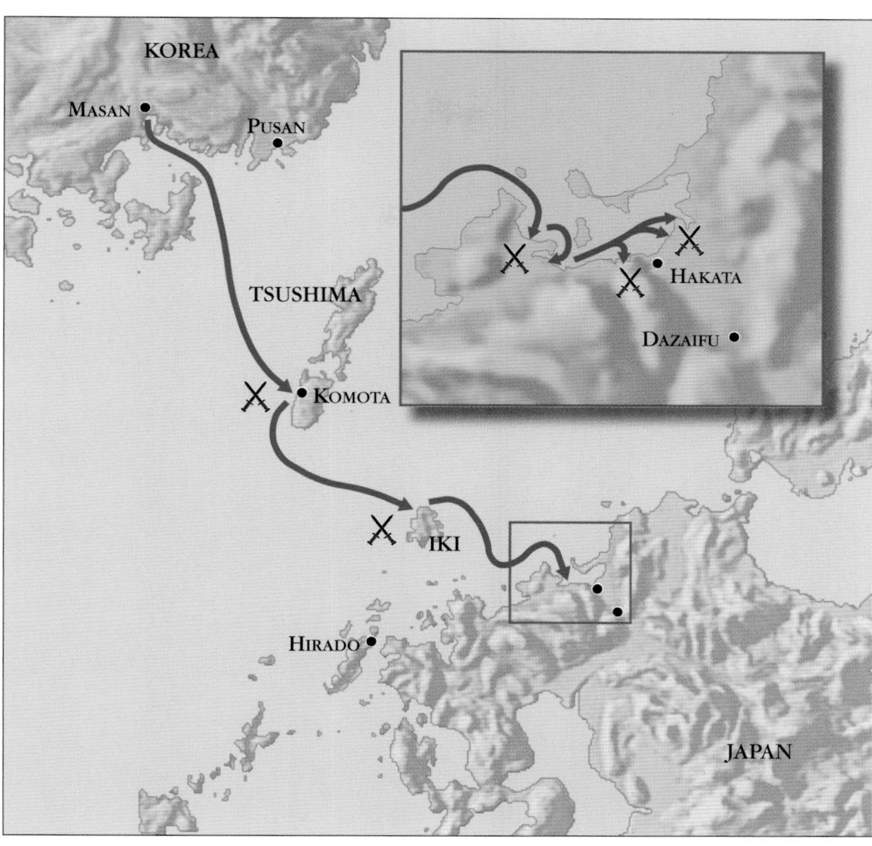

THE MONGOL INVASION of Japan in 1274 consisted of a number of different phases, including the capture of the islands of Tsushima and Iki before the main landing at Hakata Bay. There, they fought a number of exhausting battles with the Japanese, before eventually withdrawing after low supplies and storm damage seriously depletes their force.

war with Japan as early as 1268, but an expedition was not launched until 1273, when it was forced to return from Korea without ever reaching Japan itself. A year later, the Mongols had assembled a battle fleet estimated at up to 900 vessels capable of carrying approximately 23,000 Mongol warriors along with Chinese and Korean soldiers. In command were the Mongol general Hol Don and Korean leader Kim Bang-gyong.

Kublai Khan's fleet sailed from the vicinity of Pusan in October 1274, crossing the 177km (110 miles) of open water known as the Strait of Tsushima. Along the way, the islands of Tsushima and Iki were captured. Then, on 19 November, the Mongol force reached the port of Hakata, near the site of the modern city of Fukuoka. Reportedly, the Japanese had assembled a relatively small contingent of 6000 warriors, including samurai and other defenders of the shogun known as *gokenin*. The Mongols had already endured a stormy crossing. However, they seized the initiative upon landing.

The Battle of Bun'ei, 1274

During the battle of Bun'ei, also referred to as the battle of Hakata Bay, the first clash of arms between the Mongols and the soldiery of feudal Japan occurred on 20 November 1274. As the battle developed, the tactics of the Mongols soon proved superior to those of the Japanese. Fighting in tight infantry formations resembling somewhat the phalanxes of ancient Greece, the Mongols also unleashed a veritable storm of arrows from their bows, striking the Japanese repeatedly from a distance.

For years, historians have disputed one aspect of the battle which has been immortalized in Japanese tapestries depicting the fighting. The Korean sailors aboard the Mongol ships had supposedly fired rudimentary catapults, sending both hollow bombs and projectiles filled with shrapnel into the Japanese lines. Recently, archaeological work has confirmed the presence of such weapons and, therefore, possibly the first known use of gunpowder against the Japanese. Undoubtedly, the Japanese were stunned by such a weapon, while their horses were terrified.

SMOKE BILLOWS FROM SHIPS OFFSHORE, *and torrents of arrows fly as Mongol troops and defending samurai clash in this romantic portrayal of the Mongol invasion of Japan. The first incursion against the island nation occurred in 1274.*

While the Mongols employed new technology and fought as cohesive units, the Japanese samurai were unfamiliar with such tactics. Samurai doctrine dictated that a warrior confront an enemy of similar rank and engage in single combat. Moreover, it had been at least 50 years since a major battle had taken place on Japanese soil. The Japanese commanders were inexperienced in fighting on such a grand scale, and badly needed reinforcements were slow in arriving on the field.

Author Stephen Turnbull described the Mongol advantage at Bun'ei: 'Accounts of the first Mongol invasion of Japan in 1274 give a very good idea of how the Mongols combined naval and land activities in what may be termed amphibious operations. When the Mongol ships drew close to the beach, traction trebuchets launched a "shore bombardment" of iron-cased exploding bombs. This was the first experience that the Japanese had of gunpowder weapons, and it created such an impression on one leader of samurai that he deliberately included a picture of one in the painted scroll he commissioned. The nature of the weapon as a fragmentation bomb is clearly shown. When the Mongols came ashore they fought dismounted and in dense phalanxes, loosing clouds of arrows. This point of detail was no doubt added to the Japanese account because of the puzzling contrast it presented to the traditional Japanese way of fighting, which preferred single combat above all. There was no shortage of hand-to-hand fighting, however, during the hours that followed.'

The samurai no doubt displayed incredible bravery and inflicted severe casualties on the invaders. One of them, Takezaki Suenaga, was eager to rush into battle, fight heroically and thereby increase his prestige and honour. Under strict orders to retire in the face of the marauding Mongols and await further orders, he was said to have shouted: 'Waiting for the general will cause us to be late to battle. Of all the warriors of the clan, I Suenaga will be the first to fight from Higo.'

Under mounting pressure, the Japanese fell back to their defences at Dazaifu, the capital city of Kyushu. The Mongols burned the town of Hakata and carried off what booty they could gather. However, aware that strong enemy reinforcements might well be just over the next hill, and, unfamiliar with the surrounding terrain, they did not pursue the retiring Japanese. As the daylight began to fade, both sides had suffered tremendous casualties, although the exact number of dead and wounded is unknown.

'Whoever obeys us remains in possession of his land, but whoever resists is destroyed. We send you this order, so if you wish to keep your land, you must come to us in person and thence go on to him who is master of all the earth. If you do not, we know not what will happen. Only God knows.'

— *Genghis Khan*,

The Secret History of the Mongols

Advance and Retreat

The dark clouds of a storm had begun to gather and, given the circumstances, the Mongols decided to withdraw from the island of Kyushu. Theory and conjecture surrounds the decision. While some historians assert that the Mongol invasion had merely been intended as a reconnaissance in force and that the withdrawal had been a component of the overall strategy, others consider the Mongol retirement another example of their tactic of strategic retreat. Still, another faction concludes that the Korean seaman who manned the vessels of the fleet were concerned about the coming storm and wished to depart in haste rather than become marooned on a hostile shore.

Regardless of the impetus for withdrawal, the fate of the Mongol fleet was inevitable. Typhoon force winds buffeted the vessels, and as many as one-third of them capsized or sank outright. Estimates of the lives lost during the gale are as high as 13,500. The Great Kublai Khan himself may well bear part of the blame for the disaster which befell his fleet. In his haste, the Khan had ordered the fleet to be rapidly constructed and augmented with any available craft.

Apparently, many of the vessels had been intended for use on rivers and inland waterways and were unsuited for the open sea. Those with flat keels rather than curved ones were more likely to capsize in heaving oceans.

Furthermore, those ships constructed by the subjugated Koreans may well have been deliberately built to poor standards, robbing the Mongols of sturdy vessels in a subtle but highly effective form of sabotage.

The Divine Wind

Although the decimation of his fleet following the battle of Bun'ei was a blow to Kublai Khan's prestige, he, like other Mongol rulers before him, was unacquainted with the concept of defeat. By the spring of 1281, he had assembled yet another invasion fleet, this one consisting of more than 100,000 troops and 12,500 ships. By now the Japanese had bolstered their defences, including the construction of a stone wall more than 19km (12 miles) long near Hakata.

A rendezvous of Mongol vessels leaving ports in China and Korea failed to materialize. When the

Battle of Bun'ei

1274

Following a pair of failed diplomatic missions, Kublai Khan dispatched a naval force of nearly 900 vessels to Japan in the autumn of 1274. Aboard the Mongol vessels was a contingent of more than 23,000 Mongol and allied troops from Korea and China. On the morning of 20 November, the invaders landed at Hakata Bay on the island of Kyushu and were confronted by a relatively small Japanese force numbering no more than 6000. At Bun'ei, the strength of the Japanese lay in the fighting spirit and skill of their samurai warriors, while the superior tactics of the Mongols compelled the defenders to retreat. Having suffered heavy casualties, however, the Mongols did not press their advantage and retired to their ships the following day. Later, storms caused irreparable damage to the Mongol fleet, bringing to end any hope of continuing the campaign.

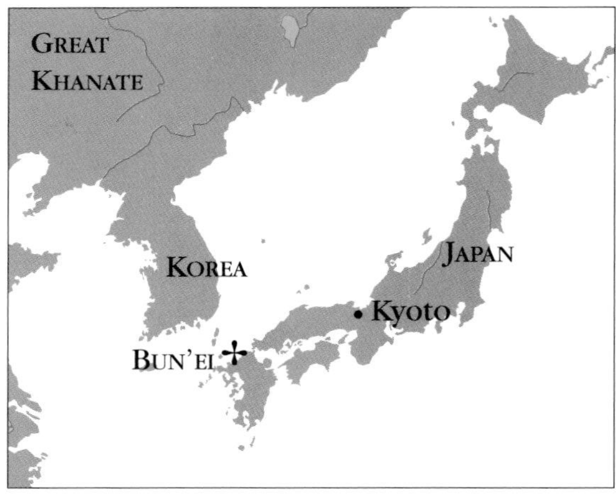

The Battle of Bun'ei, near Hakata Bay and the modern Japanese city of Fukuoka on the island of Kyushu, was the climax of the first Mongol invasion of Japan. Failure to establish a permanent bridgehead there ended the Mongols' ambitions to invade Japan in 1274.

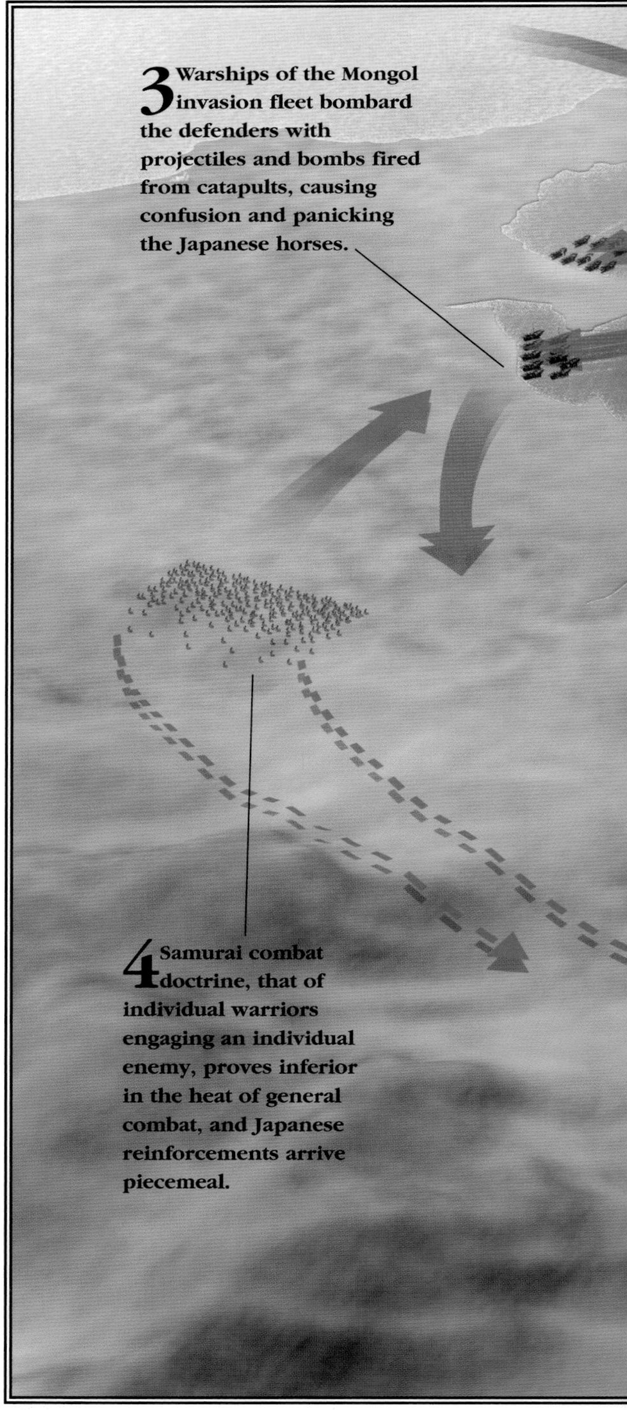

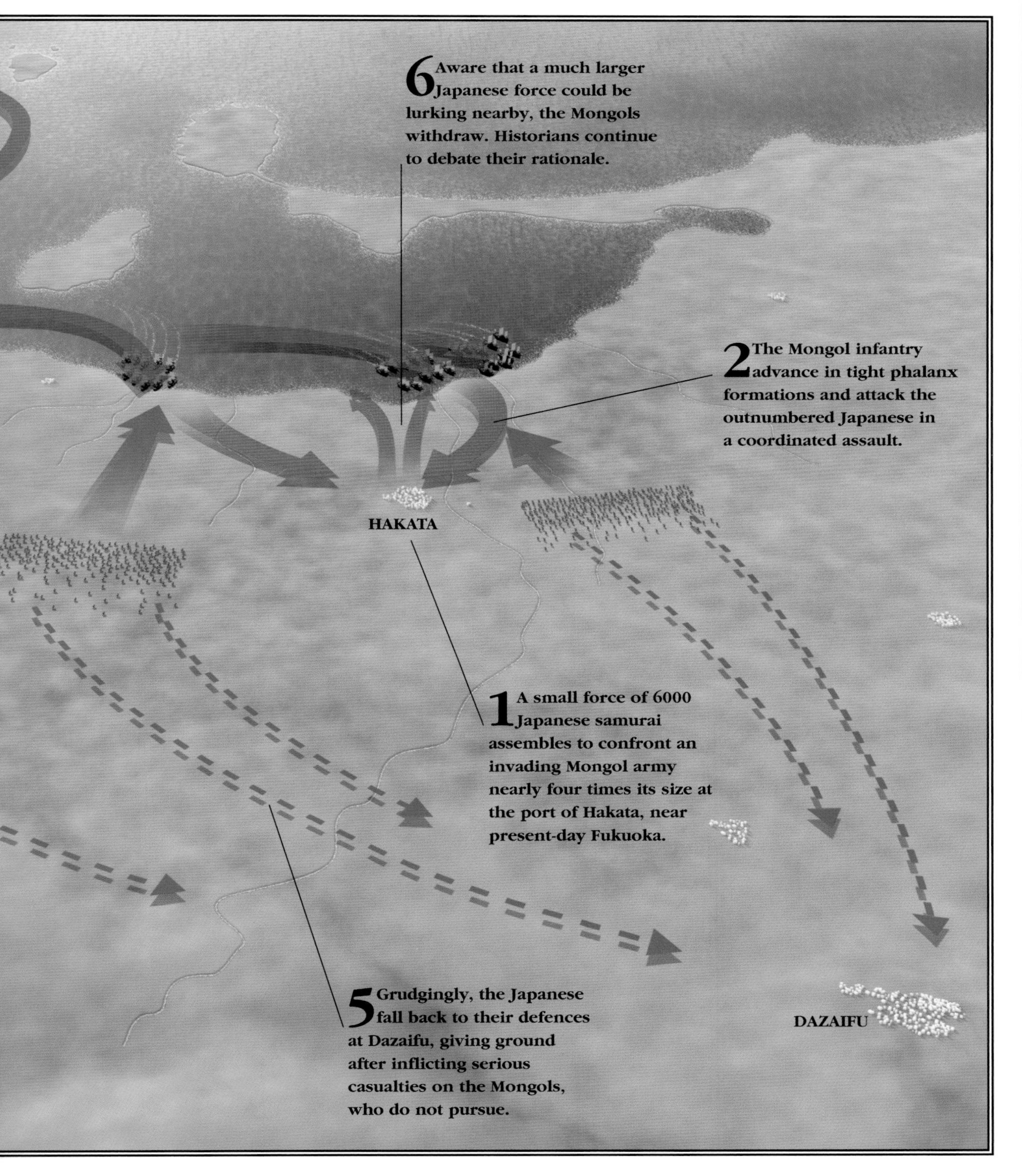
6 Aware that a much larger Japanese force could be lurking nearby, the Mongols withdraw. Historians continue to debate their rationale.
2 The Mongol infantry advance in tight phalanx formations and attack the outnumbered Japanese in a coordinated assault.
HAKATA
1 A small force of 6000 Japanese samurai assembles to confront an invading Mongol army nearly four times its size at the port of Hakata, near present-day Fukuoka.
5 Grudgingly, the Japanese fall back to their defences at Dazaifu, giving ground after inflicting serious casualties on the Mongols, who do not pursue.
DAZAIFU

first Mongol soldiers were landed by an understrength naval force which had sortied from Korea, their progress was stymied by the stone wall. They withdrew to the island of Shikanoshima in Hakata Bay, and a melee among Japanese coastal defence craft and the ships of Kublai Khan developed. Forced to withdraw, the surviving vessels took no further part in the battle.

The second force, which had sailed from China, landed approximately 48km (30 miles) south of Hakata at Imari Bay. Weeks of indecisive fighting ensued, but a second major storm assailed the Mongol fleet at the end of July. The losses in Mongol and allied lives were said to be in the tens of thousands. One Japanese observer recalled: 'A person could walk across from one point of land to another on a mass of wreckage.' The storm, which caused Kublai Khan to cancel plans for yet another invasion of Japan scheduled for 1286, has been known to history as *kamikaze,* or 'Divine Wind'.

The reign of Kublai Khan lasted more than three decades. Although the empire reached its territorial zenith during that tenure, the legacy of Kublai Khan is commingled with internal strife and significant military reverses which hastened the decline and dismemberment of the largest domain in history to be ruled by a single monarch.

The Deokpoong Buwongun

Like Nelson at Trafalgar two centuries later, Admiral Yi Sun-sin (1545–98) died in battle, the victim of an enemy bullet. At the battle of Noryang in November 1598, the greatest military hero in Korean history was fighting the Japanese, as he had for most of seven long years. At the height of his power, Yi's navy crushed the enemy, destroying 200 warships, or 40 per cent of their strength.

Wan, along with his eldest son, Hoe, bore the Admiral's lifeless body back to his cabin and then returned to the fight. As he had directed, the men kept the admiral's death a secret until victory was won.

Posthumously, Yi Sun-sin was given honorary titles including Yeongijeong, or Prime Minister, and Deokpoong Buwongun, the Prince of the Court from Deokpoong. Today, a statue of the great naval commander stands in the centre of Seoul, the capital of South Korea. Although he is a somewhat obscure figure in the west, Admiral Yi must be counted among the greatest naval tacticians in history. He is said to have commanded Korean naval forces in more than 20 battles and never to have lost.

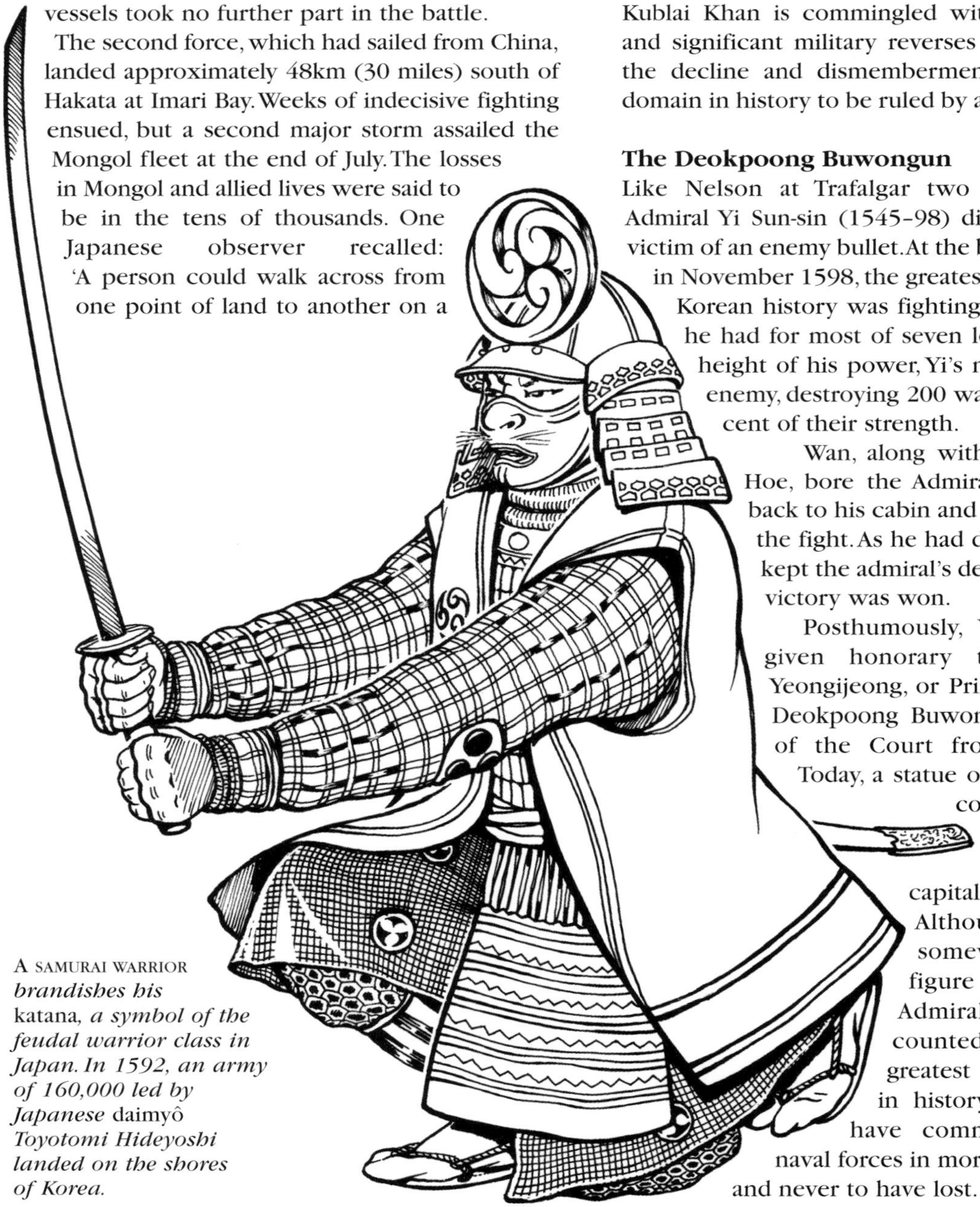

A SAMURAI WARRIOR *brandishes his* katana, *a symbol of the feudal warrior class in Japan. In 1592, an army of 160,000 led by Japanese* daimyô *Toyotomi Hideyoshi landed on the shores of Korea.*

The Japanese Menace

In 1592, known to Koreans as the Year of the Water Dragon, Japanese *daimyô* Toyotomi Hideyoshi led an army of 160,000 Japanese samurai and other soldiers on a mission of conquest. Their initial objective was to subjugate Korea. This would be followed by the rapid defeat of Ming China, and then the whole of Asia would fall under the dominion of Japan. With a Japanese landing at Pusan in May of that year, the Imjin War began.

The Japanese defeated Korean forces and marched resolutely on Seoul. When word of the Japanese approach reached the capital, the Korean government fled in panic to Pyongyang, sparking riots in the streets of Seoul. Within three weeks, the vanguard of the Japanese army marched through the gates of the capital city. Moving northward, Hideyoshi's army defeated the Koreans at the battle of the River Imjin and chased King Seonjo, his family, and government officials to the city of Uiju, on the River Yalu and the border with China. However, while the Japanese advanced inexorably on land, all was not as well as Hideyoshi believed.

The Japanese *daimyô* had made one grave miscalculation in planning his adventure on the Asian continent. Command of the sea seemed at best an afterthought and at worst a glaring oversight. For the Koreans, it was their saving grace. Stephen Turnbull wrote:

'It is a notable feature of the Japanese invasion that almost no planning went into the naval side of the operation. To Hideyoshi the fleet he impressed from the *daimyô* of western Japan was simply there to provide transport. The ships had no protection other than that provided by the samurai who sailed on them during the short journey from Tsushima to Pusan. The thought that the Korean navy might attack them never seems to have entered anyone's mind, yet we do know that Hideyoshi had tried to obtain two Portuguese ships for use in the operation, a request that was politely declined. However, the experience of the unopposed crossing only served to confirm this optimism.'

Meanwhile, Admiral Yi was engaging the Japanese at sea within two weeks of their establishment of a foothold on the Korean peninsula. Accounts of his early offensive actions are numerous and provide a glimpse of the greatness he was to display during the protracted conflict with the Japanese. His leadership and daring were to prove the decisive factors in the outcome of the Imjin War as control of the Japanese transportation and resupply routes was vital to ultimate success or failure.

> *'In an instant, our warships spread their sails, turned round in a "crane wing" formation and darted forward, pouring down cannon balls and fire arrows on the enemy vessels like hail and thunder. Bursting into flame and with blinding smoke, 73 enemy vessels were soon burning in a red sea of blood.'*
>
> – *Yi Pun, at the Battle of Hansando*

Yi's Engines of War

With the outbreak of the Imjin War, by far the most common warship of the Korean navy was a vessel called the *panokseon*. Developed initially as a defensive measure against marauding Japanese pirates, the *panokseon* was of traditional Korean construction, heavily influenced by Chinese shipbuilding techniques. By Western standards, the ships were extremely broad and appeared somewhat ponderous in motion. With crews of 125 sailors, oarsmen were separated on a lower deck from the fighting men above. Standard length was from 15–21m (50–70ft), although *panokseon* as long as 33.5m (110ft) or more are known to have been constructed. The primary features of the *panokseon* included raised gunwales for the shelter of the sailors and a superstructure which included something of a bridge or forecastle-style area for the commander. The *panokseon* was also characterized by a roof of boards which provided protection against plunging fire from enemy arrows or spears and deterred boarding during close combat. Period artwork depicts large curved hull design at the stern and decorative renderings of dragons along the sides. Sails were utilized for propulsion in open water, while the oars offered better control of the ship in harbours and inland waterways.

Despite the prevalence of the *panokseon,* it is the *kobukson* ('turtle ship'), the most celebrated Korean fighting ship of the era, and indeed in all of the nation's history, that is best known. Based on

ABOVE: A FIERCE LIKENESS OF ADMIRAL YI SUN-SIN *glowers from its perch in the heart of Seoul, South Korea's modern capital city. Admiral Yi, often called the Korean Nelson, devastated Japanese naval assets and saved his nation from defeat.*

an earlier design known as the *kwason,* or 'spear ship', the original turtle ships appeared during the mid-fifteenth century and underwent periodic revisions to the extent that the turtle ships employed by Admiral Yi were quite different from those of earlier times. Although the turtle ship had been in existence for more than a century before the admiral's rise to prominence, he did advocate a number of improvements.

At the time of the Japanese invasion of the Asian mainland, the turtle ship could best be described as a traditional wide vessel with a characteristically flat keel, high gunwales for the protection of the combatants, and a covered roof reinforced with numerous spiked appendages

BELOW: FASHIONED FROM LACQUERED LEATHER, *this Korean helmet dates from the sixteenth century. Note the added protection of leather attachments to the sides and rear along with the ornamental crown and thongs for tying beneath the chin.*

which could easily impale enemy personnel who attempted to board the vessel. With iron-reinforced decks and sides, the ship certainly did resemble a gigantic turtle. At the prow, an ornate dragon head faced forward, and inside could be mounted at least one medium cannon. The sides included 12 gunports, through which cannon could be mounted or spearmen might confront the enemy at close quarters. Masts and sails were used in open water and could be removed when the vessel was powered by the oarsmen in tighter waterways and during combat operations.

'Another special war vessel has been developed ... The whole surface of the ship, other than [these] passageways, is covered with spikes so that no enemy can walk over it. At the bow is a dragon head from which cannon can be fired ... It is called a turtle ship from its appearance.'

– Yi Pun, Admiral Yi's nephew

Capable of ramming an enemy ship, firing from a standoff position, or engaging in close combat, the turtle ship was truly a formidable weapon. Yi's modifications included the addition of an arched roof which more easily deflected enemy shot or arrows and the mounting of numerous small cannon capable of delivering close broadsides. The fact that a relative few of these were present with the Korean fleet during the Imjin War did not seem to handicap Admiral Yi's operations. His intimate knowledge of the Korean coastline, its shallows and shoals, and his audacity in command were only augmented by their presence.

Korean Naval Weapons

Chinese gunpowder technology was rapidly imported by the Koreans during the fourteenth and fifteenth centuries. By the time of Admiral Yi, five principal cannon were deployed on warships of the Korean navy. Ranging in size from smallest to largest, these were identified as: *sungja*, or victory, a diminutive weapon fired from the upper deck; *hwangja,* yellow cannon; *hyonja,* black cannon; *chija,* earth cannon ; and *chonja,* the heaven cannon, which weighed 300kg (660lb) and with its 140mm (5.5in) bore could hurl a projectile several hundred yards. Fragmentation-type iron shot, stones, and a form of buckshot, were often fired from these weapons.

Additionally, giant wooden arrows with iron tips were launched from the heaven cannon and could inflict tremendous damage on enemy vessels. The arrows were up to 3m (9ft) long and were propelled at high velocity against the wooden hulls of warships, often shattering the planks into deadly splinters. These arrows were sometimes tipped with flame and intended to set enemy vessels on fire.

Initial Encounter

The year before the Japanese invasion, Admiral Yi was promoted to command of the Korean naval contingent based along the coast of Cholla Province, west of Kyngsang Province. These two southernmost Korean provinces had been divided into two naval districts. Yi's counterpart, Admiral Won Kyun (1540–97), commander of the Kyngsang fleet, was the first to encounter the Japanese at sea.

The third commander of this Korean flotilla, Won Kyun was characteristically hesitant, and the battle was a disaster for the Koreans, who were nearly wiped out.

Undeterred by the defeat of Won Kyun, Admiral Yi boldly assumed the offensive. The Japanese had landed at Pusan on May 25, 1592, and their supply vessels remained in the vicinity. The intrepid Admiral Yi fell upon the Japanese ships anchored near Okp'o, adjacent to Koje Island, raking them with cannon and arrows from both sides. Flaming arrows set several Japanese vessels alight, and a number of Japanese sailors panicked, cutting their anchor lines in an attempt to escape the Korean onslaught. Only a handful managed to get away.

Yi's victory was overwhelming. Twenty-six Japanese ships were sunk or set on fire, while the only casualty sustained by the victors was a single flesh wound to the arm of a sailor. On the same day, the Koreans continued eastward in search of the enemy and annihilated a small Japanese squadron as darkness fell.

Yi's warships continued to seek and destroy Japanese vessels during the following week. The morning after his initial victory, Yi sailed into the shipping lanes between Tsushima Island and Pusan, where he encountered a large number of supply ships and their naval escorts. As the running fight continued for hours, Yi's force destroyed the enemy and took a number of Japanese ships as prizes. Without losing any of their own warships, the Koreans had either sunk or captured more than 40 vessels. Yi then moved his fleet westward to a more advantageous location near the peninsula's southern coast.

While the situation continued to deteriorate on land and the Korean government appealed to neighbouring China for military assistance, Admiral Yi embarked on a second naval campaign against the Japanese in the early summer. At SaChon, Japanese troops were building coastal fortifications to guard the harbour and protect shipping anchored within. Yi knew that the confines of the harbour and the ebbing tide offered no advantage to his force. Therefore, he ordered his fleet to present the Japanese with the spectacle of a confused retreat. Drawing the enemy into open water, Yi then turned on his adversary and charged into their midst. The Japanese ships were defeated in short order, but Yi intentionally allowed a few of them to remain intact in order to evacuate the soldiers whom he expected to abandon their construction project ashore. Fearing that they might be marooned, the Japanese did just that.

By this time, Hideyoshi had become painfully aware of the reverses his forces had suffered at sea and was beginning to feel the effect of dwindling stores and overextended supply lines. Meanwhile, the Korean navy under Admiral Yi continued to inflict damaging and embarrassing defeats on the Japanese. Near Tangp'o, he reportedly attacked a

THIS CUTAWAY VIEW *of the Korean* panokson *fighting vessel reveals the protection afforded to those who powered the ship with oars below decks, the positioning of archers and cannoneers, and the vantage point of the commander.*

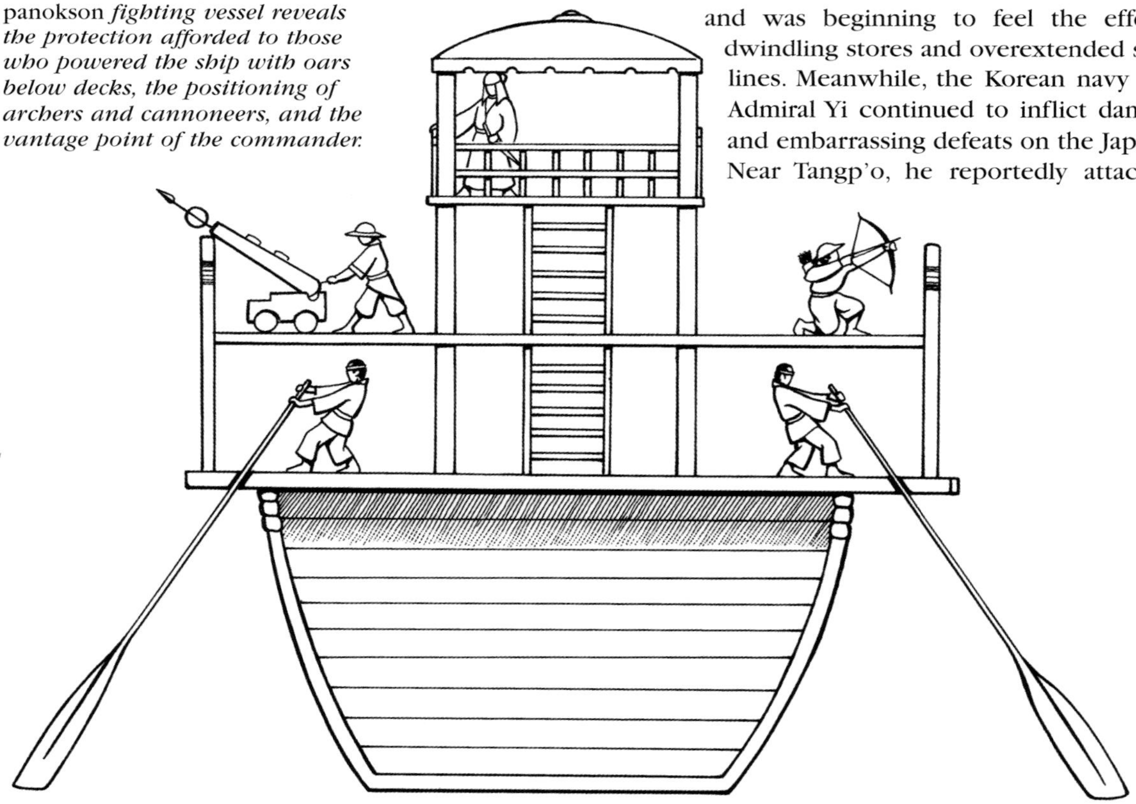

convoy of 25 Japanese supply ships bound for the Korean coast and sank every one of them. Several days later, he employed the false retreat once again and attacked a pursuing Japanese force on both flanks. Only one enemy ship survived the onslaught, and it was hunted down and sunk the next morning.

Battle of Hansando, 1592

During less than two months of war at sea, Admiral Yi had inflicted a series of stinging defeats on the Japanese. More than 100 enemy ships had been sunk or captured and thousands of casualties had been inflicted. In sharp contrast, the Korean losses in these engagements are estimated at only 11 killed and 26 wounded. Since the defeat of Won Kyun in the opening days of the war, no Korean ships had been sunk or seriously damaged.

In response, Hideyoshi had become determined to settle the score and now raised a powerful armada to wrest control of the vital shipping lanes through the Yellow Sea. Hideyoshi ordered Admiral Wakizaka Yasuharu (1554–1626) to rendezvous with two other naval squadrons before mounting his expedition. Wakizaka, however, grew impatient and decided to set out on his own with 73 ships. Of these, an estimated 36 were large ships called *atakebune* with multiple decks and numerous guns which were probably of smaller calibre than the heavy Korean weapons. Another 24 were *sekibune,* medium sized warships, and 13 were still smaller scout vessels known as *kobaya.*

Forewarned by a local farmer that a Japanese naval force was headed toward him, Admiral Yi learned that the enemy had anchored temporarily north of Kyonnaeryang Strait, near the island of Hansando. Once again, Admiral Yi assumed the offensive and formulated a plan to lure the enemy into waters which were favourable to his own warships. Having combined with two smaller naval contingents, commanded by Admirals Kyun and Yi Eok Ki, the aggregate Korean strength totalled about 54 vessels with only a few – some say but two or three – of the famed turtle ships available for action.

Just as the famed battle of Salamis fought between the Greeks and Persians in 480 BC shaped the future of western civilization, so, too, the battle of Hansando can be described as a watershed in the history of East Asian warfare. On the morning of 14

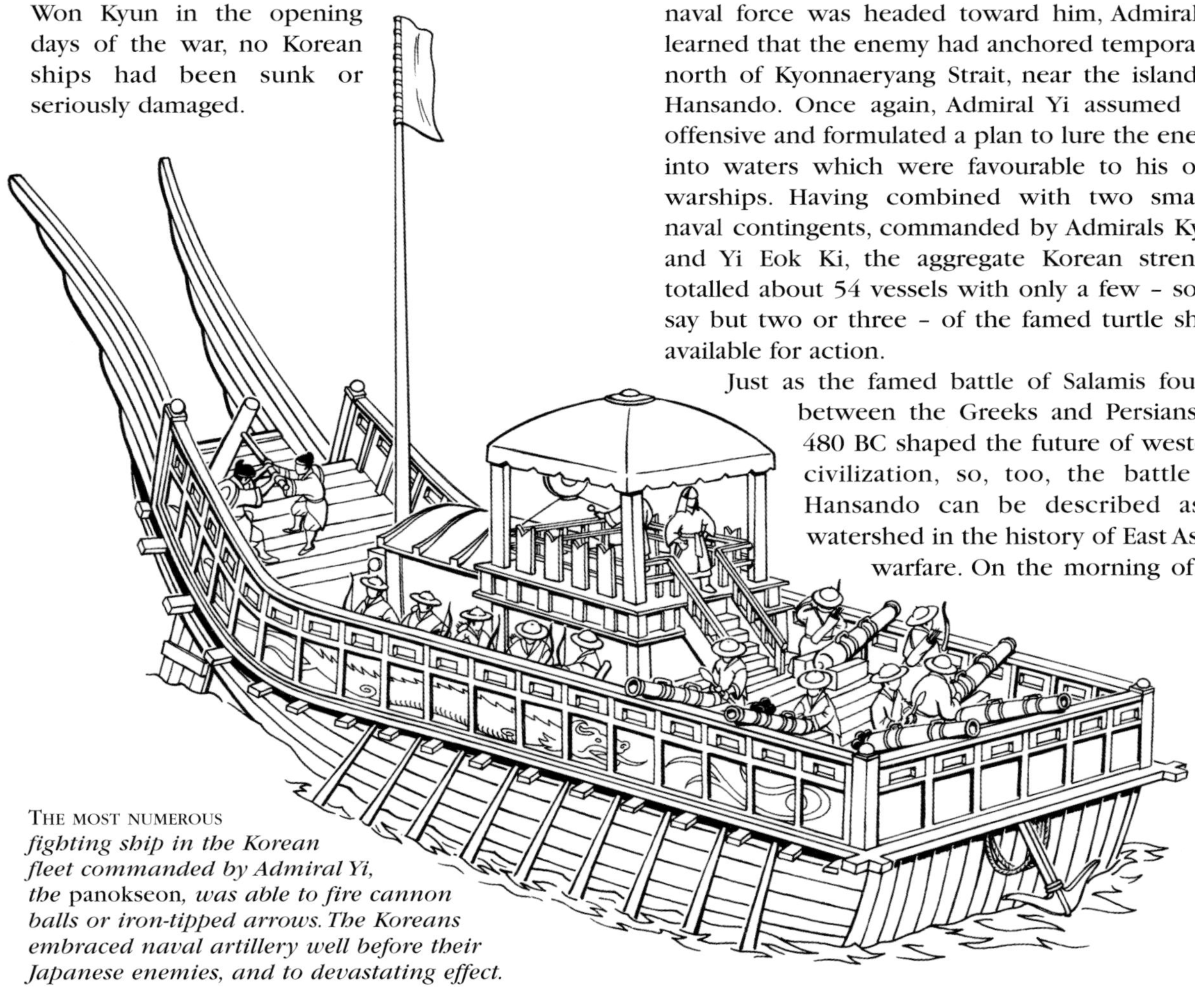

THE MOST NUMEROUS *fighting ship in the Korean fleet commanded by Admiral Yi, the* panokseon, *was able to fire cannon balls or iron-tipped arrows. The Koreans embraced naval artillery well before their Japanese enemies, and to devastating effect.*

Battle of Hansando

1592

Warned that a powerful Japanese naval force was approaching, Admiral Yi marshalled the forces of the Korean navy north of the Kyonnaeryang Strait to protect the shipping lanes in the Yellow Sea. Admiral Yi was aware that the waters where the Japanese have paused are too shallow for his own warships. Therefore, he devises a scheme to lure the enemy away from the shoals surrounding Hansando Island and into the open sea, where his heavier *panokseon* ships and artillery could wreak havoc on the lighter Japanese craft. Of the 73 Japanese ships deployed, 59 were damaged or sunk, while the Koreans suffered only minor damage to a number of ships. This crushing victory proved the superiority of Korean naval weaponry, especially the *panokseon* ships, as well as Korean tactics. The defeat also ended Hideyoshi's dreams of conquering Ming China. The supply routes through the Yellow Sea had to be open in order for his troops to have enough supplies and reinforcements to invade China, and this could no longer be guaranteed.

• Beijing
KOREA
MING CHINA
HANSANDO
JAPAN

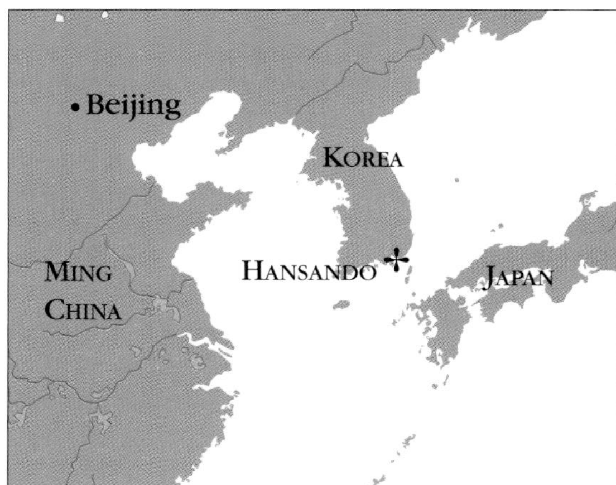

If the Koreans were going to repel the Japanese invasion of their country, they would have to destroy the Japanese navy. Control of the coast would severely limit supply of troops and provisions to the Japanese land forces.

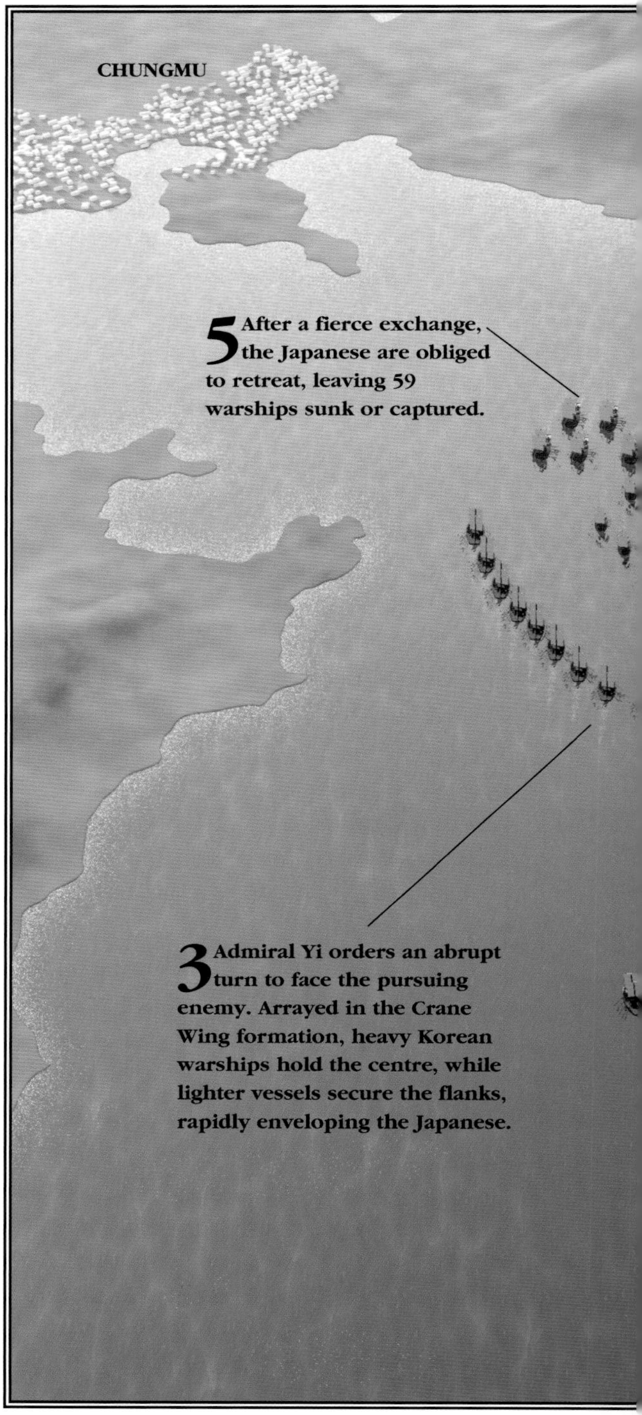

CHUNGMU
5 After a fierce exchange, the Japanese are obliged to retreat, leaving 59 warships sunk or captured.
3 Admiral Yi orders an abrupt turn to face the pursuing enemy. Arrayed in the Crane Wing formation, heavy Korean warships hold the centre, while lighter vessels secure the flanks, rapidly enveloping the Japanese.

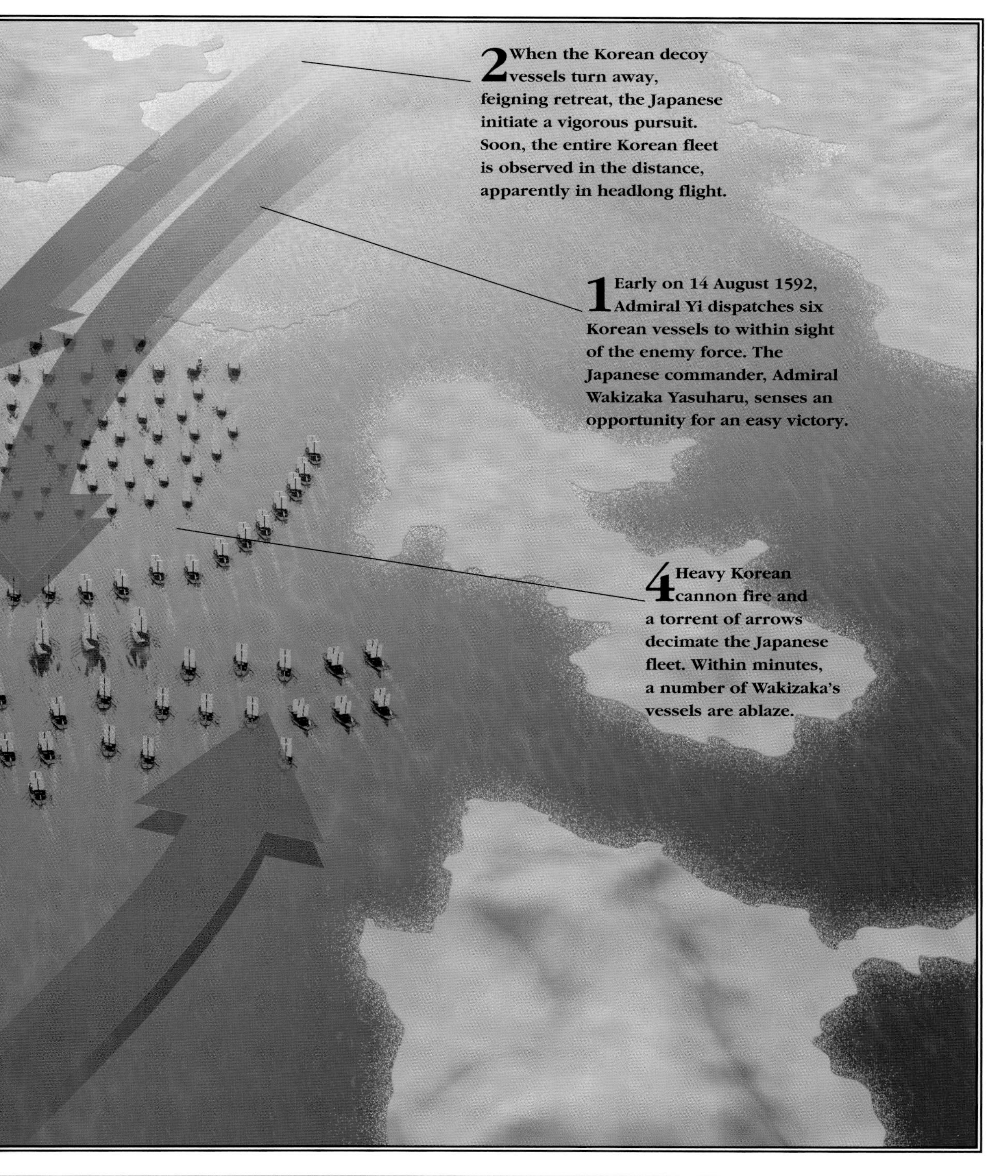
2 When the Korean decoy vessels turn away, feigning retreat, the Japanese initiate a vigorous pursuit. Soon, the entire Korean fleet is observed in the distance, apparently in headlong flight.
1 Early on 14 August 1592, Admiral Yi dispatches six Korean vessels to within sight of the enemy force. The Japanese commander, Admiral Wakizaka Yasuharu, senses an opportunity for an easy victory.
4 Heavy Korean cannon fire and a torrent of arrows decimate the Japanese fleet. Within minutes, a number of Wakizaka's vessels are ablaze.

August 1592, Admiral Yi set his battle plan in motion. Yi Pun's succinct but somewhat romanticized version of the battle of Hansando is nevertheless representative of Admiral's Yi's flawless execution of the overall plan of attack. Once again, he baited the enemy with the tactic of false retreat. Sending six of his *panokseon* through the narrow channel between Koje Island and the mainland of Korea, he tantalized Wakizaka with thoughts of an easy victory. Swiftly, the Japanese set sail in hot pursuit of what appeared to be a frightened and inferior Korean force.

Similar to the more familiar western tactic of 'crossing the T', the crane wing formation resembled a gigantic U shape, inviting an enemy force to actually envelop itself while the massed firepower of the Korean ships devastated the adversary from the front and both flanks. Admiral Yi had previously utilized line ahead or rolling tactics to engage the Japanese in the open sea. However, on this occasion the crane wing proved the perfect complement to the false retreat.

'Admiral Yi said, "Here, the sea is narrow and the shallow harbour unfit for battle, so we must lure them out into the open sea to destroy them in a single blow." He ordered his warships to pull back with feigned defeat until the jubilant enemy vessels pursued our fleet as far as the sea off Hansando…'

– Yi Pun, at the Battle of Hansando

Battle Formations

Yi placed his more heavily armed vessels in the centre of the formation, with his lighter, faster ships on the flanks. A tactical reserve was stationed behind the formation to fill in gaps which could be created during the confusion, smoke and flame of combat. Wakizaka unwittingly did exactly as the Korean admiral had hoped by attempting to employ the tactics most favoured by the Japanese at the time: closing swiftly then grappling, boarding and subduing the enemy during hand-to-hand fighting.

In the event, the Japanese did pursue the fleeing *panokseons* into the teeth of the crane wing. The greater range and heavier firepower of the Korean cannon and arrows began to take its toll immediately. The Koreans fired heavy broadsides while the inferior Japanese weaponry could make only a token response. Several of Wakizaka's ships were set on fire immediately. Interlocking fields of fire concentrated the Korean blows against their enemy from different directions simultaneously. Inevitably, the Japanese force was contained like herring in a fisherman's net.

Fighting continued into the late afternoon, and Wakizaka's impetuousness cost him dearly. Two of his lieutenants were killed in action while a third committed *seppuku*, or ritual suicide. Wakizaka was struck by several arrows during the fighting but body armour prevented a fatal wound. As the sun began to set, 59 Japanese ships had been sunk or captured by Korean boarding parties. Wakizaka eventually transferred his flag to a smaller vessel and escaped with less than 20 per cent of his original force intact.

Admiral Yi maintained a meticulous diary of his wartime exploits and reported regularly to the royal court on his progress. Of the great victory at Hansando, he wrote:

'First, I ordered out five or six board-roofed vanguard ships to make chase, feigning a surprise attack. When the enemy vessels under full sail pursued our ships, they fled from the bay as if returning to base. The enemy vessels kept pursuing ours until they came out to open sea. Immediately, I commanded my ships' captains to line up in the crane wing formation so as to surround the enemy vessels in a semicircle. Then, I roared, "Charge!" Our ships dashed forward with the roar of cannons "Earth", "Black", and "Victory", breaking two or three of the enemy vessels into pieces. The other enemy vessels, stricken with terror, scattered and fled in all directions in great confusion. Our officers and men and local officials on board shouted, "Victory!" and darted at flying speed, vying with one another, as they hailed down arrows and bullets like a thunderstorm, burning the enemy vessels and slaughtering his warriors completely.'

In a magnanimous gesture, Yi rounded up survivors and turned them over to local authorities, ordering them to render aid as necessary and send the prisoners back to Japan whenever peace should be restored. Overall, Japanese casualties were estimated at an appalling 9000, while the Koreans lost only 19 dead and 114 wounded.

Epilogue to Hansando

In the wake of the resounding Korean victory at Hansando, it became apparent that the limit of the Japanese northward advance on land would be the city of Pyongyang. The only alternative to receiving supplies and reinforcements from the sea was an arduous trek overland, and even then the levels received could not sustain a protracted offensive. Hideyoshi was compelled to order his naval forces to avoid direct confrontations with the Korean navy, which dominated the waters off the country's coastline. Hideyoshi's grand vision of conquest, which included Ming China, the Philippines, and dominion over Southeast Asia as well, had been thwarted, largely by his failure to grasp the strategic importance of command of the sea.

A former vice admiral of the British Royal Navy, George Alexander Ballard (1862–1948), assessed the importance of Yi's great victory at Hansando in his book, *The Influence of the Sea on the Political History of Japan:* 'This was the great Korean admiral's crowning exploit. In the short space of

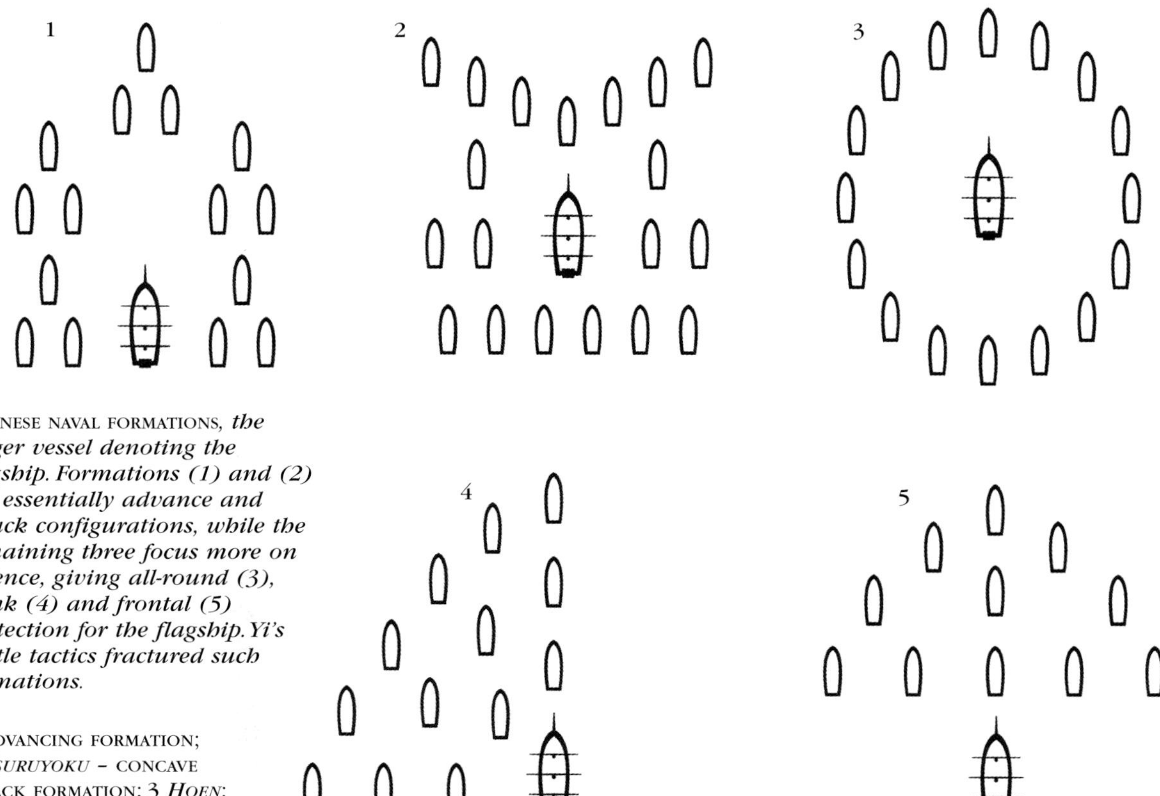

JAPANESE NAVAL FORMATIONS, *the larger vessel denoting the flagship. Formations (1) and (2) are essentially advance and attack configurations, while the remaining three focus more on defence, giving all-round (3), flank (4) and frontal (5) protection for the flagship. Yi's battle tactics fractured such formations.*

1 ADVANCING FORMATION; 2 *TSURUYOKU* – CONCAVE ATTACK FORMATION; 3 *HOEN*; 4 GANKO; 5 *GYORIN*.

six weeks he had achieved a series of successes unsurpassed in the whole annals of maritime war, destroying the enemy's battle fleets, cutting his lines of communication, sweeping up his convoys… and bringing his ambitious schemes to utter ruin. Not even Nelson, Blake, or Jean Bart could have done more than this scarcely-known representative of a small and cruelly oppressed nation; and it is to be regretted that his memory lingers nowhere outside his native land, for no impartial judge could deny him the right to be accounted among the born leaders of men.'

Although his command prowess had saved Korea, Admiral Yi subsequently fell into disfavour with the king after refusing an order to attack the Japanese. Yi knew that the order had been prompted by false information provided by a Japanese spy. Yi was imprisoned and narrowly escaped the death penalty, while his rival, Admiral Won Kyun, assumed command of the Korean fleet and once again led it to disaster. In desperation, the Korean government reinstated Yi, who could muster only 13 warships and 200 sailors. 'I still own 13 ships,' he pledged, 'As I am alive, the enemies will never gain the Western Sea.'

> *'On the nineteenth at dawn Ch'ungmu-kong [Admiral Yi] plunged his entire fleet into a final battle with the enemy, thundering, "Charge!". Suddenly, a stray bullet from the enemy vessel struck him. "The battle is at its height; do not announce my death!" With these words, he died.'*
>
> *– Yi Pun, Admiral Yi's nephew*

True to his word, Admiral Yi prevailed against overwhelming odds at the battle of Myeongnyang. A year later, the great admiral sailed for Noryang to do battle with the Japanese once more and rendezvous with his destiny.

The Wane of Hideyoshi

Toyotomi Hideyoshi, the son of peasants, had succeeded in unifying Japan and bringing to an end a century of civil war in the island nation, which is remembered as the Sengoku period. Although he had contributed greatly to stability at home, his peasant lineage prevented Hideyoshi from ever attaining the title of shogun, and his ill-fated campaign of military conquest against Korea and China proved costly in terms of lasting peace. Further, Hideyoshi was unable to truly consolidate power in Japan before he died in 1598 at the age of 61. Nominal power passed to his young son, Hideyori, who was but five years old at the time of his father's death.

The simmering rivalries among the powerful feudal warlords, known as *daimyô*, were a recipe for renewed civil strife. The so-called Council of Five Elders squabbled internally and was marked by infighting. Unsurprisingly, a pair of large factions emerged in Japan during the final years of the sixteenth century. One, headed by Tokugawa Ieyasu, *daimyô* of the Kanto Plain, was centred in the east at Edo, the site of modern day Tokyo. The other, led by one of Hideyoshi's closest and most able allies, Ishida Mitsunari (1560–1600), was strongest in the west. Mitsunari was sworn to defend Hideyori, the heir-apparent to his father's preeminent position.

Both Ieyasu and Mitsunari sought the support of various *daimyô* as the two sides prepared for a showdown. By the time of Hideyoshi's death, Ieyasu had a distinguished military record and had fought with the great Oda Nobunaga at Nagashino in 1575. He was said to be prone to episodes of rage but managed to hold onto the loyalty of his lieutenants. Conversely, Mitsunari had risen through political and diplomatic service in the court of Hideyoshi but had gained the enmity of several *daimyô* during his political manoeuvring. As a result, the commitment of his subordinates was tenuous at best and would prove his undoing on the battlefield at Sekigahara in October 1600.

According to authors Thomas A. Stanley and R.T.A. Irving, the military confrontation which

TOYOTOMI HIDEYOSHI, WHO ROSE *from peasant stock to lead the unification of Japan, arouses his soldiers with a blast on the great war trumpet prior to his victory in the battle of Shizugatake in 1583.*

decided the future course of Japanese rule was preceded by intrigue, posturing, promises, and treachery:

'The campaign which culminated at Sekigahara had begun with political manoeuvrings some months earlier. In July, Tokugawa Ieyasu was drawn from the Regent's Council at Osaka to defend his eastern domains from potential threat by a neighbour allied to the Ishida faction. Ishida promptly called his western allies to arms to mount a surprise attack on Ieyasu from the rear. Ieyasu was not to be fooled, however, and his elaborate network of spies kept him informed of exactly what was going on. Having used his time to draw together his own allied forces, he set off from Edo in August to feign attack on his neighbour to the north. He then turned west with his main body of troops to thwart Ishida's intended line of campaign.'

Two castles, Gifu and Konosu, fell to Ieyasu, who maintained control of the primary route to his capital at Edo and threatened to march on Ishida's base at Sawayama Castle. Taken aback by the swiftness of Ieyasu's forces, Ishida realised that the momentum of his adversary could sweep quickly to the east and even place young Hideyori in danger. Ishida abandoned his defensive position at Ogaki Castle and moved his forces to Sekigahara, where the mountainous terrain was sliced by a somewhat narrow pass.

Unwittingly, Ishida's decision actually benefited Ieyasu, who was well schooled in the finer points of warfare and much preferred the battlefield to siege warfare.

Clandestine Deal

Ishida set out for the pass at Sekigahara on 20 October while throughout the night a steady rain pelted down on his army. Although he was probably acutely aware that his armed coalition was fragile, he did not know that the wheels of betrayal had already been set in motion. During his rise to political prominence, Ishida had alienated quite a few of the *daimyô* across Japan. While some of those who joined him had done so out of obligation to Hideyoshi, their former leader, others were far less committed to the preservation of the status quo, particularly if Ishida stood to benefit.

Among these was 17-year-old Kobayakawa Hideaki, who, in 1597, had been appointed commander of Japanese forces in Korea during the waning months of that adventure. Internal strife among the Japanese generals contributed to the undoing of the campaign and, in the eyes of Ishida, Hideaki was to blame. Ironically, Ieyasu had been Hideaki's defender, and to compound the mixed emotions the young commander felt, his mother had urged that Hideaki join Ieyasu in the coming struggle for power.

Also among those ostensibly with Ishida was Kikkawa Hiroie, who became disenchanted over a disagreement between Ishida and Mori Terumoto, a *daimyô* whose troops were subsequently to remain out of the coming fight. Hiroie sent a secret message to Ieyasu that Terumoto would not be participating in the battle.

The Battle of Sekigahara, 1600

Covering the 19km (12 miles) from Ogaki Castle to Sekigahara, Ishida arrayed his rain soaked soldiers to block the Nakasendo, one of the major roadways in Japan, which connected Edo and Kyoto. Two other routes, the Ise and Hokkoku roads, running north-south and north-west respectively, were possible avenues of approach, and Ishida was obliged to cover all three.

Stanley and Irving described his dispositions:

'Setting his own camp on the northern flank of the valley, just above the Hokkoku road, his battle line extended southeast across the valley floor to cross the Nakasendo at Fuwa, the site of the old seventh century barrier station. It was here, on a mound overlooking Fuwa from the north, that he placed one of his most trusted and battle hardened divisions, under the command of Otani [Yoshitsugu]. It was also here, within the next few hours, that the fighting would be at its fiercest and where the most crucial stages of the battle would be fought.'

If, in fact, Ishida had any idea that there were turncoats among his *daimyô,* he may have given evidence of his concern by placing Kobayakawa in command of the right wing of his line facing Ieyasu. In this southern position, Ishida believed that Kobayakawa was less likely to be involved in the decisive fighting. This proved to be a grave miscalculation on the part of the less experienced commander.

Ieyasu reached the field at Sekigahara in the early hours of 21 October. He chose to maintain those forces under his direct command in reserve to the east of the village itself and to place his command post on a small hill adjacent to the Nakasendo road. From this vantage point, he could watch the battle take place and exert more effective command and control than would be possible in the midst of the fray.

In terms of manpower, the forces of Ishida and Ieyasu were evenly matched at approximately 85,000 each. However, Ieyasu's position was precarious for most of the day, owing in part to the absence of his son, Hidetada, at the head of 38,000 soldiers who had been stymied in their effort to reduce Ueda Castle by siege. Actually, a large number of Ieyasu's troops failed to take part in the battle at all. The victory turned out to be such a near thing for Ieyasu that he later had to be restrained from executing Hidetada for his tardiness. Ishida had seized the advantages the terrain offered, placing his forces in command of the high ground overlooking the valley. He hoped that Ieyasu would act impetuously and attempt a frontal assault against the centre of his line, allowing forces on both his wings to attack the exposed enemy flanks and assail Ieyasu from three sides simultaneously.

Battle Joined

As the grey light of day peeked through the overcast of 21 October 1600, the opposing armies were no more than 400m (1310ft) apart. The soldiers of both sides had been drenched by the incessant rain, and the fog and mist were slowly burning off. The opening action of the battle occurred at about 8 a.m. and it is very likely that Ieyasu was unaware the fighting had even begun until it was well underway.

Despite the fact that the inclement weather had turned the valley floor into a quagmire, the superb shock cavalry of Ii Naomasa splashed forward, directly into the centre of Ishida's line. Though their vanguard numbered only about 30 horsemen, Naomasa's warriors were ferocious. Known as the Red Devils, they wore red-lacquered armour from head to toe and carried red spears.

Seeing the movement by Naomasa, the soldiers of Fukushima Masanori joined in. Quickly, up to 20,000 of Ieyasu's troops were engaged in the centre and to the north, pushing steadily against the defensive rings which surrounded Ishida's command post.

To the south, Otani stood firm, denying further progress along the Nakasendo to Ieyasu. Casualties were heavy on both sides, and Ishida sent repeated signals and dispatched couriers to Kobayakawa, imploring him to engage the enemy and relieve some of the pressure on Otani. Neither Kobayakawi nor Kikkawa was moved by these communications, and compounding the problem for Ishida was the fact that approximately 1500

Samurai, campaign dress (1600)

Samurai warriors at the battle of Sekigahara were often protected by extensive use of armour, including a helmet of iron or steel covered by heavy lacquer and possibly some adornment indicating their family and status. The interior of the helmet was generally lined with woven hemp for the comfort of the wearer. Plates, covering the neck, extended from the edge of the helmet. Thongs of silk or hemp were tied under the chin to secure the headwear. Custom body armour consisted of iron or steel plating, sometimes reinforced by chain mail, complementary shoulder and forearm protection and leggings. The armour was initially constructed to be lightweight out of necessity: a samurai engaged in close quarter hand-to-hand combat would be restricted by heavy armour. However, the advent of the firearm during the latter part of the sixteenth century required armourers to construct heavier protection against low-velocity musket balls.

This man carries an arquebus. Although inaccurate, it did allow for sustained and damaging firepower when employed at close range. By the end of the sixteenth century, the Japanese were producing such firearms in quantity, and these weapons were common. While this warrior is seen carrying a matchlock weapon, it is the short sword, or katana, *for which the samurai are best known.*

Battle of Sekigahara

1600

Following a century of civil war in Japan, Toyotomi Hideyoshi had succeeded in unifying the island nation near the end of the sixteenth century. However, his untimely death in 1598, along with the tender age of his son, the five-year-old Hideyori, precipitated a struggle for preeminence between two of Hideyoshi's former subordinates. On 21 October 1600, the Battle of Sekigahara pitted the forces of Ishida Mitsunari, loyal to the young Hideyori, against those of Tokugawa Ieyasu, *daimyô* (lord) of the Kanto Plain. Secret negotiations between Tokugawa and several of Ishida's lieutenants, including Kobayaka Hideaki, resulted in their defection to the Tokugawa faction and contributed to eventual victory. The outcome of the battle facilitated the establishment of the Tokugawa shogunate, which would rule Japan for the next 250 years.

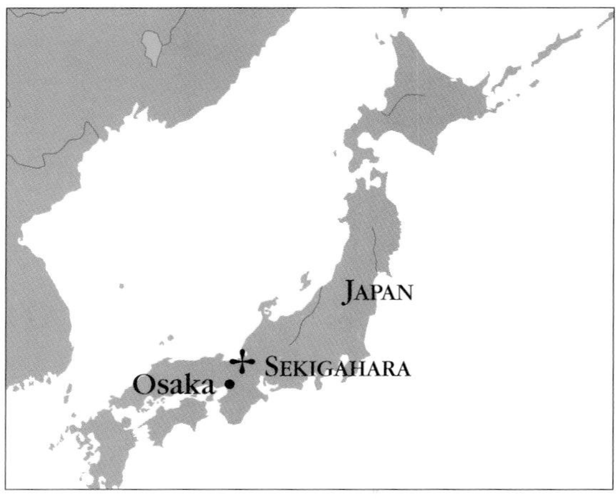

At Sekigahara, a narrow pass to the west of Edo, the site of modern day Tokyo, forces under Tokugawa Ieyasu gained a tactical victory over those of Ishida Mitsunari, giving rise to the powerful Tokugawa shogunate.

4 A general engagement develops, with Tokugawa's troops sluggishly advancing against Ishida's centre and northern flank. Defenders on the southern flank, however, hold firm.

SEKIGAHARA
3 The battle begins as the cavalry of Ii Naomasa, known as the Red Devils, assail the centre of Ishida's line, anchored on adjacent high ground.
5 Turning on their commander, Ishida's former allies under Kobayaka Hideaki strike in favour of Tokugawa and collapse the southern flank of Ishida's army.
6 As his battle line disintegrates, Ishida flees to temporary safety on Mount Ibuki.
2 Through driving rain, Ishida assumes defensive positions astride the Nakasendo, a major thoroughfare between Edo and Kyoto. Meanwhile, Tokugawa arrives on the field from the west the following day.
1 Startled by the swiftness of the forces of Tokugawa Ieyasu, Ishida Mitsunari rushes to block his enemy's line of advance at the narrow pass of Sekigahara.

ASHIGARU LOYAL TO TOKAGAWA IEYASU *rally around the famed Boar's Eye standard which flew during major engagements with rivals for dominion over the Japanese nation. The Tokugawa Shogunate endured for nearly three centuries.*

men under Yoshihiro Shimazu also remained inactive. The reason for Shimazu's failure to act was yet another perceived slight, this time by a representative of the overall commander of the Western Army.

In the predawn darkness of the eve of battle, Shimazu had advocated an attack against the gathering Eastern Army under the cover of darkness. Another of Ishida's lieutenants had called such an attack an act of cowardice. When Ishida did not speak up on his behalf, Shimazu determined that his forces would remain unengaged until late in the battle. By then, Shimazu was fighting for his own life as much as he was for Ishida.

Turning Point

By noon, the outcome of the battle was hanging in the balance. As Ishida was urging Kobayakawa to come to his aid, Ieyasu was fully expecting the 17-year-old commander to follow through on their treacherous arrangement. Apparently exerting little tactical control over unfolding events after all, Ieyasu could wait no longer. He ordered his arquebusiers, armed with European-style matchlock muskets, to fire a volley in the direction of Kobayakawa's ranks to get the soldiers moving.

Stanley and Irving wrote: 'When Kobayakawa finally made his charge down the hill, prompted by a volley fired at him on Ieyasu's orders, it was against the brave Otani rather than the Eastern Army that his men's bloodlust was directed. Otani still held out for a while longer, despite being heavily outnumbered. Other defections to the Eastern Army eventually proved too much though, and Otani, now faced with defeat, performed his last action that day by ripping his stomach open in the ritual manner.'

With the collapse of his southern wing and several of his commanders lying dead on the field, Ishida abandoned the fight and fled to temporary safety on nearby Mount Ibuki. The last vestige of an organized resistance by the Western Army was Shimazu's force, caught in a desperate struggle with the Red Devils and their fiery commander Naomasa, who had sustained a gunshot wound in his arm.

As the rest of Ishida's battle line disintegrated around him, Shimazu was cut off. It became apparent that the only way out was a desperate charge through the centre of Ieyasu's line. If he could reach the Ise road, he could possibly find an avenue of escape.

Shimazu and a nephew exchanged helmets, and the commander sallied forward with 200 of his remaining warriors. Shimazu and 80 survivors ultimately reached safety on the island of Kyushu. However, his nephew stood against the marauding Red Devils. Delaying their pursuit just long enough, the nephew's heroic gesture cost him his life. His head was cut off and later displayed with many others as a war trophy.

Ishida was on borrowed time. He remained at large for only three days before being captured on Mount Ibuki. Along with a number of other leaders of the former Western Army, he was executed at Kyoto. With his hard won victory at Sekigahara, Tokugawa Ieyasu was now in position to eliminate the only remaining threat to his consolidation of power, the young Hideyori, who remained alive and ensconced in Osaka Castle. Despite their willingness to fight for Ieyasu, some of the eastern *daimyô* hesitated in the final act of the drama. They prevailed upon Ieyasu to spare young Hideyori, who was allowed, for a time, to remain in Osaka with nominal rule over three of the western provinces.

Following the battle of Sekigahara, Ieyasu reportedly seized the land and other holdings of 90 families in Japan, redistributing large tracts to those who had been loyal and fought with him. Interestingly, he also rewarded some of those *daimyô* who had offered tacit support to Ishida or essentially remained neutral. Perhaps this was a pragmatic, politically motivated attempt to defuse a future confrontation with those who might otherwise prove to be powerful enemies in the days ahead.

In 1603, Tokugawa Ieyasu reached the zenith of his power when Emperor Go-Yozei bestowed the title of shogun upon him. Not until 1615 did Ieyasu move against Osaka Castle. At the age of 23, Hideyori committed ritual suicide, and the influence of the Toyotomi family was at last extinguished. The Tokugawa shogunate, born of blood and battle at Sekigahara, was to rule the island nation of Japan for more than 250 years.

Conclusion

While certain influences, such as weather, failure of technology, or even an accident may shape the course of a battle or campaign and will forever remain beyond the influence of a military leader, command and control are crucial to the success of any military undertaking. Regardless of culture, allegiance, or happenstance, an undisputed maxim of warfare is simply that the combatant who exerts effective command and control wins the day. The commander bears ultimate responsibility for that which his troops do or fail to do. Therefore, the force of personality, presence of mind and military skill demonstrated by Subutai and Jebei at the River Kalka, Admiral Yi at Hansando and Tokugawa Ieyasu at Sekigahara shaped the course of history not only in Asia, but also on a global scale.

A TRIO OF ASHIGARU *prepare to load their arquebuses. The* ashigaru, *of lower class than Samurai, provided a broad base of manpower and allowed the* daimyô *to raise larger armies.*

CHAPTER 4

SIEGE WARFARE

For many centuries, China considered itself the only civilization of any consequence in the entire world. It was *zhongguo*, the 'Middle Kingdom,' the centre of all culture and the civilized arts. And this civilization needed to be protected from the 'outer barbarians' (non-Chinese) that surrounded it and threatened to engulf it. The Great Wall of China was the supreme example of this protective mentality.

However, once cities were fortified, the creation of means of taking them soon followed. During periods when China was politically fragmented, a leader would never know if a neighbouring kingdom might invade with a conquering army. Even in later periods, a scheming official or ambitious general might start a revolt and seize a provincial capital. Techniques were developed not only to defend cities, but also to take them.

THE GREAT WALL OF CHINA *is one of the greatest engineering marvels of the world. The wall as it exists today was built during the Ming Dynasty (1368–1644). It was built in an attampt to keep out Mongol armies along China's northern borders.*

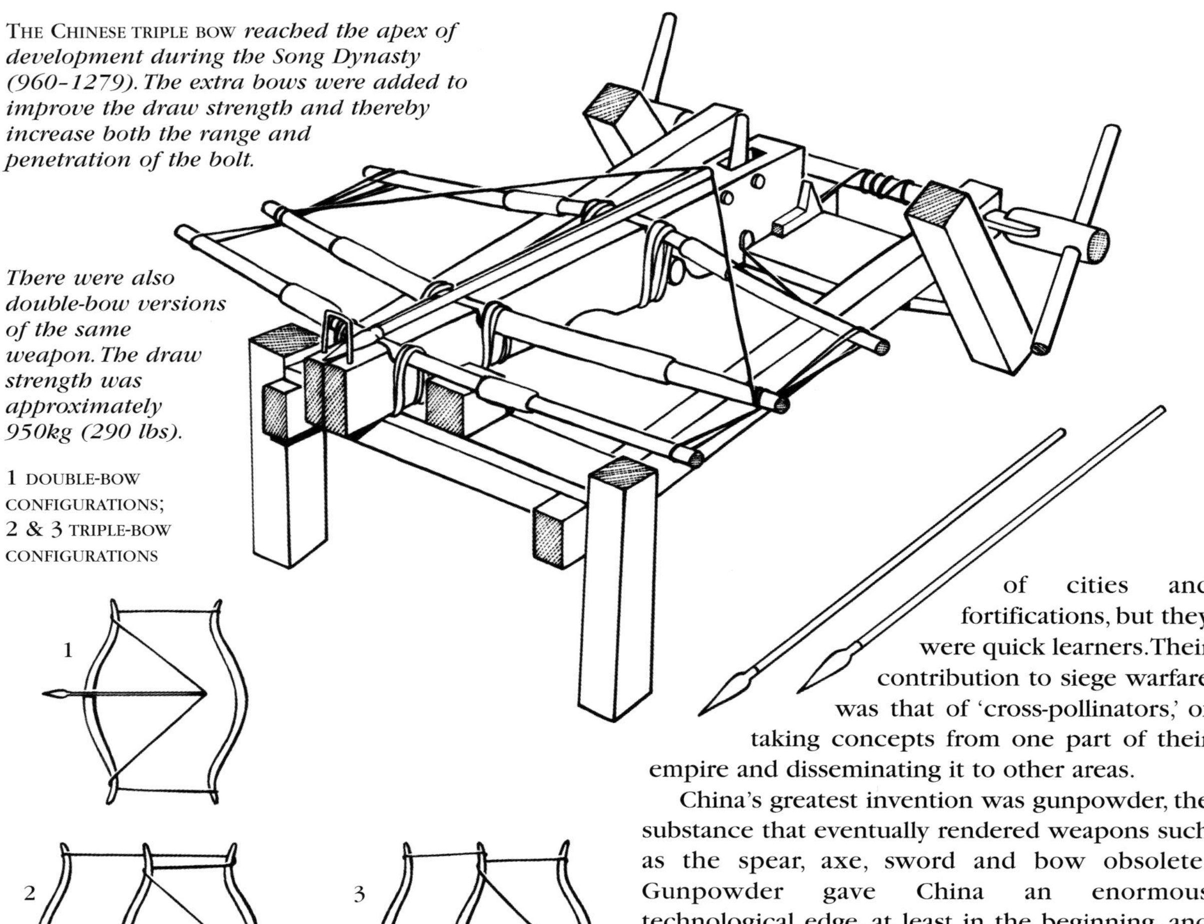

THE CHINESE TRIPLE BOW *reached the apex of development during the Song Dynasty (960–1279). The extra bows were added to improve the draw strength and thereby increase both the range and penetration of the bolt.*

There were also double-bow versions of the same weapon. The draw strength was approximately 950kg (290 lbs).

1 DOUBLE-BOW CONFIGURATIONS; 2 & 3 TRIPLE-BOW CONFIGURATIONS

China developed a wide range of siege engines, many of which were also independently developed in the west. But the Chinese introduced the traction trebuchet, which was copied in the Middle East and Europe during the Middle Ages, then reintroduced to China in a new and more powerful form, the counterweight trebuchet.

The Spread of Ideas

The Mongols were a nomadic people who carved out the greatest empire the world has ever seen, stretching from China to the edges of Poland in Europe. These hard-riding horsemen knew little of cities and fortifications, but they were quick learners. Their contribution to siege warfare was that of 'cross-pollinators,' of taking concepts from one part of their empire and disseminating it to other areas.

China's greatest invention was gunpowder, the substance that eventually rendered weapons such as the spear, axe, sword and bow obsolete. Gunpowder gave China an enormous technological edge, at least in the beginning, and the Middle Kingdom soon invented handguns and many types of cannon.

Tiny Choson (Korea) was a vassal in the Chinese tributary system, and adopted Chinese guns and siege techniques without hesitation. But in the fifteenth century the Koreans developed their own artillery that was an improvement over Chinese models. They also revolutionized naval warfare by the use of turtle ships *(kobukson)*: cannon-firing armoured warships that were designed to blast an opponent into submission, not grapple and board him.

The islands of Japan were relatively isolated, but even here Chinese technology seeped through over the centuries. The samurai lords had already adopted some crude Chinese guns, but responded enthusiastically when the Portuguese introduced modern arquebus muskets in 1543.

The Japanese invaded Korea in 1592 and in the process helped spread European-style weapons into the Korean peninsula. European contact though trade also brought European artillery and firearms into China. For a brief period during the late Ming dynasty, European military technology was in vogue, in part because it was superior to anything the Asian countries had in their arsenals.

Chinese City Fortifications

China was the dominant culture though much of Asian history and this is particularly true in military matters. Chinese building techniques were admired and its military architecture widely copied. Countries like Korea borrowed heavily from the Middle Kingdom, even though they added their own variations to suit their native culture, climate and geography

The Chinese came naturally to wall building, and in fact the Chinese character for 'city' and 'wall' is the same. Wall building has a long history in China, and references can be found from the earliest times. At the site of Ao, which flourished in Shang dynasty times (fifteenth century BC), large walls were constructed that were 25m (65ft) thick. Many Chinese cities were planned and these sites tend to have rectangular or square walls. Sometimes, they might feature several walls, all arranged in a concentric manner. They had crenellations to protect the defenders from enemy missile fire and in later times were pierced by gunports to allow cannon to fire at the enemy.

The main city gate usually faced south and was crowned by a wooden pagoda-roofed gatehouse that stored weapons. The main entrance was a tunnel just under the gatehouse, but getting through this passageway was never easy. Thick wooden doors barred the way and there was normally at least one portcullis.

Towers were placed at intervals along the wall, especially at the corners. Since the towers stood out from the main wall, archers stationed in them could provide enfilade fire with flights of arrows. Towers also served as high vantage points to watch enemy movements.

Walls could be made with a variety of materials, depending on local conditions. Compressed earth, brick, or stone could be used. Peasant farmers might be brought in from the country as forced labour on the bigger projects.

The Great Wall

The Great Wall is the most famous defensive work in Asia and a symbol of Chinese civilization. There have been many walls built over the centuries, and many were not even continuous. It is estimated that if all the walls built between the fifth century BC and the sixteenth century AD were joined together in terms of length, the total would be something like 56,000km (35,000 miles).

Today's Great Wall is of relatively recent construction (at least by Chinese standards) and dates to Ming Dynasty times, around 450 years ago. But the Great Wall had always been impressive, whatever its incarnation. As the crow flies, it is 2700km (1600 miles) long, but there are many twists, turns, and double loops so that it is, in fact, much longer. Counting all the bends, the total is more like 6440km (4000 miles), which makes it the longest human-made

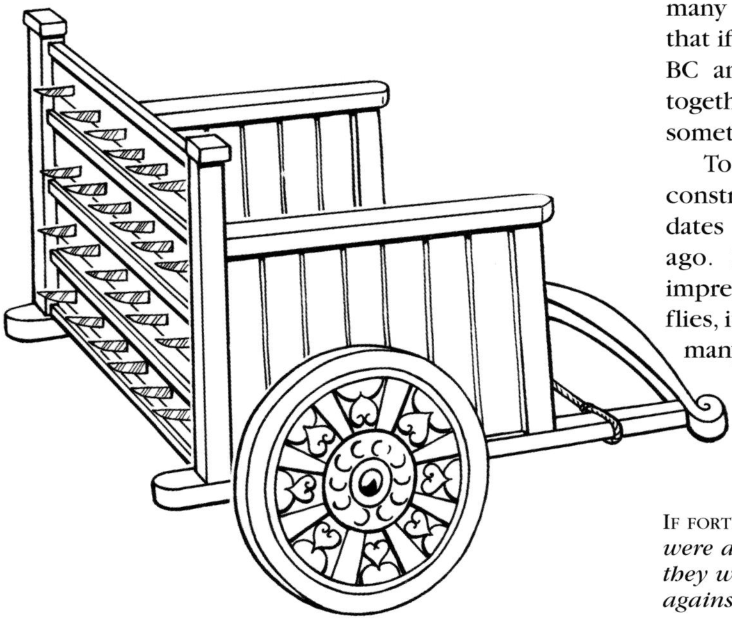

IF FORTIFICATIONS WERE BREECHED *these dagger carts were a literal 'stop gap' to restore defences. Often, they were used to block up entrances or doorways against attackers.*

structure ever built. The first really famous wall was built by Qin Shi Huang Di (259BC–210BC), the fabled First Emperor of China. He was a superb administrator and organizer but also a ruthless tyrant. He unified the country but recognized that nomadic tribes – China's perennial problem – were a serious threat to his expanded realm. The Qin Great Wall was finished in only 12 years, at the cost of much hardship and death for the workers who laboured on it. One of the reasons it was completed so quickly is because it was not made of stone like the later Ming version, but of packed, 'rammed' earth.

'Build high walls, stock up rations.'

– Zhu Sheng, advisor to the first Ming emperor Hongwu

It is a technique that is used in the Chinese countryside to this day by peasant farmers. Wooden planks are placed parallel to one another to form the mould or framework of the wall. The space in between is filled with earth sprinkled with water, and pounded down by human muscle. Because rammed earth walls are less labour-intensive they are much easier, and quicker, to build.

The Ming dynasty created the Great Wall we see today. The Mongols had been expelled from China in the mid-fourteenth century but still remained a potential threat. And even when the Mongols were out of the picture, there was seemingly no end to the nomadic northern tribes, like the Jurchen (Manchus), who were ready to sweep south. It was decided that only stone and brick walls would be an effective barrier against these invaders, and well worth the Herculean effort involved. Even today, modern engineers marvel at the construction, because at times the Great Wall seems to defy gravity itself as it marches down valleys and up precipitous cliffs.

Watch towers were placed at intervals, in the main used to store weapons, house garrison troops and provide bastions of defence. Signal towers always had a ready supply of animal dung to light fires at a moment's notice. Once ablaze, these fires could be seen for miles, and were a much faster means of communication and warning than horse and rider.

Nanjing: A Triumph of Engineering

Though comparatively little known in the west the Nanjing walls are a marvel of military engineering, just as important in their own way as the more famous Great Wall in the north. The Nanjing fortifications are massive, yet aesthetically pleasing, a marriage of Chinese military engineering, science, and art.

Nanjing, formerly 'Nanking,' is located along the banks of the great River Yangtze, in the very centre of China's heartland. Its strategic location assured it a prominent place in China's rich and turbulent past. Though Nanjing has been important since at least the second century BC, the city really came into its own during the reign of Hongwu, the first emperor of the Ming dynasty.

Hongwu decided to make Nanjing his capital, and lost no time in building projects that reflected its new imperial status. In fact, the very name 'Nanjing' means 'southern capital' in Chinese, though in Ming times it was named 'Yingtianfu' ('responding to heaven'). The Chinese had just emerged from nearly a century of alien Mongol rule, and the new emperor was determined not to repeat the experience. Hongwu's engineers were summoned to improve the natural defensive advantages Nanjing already enjoyed.

Originally the new walls were just going to be an extension of previously existing defences, but plans changed. An area to the northeast called Lion Hill was going to be included in the new additions, thus doubling the perimeter of the new walls. A rammed earth wall was also constructed just to the south, but this was a pale reflection of the main stone and brick works.

No less than 200,000 workers toiled on the walls, moving 0.2 million cubic metres (7 million cubic feet) of earth during the course of construction. Normally, the Chinese preferred to follow a rectangular shape for city defences, but in this case their plans were influenced by Nanjing's unique topography. The River Yangtze flowed nearby, a natural moat that guarded the city's

LEFT: THE ZHONGHUA GATE, NANJING. *The massive gate is not so much a gate in the conventional sense but rather a mini fortress attached to the main wall. It provided a first line of defence for the city.*

BELOW: THIS PORTION *of the Nanjing Wall runs past Lake Xuanwu. Ming engineers incorporated natural features into the overall design to improve the defences.*

western flank. Nanjing also nestled at the foot of Zijin Shan (Purple Mountain), the tree and bamboo-shrouded slopes blocking approaches from the east. Further natural protection was provided by Lake Xuanwu, another 'moat' that also provided the city with drinking water.

Taking advantage of this local topography, the walls assumed a meandering course, snaking around natural landmarks but always complementing them. Even today the Nanjing walls are unobtrusive, very much part of the total urban environment.

The fortifications took 20 years to complete, from 1366 to 1388, and about two-thirds still survive. When completed, the wall was 33km (20 miles) in length and encompassed about 60 square kilometres (23 square miles) of land. It was said to be the longest wall in the world at the time of its completion.

Basic construction techniques were similar to the ones used on the Ming-era Great Wall. First, a foundation course was set down, mainly in granite or limestone. Then, the inner and outer brick walls were raised, using a special kind of mortar that not only included lime but glutinous rice. Though odd-sounding to western ears, the rice was a wonderful binding agent that made for a particularly stable and strong mortar. The space between the inner and outer walls was filled with rubble of various kinds, including broken brick, gravel, and yellow earth. The Chinese engineers showed their genius by using this technique. Masonry would have been even more time consuming, not to mention prohibitively costly.

Though thousands of artisans and ordinary workers were involved, there was some attempt to achieve a level of personal responsibility as well as quality control. Even today, visitors are surprised to see that every few feet some bricks are covered in Chinese characters. This is not modern graffiti but is actually stamped into the brick surface. The stamped bricks are 'signature' bricks, bearing the name of the kiln where that section was fired, the date of the manufacture, and sometimes the name of the maker himself.The brick makers thus had to stand behind their work, and not hide behind the anonymity that usually would come with a project of this scale.

It was said that inspectors would examine the new bricks, checking carefully for shoddy workmanship or materials. The inspector would ask two soldiers to hold the bricks and hit them. If they 'rang true' and didn't break, all was well.The consignment would then be passed.

The Nanjing walls are around 21m (69ft) thick and 14m (39ft) high, strong enough to resist most catapult fire. They might have been more vulnerable to heavy counterweight trebuchets, but the city's geographic location - the river on one side and the mountain on the other - made it difficult to properly deploy such weapons. Cannons were relatively puny at the time and not regarded as a major threat.

ONCE WHEELED INTO POSITION *these mobile walls - in effect shields - offered attacking troops protection from enemy missiles. Here, early Chinese musket men provide covering fire for assault troops.*

Zhonghua Gate

The city wall was pierced with 13 gates.The most spectacular surviving example is the Zhonghua gate which guarded the city's southern approaches.This gate had to be particularly strong because there was no 'help' from local topography: no thick woods, lakes, rivers or mountains. Zhonghua gate is not so much a gate in the conventional sense but rather a fortress attached to the main wall. After one passes through the massive main gate you are confronted by three more smaller passageways, each leading to an enclosed courtyard. The three inner gates featured an arched gateway 'tunnel' surmounted by a wooden gatehouse.

Each gate had a portcullis and two strong wooden doors. It was a layered defence, because if an enemy broken though the main gate - an unlikely event - they would find themselves confronted with another gate, and then another. While they stood there trying to decide what next to do, soldiers would pour out from side rooms that opened into the enclosed courtyards and slaughter them.

The gate was also honeycombed with long tunnel-like chambers sometimes referred to as 'caves' in English. These tunnels could be used for storage, but mostly they were used as barracks. Altogether the Zhonghua gate could house a 3000-man garrison. The Zhonghua gate 'citadel' is flanked by two long ramps. About two-thirds of the ramp has flights of stairs, the remaining section is smooth. The stairs were presumably for foot soldiers, while the smooth ramp could provide traction for draft animals, horses, supplies, and siege engines to reach the defenders stationed on the wall.

Chinese Siege Engines

Siege warfare had but one primary goal: to capture the enemy's fortified position by capitulation or direct assault. If the enemy refused to yield by negotiation or intimidation, more direct means were employed.

A city's walls, towers, and gates were formidable barriers, but an attacking force usually employed a number of siege engines to achieve the desired result.

One of the earliest siege engines was the catapult. Technically, any siege engine that uses an arm to fling projectiles great distances is a catapult, so the definition includes the trebuchet. In popular terms, however, a catapult would be something that had a single stout arm fixed to a frame. The arm was attached to a twisted skein of ropes, and when the arm was forced down by a windless, the rope produced a lot of tension - straining to be released. The end of the arm usually had a sling attached to it to hold the projectile. In the popular mind, reinforced by the cinema, the arm would have a kind of scoop or cup attached to it, but that was rarely if ever used. When the arm was released, the arm sprung forward, simultaneously releasing the sling and hurling the projectile.

By contrast, the Chinese invented the traction trebuchet around the fourth century. In this concept a throwing arm was unequally balanced on a fulcrum point, with the long part of the arm holding the sling and projectile. The hurling power was provided by a team of men pulling ropes attacked to the arm's short end, usually two men to each rope.

The traction trebuchet had ranges of 30–60m (100–200ft) when casting projectiles weighing upwards of 113kg (250lb). The traction trebuchet idea spread west, probably though the trade routes like the fabled Silk Road. Certainly, the Byzantines knew the concept in the sixth century.

But within a few centuries the Europeans had transformed the trebuchet into a heavy siege engine capable of throwing far greater weights. Evidence is disputed, but it seems Europeans created this new form, the counterweight trebuchet, around the time of the Crusades. At the siege of Acre in 1191, Richard the Lionheart assembled two counterweight trebuchets named 'God's Own Catapult' and 'Bad Neighbour'.

THE CROUCHING TIGER TREBUCHET (Hudun Pao) *was originally a traction machine that relied on the muscle power of a pulling crew. After the Mongol conquest these were converted into counterweight machines, as seen here.*

It was the Mongols who introduced this transformed machine back to China, where it was received with surprise. The counterweight trebuchet uses a heavy weight instead of men pulling on ropes to provide the motive power. It is capable of throwing heavier weights much greater distances than the traction trebuchet. Up to that time the Chinese had stuck to their traction trebuchets, and had developed a bewildering variety of them. Many had colourful names, such as *cijiao pao,* or 'Four Footed Catapult,' and the *hudun pao,* or 'Crouching Tiger Catapult'. They were all variations on a common theme. Teams of 15 to 45 men pulled on the ropes, with muscle power providing the force.

Various explanations have been advanced as to why the Chinese never bothered to develop the heavier, counterweight version of their machine. One suggestion is that Chinese military philosophies stressed victory without massive destruction, and that pulverizing a city to rubble seemed counter-productive to the Chinese mind.

China enjoyed long periods of relative peace, and so there was no real need for the brute force that the counterweight trebuchets provided. And even during times of warfare, the emphasis was negotiation, or, failing that, some sort of trickery or ruse. Brute force was used only as a last resort.

Some historians also point out that China did not have the numerous fortresses and castles that dotted Europe and Japan. Fortresses and protective constructions were generally along the frontier and coastal areas, guarding against external threats such as Japanese pirates or nomad raiders. Major interior cities such as Beijing did have walls, but the mandarins who served the emperor were not a military caste, and not feudal lords like those found in Europe or Japan. The Chinese invention of gunpowder may have played a significant role. By the time the counterweight trebuchet reached China, gunpowder weapons were well established. Small bombs and other devices actually gave the traction trebuchet a brief, if significant, renaissance. It was much easier to hurl small bombs and grenades from them than from their heavier, more cumbersome, cousins.

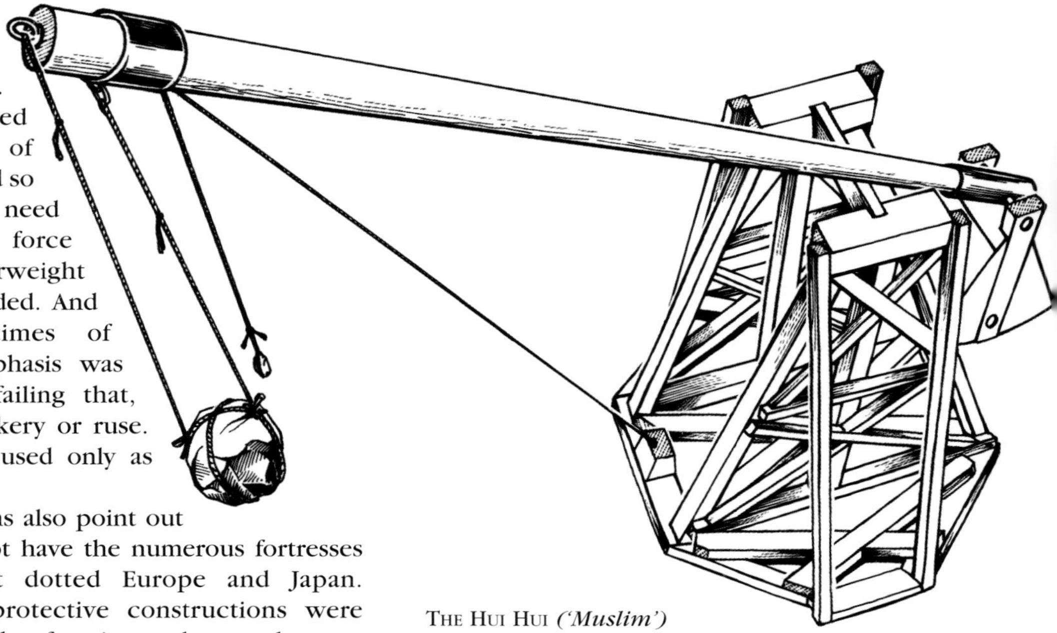

THE HUI HUI *('Muslim') trebuchet was a hinged counterweight siege engine. It played a major role in the Mongol conquest of Xiangfeng in 1273.*

Towers and Ladders

The Chinese also developed siege towers, both static and mobile. The static tower was essentially an observation post, useful during longer, more static sieges. The mobile siege tower was also known, and came in many names and varieties. In Europe such machines were called belfries and their concept goes back to Hellenistic times.

The Chinese and European devices were similar in nature. The mobile siege tower, which is sometimes several stories high, is rolled up to an enemy wall. A drawbridge or ramp comes down and soldiers rush forward from within the tower to engage the defenders. It's a familiar concept, reinforced by its many depictions in the cinema.

The Chinese had siege towers at least as early as the fifth century AD, which went by various generic names. There was the *lin che* (overlook cart), *long che* (high cart), and the *chong che* (assault cart). Sometimes a siege tower would have its own special appellation.

In 783 the town of Fengtian was put under siege. It was the middle of the Tang dynasty and engine designs were quite well developed. A *chong che* siege engine nicknamed 'Cloud Bridge' was employed during the siege. A typical design of the period, its sides were covered in layers of thick cowhide, giving protection against fire arrows and flaming brands. As an added precaution against fire – always a major source of concern – there were leather bags of water lined around the top. If a section did catch fire, the water would be released to run down the sides and douse the flames.

It seemed that the designers had thought of everything – except gravity and ground obstacles. The city defenders had dug a trench in front of the walls, and when the attackers tried to push 'Cloud Bridge' across it, the tower tipped over and crashed to the ground.

Protective Carts

The Chinese also developed *dongwu che,* a kind of mobile protective cart with a shed-like roof. This would be rolled up to city fortifications to provide protection for sappers trying to dig underneath and weaken a wall's foundation. One variation, the *tiao zhuang che,* is shown as a wheeled frame with a triangular, tent-like roof. A description written during the Tang dynasty speaks of the roof as being covered in rhinoceros hide and being able to shelter 10 men.

Of course, one of the most common attack methods was also the most basic – scaling ladders. They were certainly cost-effective and could be produced in large numbers. Unfortunately they provided no protection for the soldier. As he climbed he would be subjected to arrows, boiling water or oil, and boulders. If he managed to reach the top, the defenders would be there to welcome him with swords and a wide variety of pole weapons. Nevertheless the Chinese had an interesting device in the 'cloud ladder'. It was a frame that held a ladder that was hinged in the middle. During the approach the upper half was down or retracted, but when the machine got near enough poles would move the upper half to its fully extended position. Once in place against a wall, the rungs were more like a flight of stairs than a ladder in the conventional sense.

The 'cloud ladder' must have been effective, because it was widely copied in both Japan and Korea. In Korea it was called *unje,* literally 'cloud-scraping ladder'. The lower, cart-like section of the device had six wheels instead of the customary four wheels seen in Chinese models. By contrast, the Japanese version does not seem to have had any wheels at all.

Gunpowder: Catalyst of Change

China's invention of gunpowder in the ninth century ranks as one of the most epochal events in the history of the world. The very course of history was altered by this invention, which made war a far more deadly affair and influenced the destiny of whole nations.

Ironically one of the most destructive forces in the world was made by those seeking to prolong life, not to shorten it. The general academic consensus is that gunpowder was discovered by Chinese alchemists seeking an elixir of immortality. The Chinese experimented with various natural elements, hoping to find just the right combination to produce the desired effects. The effort was painstaking and could encompass decades. According to one version, once a Chinese alchemist mixed the proper ingredients, he would have to cook the 'recipe' for 50 or more years. As the experimenter grew older, no doubt the work took on an added urgency! But that was the crux of the whole matter – getting the right ingredients, then the proper amounts of those ingredients. While Chinese alchemists dabbled with various

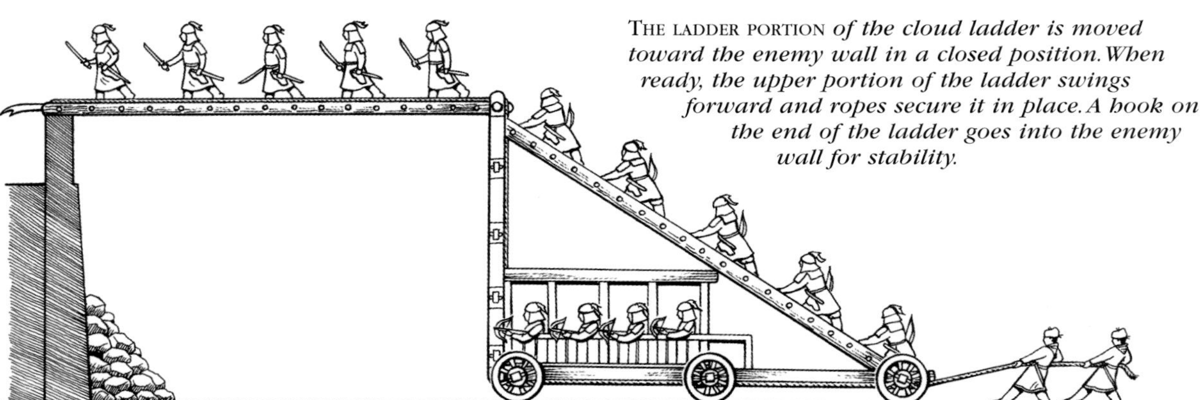

THE LADDER PORTION *of the cloud ladder is moved toward the enemy wall in a closed position. When ready, the upper portion of the ladder swings forward and ropes secure it in place. A hook on the end of the ladder goes into the enemy wall for stability.*

substances, they were learning their properties as they went along. In 142 AD Wei Boyand wrote a treatise called *Kinship of the Three,* in which he describes what happens when heat is applied to some unnamed elements. He notes the reaction is a violent one and that they 'dance and fly' in all directions.

This may mean that the Chinese invented gunpowder well before the birth of Christ, but the writings are too exoteric to provide creditable evidence. Researchers are on firmer ground with *Zhenyuan Miaodao Yaolue,* a Tang dynasty work dating from 850 AD.

Searching for immortality could be a deadly business and the writer is at pains to warn alchemists they should not fool around with certain substances. The Tang author makes it clear that if you mix sulphur, saltpetre, honey and a something called realgar, 'smoke and flames' result. As a matter of fact, he cautions that others who have applied heat to this concoction had their 'hands and faces' burnt, and 'even the whole house where they were working burnt down'.

Once gunpowder was invented the Chinese began seeking applications for this wondrous substance. It used to be thought that the earliest gunpowder was only used for benign purposes, such as fireworks displays on the lunar New Year. We now know, however, that gunpowder had military applications from the start.

The first known Chinese military usage of gunpowder was at a late Tang siege in 904. Tang engineers used catapults to hurl ignited lumps of slow-burning gunpowder into enemy positions. It was a marriage of new and old technologies and seemed to work quite well. The machines used in these operations were termed *fei huo* or 'flying fire' catapults.

Lances of Fire

The earliest known firearm was the *huo qiang* 'fire lance' (or spear), a weapon that may date to 950 or earlier. It is considered a 'proto-gun', not a true gun but definitely a remote ancestor of today's handheld weapons.

By contrast the fire lance was just that, a spear that had a hollow gunpowder-filled tube strapped to it. A soldier would carry a little iron box filled with smouldering embers to ignite the device. It was in essence a mini flamethrower, capable of spurting flames to over 3m (9ft).

The tube was about 60cm (24in) long and was constructed of strong paper, some 20 layers in thickness. Though the main purpose was to shoot flame, Chinese soldiers would sometimes fill the tube with pieces of porcelain, iron filings, and whatever else they could find. These odds and ends could do damage, and might be considered ancestral 'bullets.'

The fire lance was perfect for soldiers defending a city wall. Racks of fire lances could be wheeled about the battlements where needed, and the defenders would not have to carry embers around with them. Braziers of hot coals might be positioned at intervals along the walls as an ignition source. The weapons would be most effective at close quarters, such as when enemy troops were climbing scaling ladders.

Attackers might be set afire as they climbed, and if all else failed, the fire lance's spear point could act as a defensive backup. During a battle in 1276 a Chiang Tsai found himself attacked by two fire lance soldiers who apparently had already discharged their weapons. He was forced to dispatch them with his sword.

The earliest physical evidence of fire lances is subject to some controversy. The Dunghuang

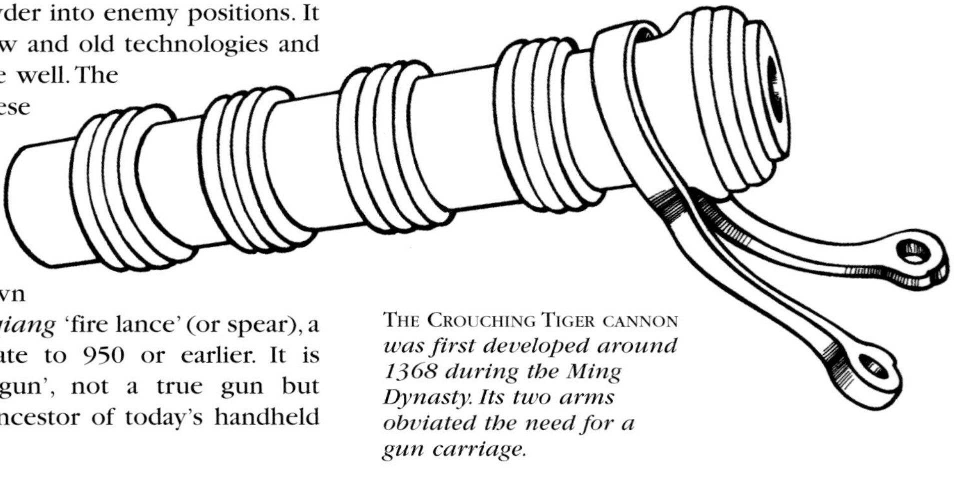

THE CROUCHING TIGER CANNON *was first developed around 1368 during the Ming Dynasty. Its two arms obviated the need for a gun carriage.*

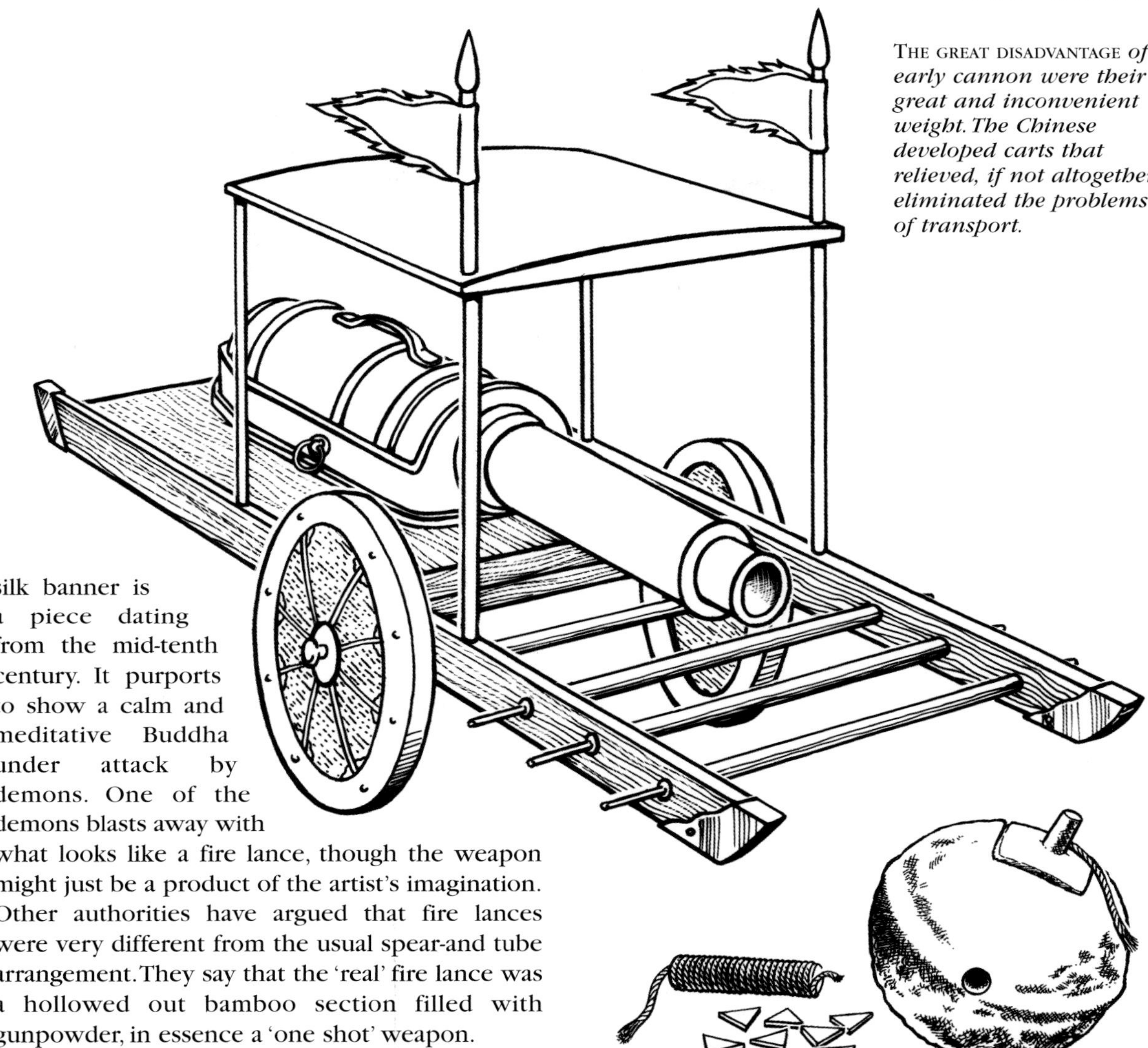

THE GREAT DISADVANTAGE *of early cannon were their great and inconvenient weight. The Chinese developed carts that relieved, if not altogether eliminated the problems of transport.*

silk banner is a piece dating from the mid-tenth century. It purports to show a calm and meditative Buddha under attack by demons. One of the demons blasts away with what looks like a fire lance, though the weapon might just be a product of the artist's imagination. Other authorities have argued that fire lances were very different from the usual spear-and tube arrangement. They say that the 'real' fire lance was a hollowed out bamboo section filled with gunpowder, in essence a 'one shot' weapon.

THE CHINESE THUNDERCRASH *bomb was essentially a kind of fragmentation device. When the fuse ignited the power inside the thin iron shell, the resulting explosion spread metal shards in a kind of 'shrapnel' effect.*

Guns and Bombs

In any case the first true gun was indeed made of bamboo and called the *tuhuojiang* or 'fire-shooting spear'. This weapon was first seen during the late Song period, in 1259. Before long handguns were being cast in iron or bronze. The oldest Chinese handgun found so far was discovered in 2004. Made of bronze, it bears in inscription date of 1271, when the Mongol Kublai Khan was consolidating his rule in China.

But perhaps of even greater significance was the Chinese invention of artillery. The oldest reliably-dated cannons are from 1356. They are of cast iron, and weigh between 60 and 270kg (132 and 600lbs) each. They generally fired stone or iron projectiles, though sometimes large steel-tipped arrows that sported leather tail fins were used.

Artillery was going to prove a vital element in siege warfare for both attack and defence. Once gunpowder was invented, there seemed to be no

THIS IS THE EARLIEST KNOWN DEPICTION *of a proto-gun called the fire lance. It is found on a painted silk banner that dates to the mid-tenth century* AD. *The demon in the red loincloth at upper right shoots a firelance to distract a meditating Buddha.*

limit to Chinese ingenuity. Rockets were developed, as well as various kinds of soft-cased and iron-cased bombs. The soft-case bomb, also called the *pi li pao* or 'thunderclap bomb', was essentially a hollowed-out bamboo tube with a fuse in the core. Thirty or so pieces of broken porcelain were packed around the bamboo, together with gunpowder and layers of paper. Once the fuse was lit, they could be hurled from a besieged city's walls with devastating effect.

The hard-cased bomb, or *zhen tian lei* ('thunderclap' or literally 'heaven-shaking thunder') was a variation on a common theme. They are usually credited to the Jin, the dynasty name of a Jurchen people that dominated northern China in the early thirteenth century. The bombs were shaped like gourds and were made of iron about 50cm (2in) thick. When detonated, the iron pieces would fly through the air like modern shrapnel.

Knowledge of gunpowder and guns travelled west via the various trade routes, including the fabled Silk Road. The Arabs were the middlemen traders for more than just luxury items such as silk and spices. They also were the conduits of new ideas, and Europeans most likely learned of gunpowder though them.

Genghis Khan

The Mongols were a group of nomadic tribes who lived on the fringes of the Gobi Desert. They were fierce warriors and splendid horsemen, but so were other groups in Central Asia. The Mongols were also beset by internal divisions, and tribal rivalries were not uncommon. In the mid-twelfth century few outsiders paid them any heed.

This was to charge with the rise of Temujin, a Mongol ruler of extraordinary talents, and a man who was going to lay the foundations of an empire. Within a few years he would be known as Genghis Khan, the universal ruler. And this title was no empty boast. By 1225 the Mongol realm was well on its way to becoming the greatest empire in world history. At its greatest extent it stretched from the borders of Poland to China. Not even the Roman Empire at its height could make such a claim.

After uniting his people Genghis began his programme of conquest. His first step was to organize unruly tribesmen into an effective fighting force. The idea was to organize his armies in such a way that would give him command and control, but at the same time not retard their innate fighting abilities. He divided his men into decimal units (one hundred, one thousand, and ten thousand), and during battle he would move them chess pieces on a grand scale.

The Mongol warrior would carry a curved sword, but his greatest weapon was his composite bow, which had a very heavy draw weight. Arrows launched from these bows were exceptionally powerful and could even pierce chain mail. A Mongol could shoot clouds of arrows at the gallop, taking care the release of the bowstring was synchronized with his horse's stride They also had stirrups, which meant stability in riding and more smooth and accurate shooting

Genghis Khan had contact with traders along the Silk Road caravan route, and it was from them that he first heard of China and its fabulous wealth. These fabulous tales aroused his curiosity and perhaps acquisitive spirit. But he also heard of great walled cities, and war engines that could 'hurl fire' great distances. If Genghis wanted to conquer this fabled land he knew it was not going to be an easy task.

It did help that the region was fragmented politically. Around 1100, the Song Dynasty ruled supreme in China. But the Songs could not stand up against the Jurchen, a Tungusic people from Manchuria who took over the northern part of China 1127. Before long, these Jurchens had established their own rulers known to history as the Jin dynasty. The Jin captured most of the Song royal family – but not all. One member escaped

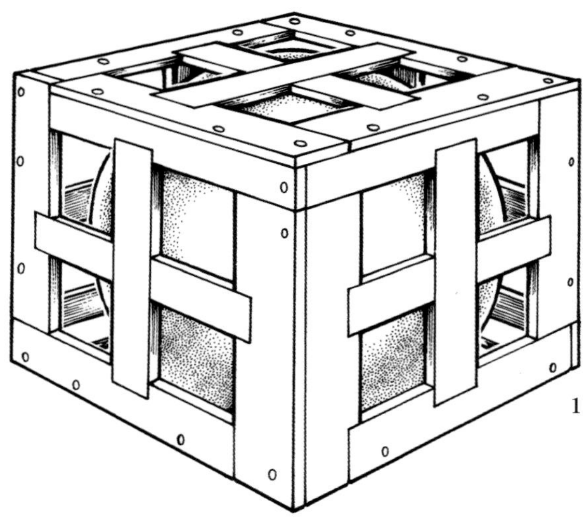

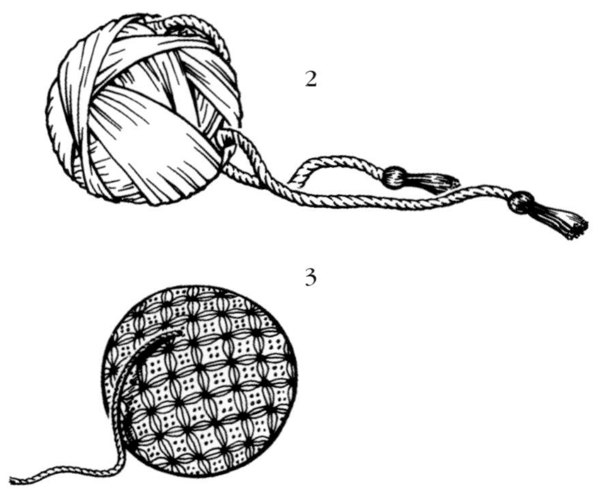

CHINESE EXPLOSIVE DEVICES: *The 'ten thousand enemies' was made of clay housed inside a wooden framework (1); the bee-swarm bomb was often thrown by catapults, and had a variety of small projectiles embedded within it (2); soft-case bombs had about 30 pieces of broken pottery inside to produce a 'shrapnel' effect when exploded (3).*

and set up the Southern Song, a dynasty that would last until 1279.

Southern Song culture flourished and it was in this period that the Chinese were making use of the first gunpowder weapons, from cannon to handguns. They also possessed some of the finest military architecture in the world. Genghis Khan knew he couldn't conquer China all at once but had to be patient, devouring it piece by piece. Conquering the Southern Song was still in the future, and was not going to be fully accomplished until Genghis was long dead. In the meantime, the Mongols' first course in the Chinese banquet was the Tangut state of Xi Xia (western Xia), a state which had cities with defences almost as powerful as those in China.

The First Mongol Siege

This, then, was going be the Mongols' greatest challenge. They were nomads with no acquaintance with cities or siege warfare. But they had to learn, and learn fast, if they had a hope of conquering the Middle Kingdom. The Xi Xia campaign was going to be a testing ground for the battle-hardened warriors to see if they could adapt to changing conditions and alien ways.

Faced with walled fortifications for the first time, Genghis Khan improvised, substituting cunning for real siege craft. He started the campaign with a real epic of human endurance. In 1207 the Mongol army crossed the blazing Gobi Desert, a distance of 965km (600 miles), drinking milk and the blood from their own horses.

Once they successfully crossed the desert, the Mongols had to besiege the Xi Xia fortress of Volohai. The Khan's plan was ingenious, but to modern ears it sounds like a chronicler's tall tale. In any case, the plan was feasible, if fantastic. The Mongols were baffled by Volohai's great walls, never having encountered anything like them. They had routed the Xi Xia's land forces, but this was a whole different problem. Genghis tried to attack the walls, but without success.

The Khan then told the Jin commander he would lift the siege if the Mongols were sent tribute or gifts. Genghis was specific – he desired one thousand cats and '10,000 swallows'. These were supplied, but once in possession of the animals the Mongols tied little cotton tufts to them and set these ablaze.

The panicked animals rushed back to Volohai, instinctively seeking shelter in their home roosts or dens. They set numerous blazes in the fortress, diverting the garrison from defence to fire-fighting. The Mongols took the opportunity to assault the walls, and Volohai soon fell.

The city of Xi Xia was another tough nut to crack. It is situated near the River Yellow on the arid Ningxia plain near modern Yinchuan, and Genghis actually tried to use the mighty waterway as an ally. The Mongols attempted to divert the River Yellow to flood

MONGOL INFANTRY ARMED *with mobile protective screens attempt to storm a defended position. These screens were effective against arrows but afforded no protection against Chinese thunder crash bombs.*

the city, but the plan proved a fiasco. The city held, but the country was laid waste by marauding Mongol horsemen. The Xi Xia emperor finally sued for peace, in effect becoming a Mongol vassal.

Genghis Khan died in 1227, but the Mongol empire continued to grow. Some accounts insist that when Genghis was on his deathbed he urged his followers to complete the conquest of China.

Mongol Campaigns in China

When the Xi Xia city of Zhengdu fell in 1215, Genghis acquired a prize greater that any treasure – Yelu Zhucai. Zhucai was a scholar and mathematician, well versed in the arts of Chinese civilization, and he had some Mongol blood. He was thus a kind of bridge between the Mongols and Chinese civilization, and became Genghis Khan's trusted adviser.

Though we cannot know how much he influenced Genghis, he is a fitting symbol of how the Mongols were going to freely adapt Chinese knowledge and use it to help them conquer the known world.

When the Khan pushed into the Khwarezm Empire, in what is now Afghanistan, Turkistan, and Iraq, he made sure he had a corps of Chinese artillerymen with him. From now on, the Mongols were going to use every kind of Chinese siege engine at their disposal, including the various kinds of traction trebuchets. Even if they never mastered the art of siege craft themselves they were going to bring Chinese engineers and artillerymen to do the job for them.

Occasionally, the Mongols resorted to the old-style techniques to achieve their ends. When Xi Xia balked at cooperation with the Mongols, the raging horsemen returned once again. When the walls at Ningxia proved too strong, the Mongols diverted water from a canal to flood the city. This time the technique was successful and Ningxia fell. Xi Xia ceased to exist as a political entity.

As the years passed the Mongols continued their inexorable march south. The Jurchen kingdom of Jin, long a source of trouble, finally fell to the Mongols in 1234. The Mongols now held northern China (roughly everything north of the River Yangtze) but against all expectations the Southern Song held out.

Song fortresses were particularly sturdy, and the Song used gunpowder weapons in their defence, including explosive grenades, firearms and cannon. In the year 1259 a Song Chinese official named Li Zengbo wrote that the city of Qingzhou was manufacturing 1000–2000 iron-cased bombs a month for use against the Mongols.

This is not to suggest that gunpowder alone was responsible for Song survival, though it did play an important part. There were other factors involved as well. The lands south of the Yangtze were unsuitable for the large-scale cavalry tactics so favoured by the Mongols. These territories also had almost none of the grazing grounds that the Mongols needed to feed their mounts.

The climate also played a significant role. The Mongol warriors and their horses found the semi-tropical heat of southern China debilitating. There were also a variety of diseases lurking in the south to which men who were used to the windswept steppes were easy prey.

But the Mongols refused to be discouraged from their path of conquest. They persisted, though for a time it looked like the Song would never be conquered. Several fortresses barred Mongol progress south, past the great River Yangtze. One of them, the small hill fortress of Diaoyucheng ('Fishing Town'), became a symbol of continued – and successful – Song Chinese resistance.

Fishing Town

Diaoyucheng is a fortified town situated on a steep hill surrounded on three sides by water. 'Fishing Town' is an historical landmark today, located about 5km (3.1 miles) from Hechuan, Chongqing, near the confluence of the Qu, Fu, and Jialing rivers. According to some accounts, Diaoyucheng experienced some 200 Mongol engagements before its walls during the period 1243 to 1279.

'Fishing Town' is a small fort, but it looms large in Song Chinese history. In 1258, the Great Khan Mongke (c.1208–59) appeared with an army of some 10,000 men. Mongke, a grandson of Genghis Khan, was nicknamed the 'Whip of God' for his ruthless soldiering. Once again, the Mongols were frustrated by the stubborn little fortress and its gallant defenders. Diaoyucheng apparently had cannon on the walls, which were used to good effect, and the tiny fortress altered the course of history when a lucky cannon shot mortally wounded Mongke. Some sources claim he died from some accident, or even illness, but it is generally accepted a cannon shot led to his demise. His passing led to something of a secession crisis but when the smoke cleared Mongke's brother Kublai Khan (1215-94) emerged as the paramount leader.

Kublai was determined to make himself ruler of all China, and made the subjugation of the southern Song a top priority. In the meantime he adopted some Chinese ways, and named his rule the Yuan (great originator) Dynasty. But no amount of window dressing could disguise the fact Kublai was an alien ruler, bent on conquering the last remaining pocket of native Chinese resistance.

The twin fortified cities of Fancheng and Xiangyang remained the final obstacles to the Mongol takeover. These fortresses were state of the art, built to withstand the conventional siege engines of the day - or at least the Chinese siege engines of the day. The defenders had siege engines of their own and cannons as well. By most accounts the siege of these two fortresses was long and protracted. Most historians claim the affair lasted almost five years, from 1268 until 1273. The Mongols finally achieved victory when they imported some Middle Eastern Moslem engineers - men who knew how to construct formidable counterweight trebuchets.

Once the walls were smashed and the Mongol hordes poured in, the outcome was never in doubt. Xiangyang and Fancheng fell to the invaders, leaving the door wide open to the vulnerable south. Though the Song held out a few more years, by 1279 the Song Dynasty was finished, replaced by Kublai Khan and Mongol rule. If Chinese military technology helped stave off the Song's inevitable end, it was also Chinese military technology, altered and improved by others, which brought the dynasty down.

Siege of Xiangyang, 1267

The siege of Xiangyang was one of the most celebrated clashes in Chinese history, an event that is commemorated even today. Though history only remembers Xiangyang, it was actually a joint siege of two cities, Xiangyang and neighbouring Fancheng. Xiangyang and Fancheng (the latter now known as Hubei) straddled the River Han, which in turn led to the Yangtze and the Song China heartland.

> *... a strong castle, and stout wall, and a deep moat ...'*
>
> – CONTEMPORARY PERSIAN CHRONICLER RASHID AL-DIN DESCRIBING XIANGYANG

The River Han was a watery highway to southern China, a path that had to be guarded at all costs. The twin fortress cities blocked the way, and had to be taken if the Mongols hoped to complete the conquest of China. The sources differ, but Xiangyang was a bone of contention for at least five years, perhaps more. Most historians say the first abortive attempts to seize the twin fortresses began as early as 1267 or 1268.

In any event these were huge and powerful fortresses, and befit the guardians of not just a river but an entire nation. Emperor Duzong (1265-74), the last reigning Song monarch, recognized their importance and appointed two loyal and distinguished soldiers to command the garrisons. Lu Wenhuan and Zhang Tianshun were both dedicated to holding the cities at all costs.

The two cities were indeed well fortified, but specific information is scanty. There are also some indications that the entrance gates had more than one wall, i.e., they had multi-layered defence gatehouses much like Nanking. Xiangyang and Fancheng were also linked by a pontoon bridge that facilitated both the movement of troops and general communications between the two posts.

The Chinese were not stupid and knew that the Mongols would appear with Chinese-style siege engines and probably northern Chinese to build and man them. That meant that the Mongols would employ Chinese traction trebuchets. To neutralize such machines the defenders widened the moat to extend the range, strengthened the already thick walls, and also lowered netting to

DEFENSIVE SIEGE WEAPONS: *the Chinese constructed windlasses that could lower troops for sorties against enemy siege engines(left); they also built striking boards that could be used against enemy troops trying to scale ladders.*

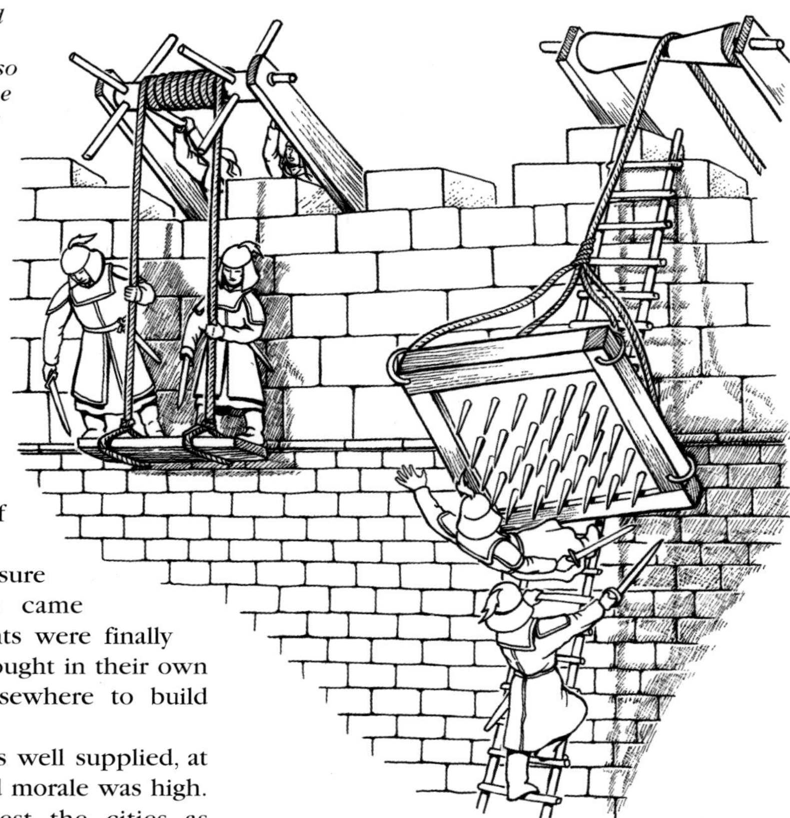

buffer and cushion the ramparts. The Song Chinese also made sure the cities were well stocked with food and other supplies, enough to last several years. It was said such foresight extended only to essentials like food and water. Eventually there were shortages of such items as salt – used for the preservation of food – and clothing.

The Emperor also made sure that fleets of supply junks came upriver, though these shipments were finally stopped when the Mongols brought in their own craftsmen from Korea and elsewhere to build patrol boats to plug this gap.

The Xiangyang garrison was well supplied, at least for the first few years, and morale was high. The Mongols did finally invest the cities as expected and built field fortifications designed to cut them off from the outside world. As the years dragged on, a kind of stalemate developed. The Mongols could not take the fortresses, but did manage to repulse numerous Song attempts to bring in relief troops and supplies.

Some sources say that the Mongols attacked at one point, probably at one of the gates, and were trapped between the gatehouse walls. Song defenders made short work of them and they were all wiped out. The Song troops were armed with crossbows and more than likely had firearms and cannon as well.

Predictably the Mongols brought up traction trebuchets, which were not as effective as hoped. The range was almost too far for the trebuchets, thanks in part to the widened moat, and few shots hit the walls. Those stones that did hit the target had little impact, thanks to the net 'padding' that festooned the walls.

After two or three years of stalemate, the Mongols were growing almost desperate. Even damming the river was tried, but this time it didn't have the desired effect. Kublai Khan was growing more angry and impatient with every passing week. Yet the Song Chinese were not having it all their own way. On occasion, the trapped defenders would stage a sortie and try to break out, without success. It was said that Song Chinese captured in this manner were beheaded without mercy.

Kublai sent Shi Shu, a northern Chinese general in Mongol service, to Xiangyang to help tighten the siege. It is ironic that his troops were also Chinese: northerners who apparently did not care that they were in the employ of foreign invaders. Shi Shu lost no time in building 500 river junks to help seal off any waterborne aid. Two forts were also built to control river traffic. In August 1269 there was a major river battle as a Song General

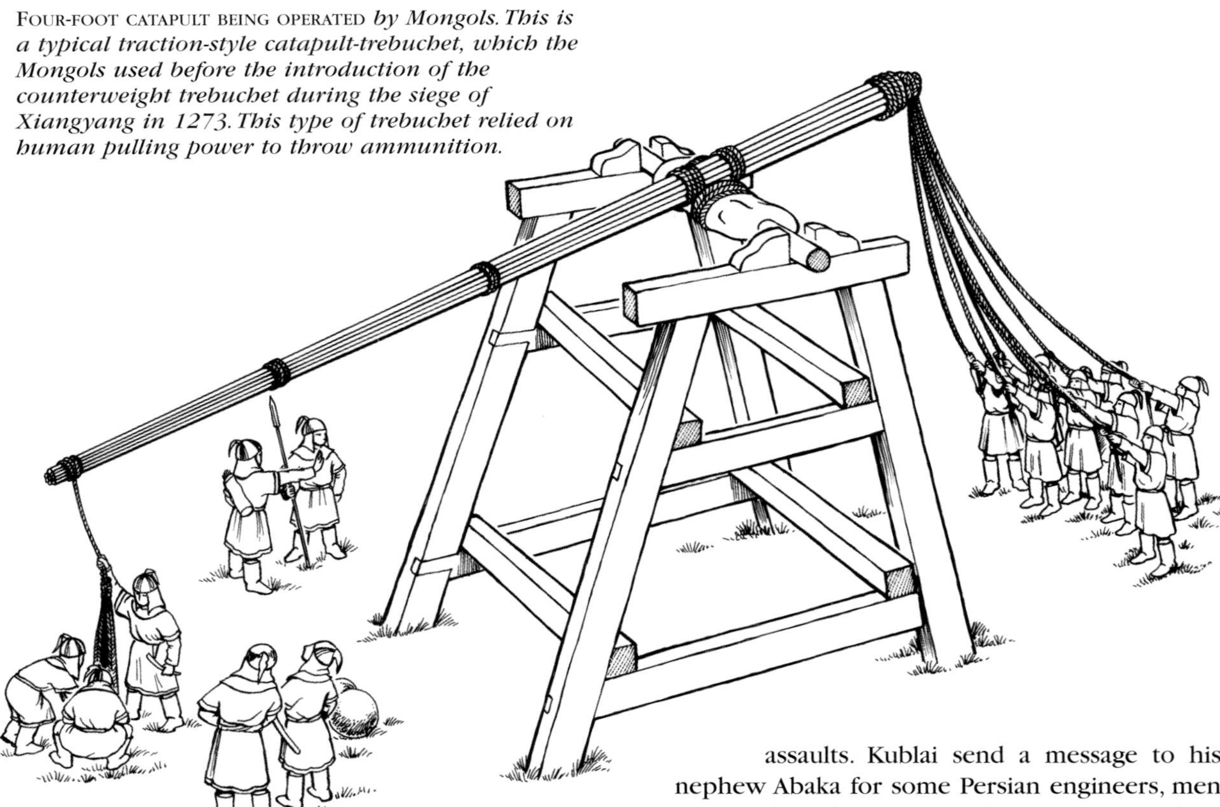

FOUR-FOOT CATAPULT BEING OPERATED *by Mongols. This is a typical traction-style catapult-trebuchet, which the Mongols used before the introduction of the counterweight trebuchet during the siege of Xiangyang in 1273. This type of trebuchet relied on human pulling power to throw ammunition.*

named Hsia Kuei tried to take one of the River Han forts with an armada of 3000 junks. There was heavy fighting and the Song fleet was defeated with a loss of 50 junks.

Thereafter the noose tightened around the twin cities. In September 1272 one relief force of some 3000 Song troops actually managed to reach the beleaguered cities, but unfortunately lost most of their supplies in the process. Worse still, they were now trapped inside. In the end, they only added 3000 hungry mouths to an already worsening situation.

Persian Engineering

Xiangyang still held out, even though food was getting scarce. Finally, Kublai Khan's patience was at an end - *something* had to be done, and quickly. He was tired of the stalemate, yet did not want to waste the lives of his troops in needless frontal assaults. Kublai send a message to his nephew Abaka for some Persian engineers, men acquainted with siege warfare as practised in Europe and the Middle East. In response Abaka sent Ismail of Hilla and Ala al-Din of Mosul to Xiangyang, who were two of the most skilled engineers in Persia. The moment they arrived they organized the building of counterweight trebuchets, the first ever seen in China. These machines were capable of hurling heavy projectiles more than 500m (1640ft). The actual weight of the projectiles are disputed, but they could handle stones of at least 75kg (165lb), and some claim as much as 300kg (660kg).

All do agree on the impact they made on the defenders, and accounts speak of the new engines as shaking heaven and earth when they were fired. Fancheng was the first city to be subjected to the counterweight trebuchets. The Song soldiers could not believe their eyes when they saw huge boulders sail over the walls and crash into city streets and houses. Then the Trebuchet gunners turned their attention to the walls themselves, which crumbled into piles of broken brick and shattered mortar. When the breaches were practicable, the Mongols and northern Chinese attacked in full force. What the Mongol army could not accomplish in almost five years was achieved in a few days.

Xiangyang held out for longer, in part because it had no other options. It was the last remaining bastion of Song power, and surrender would mean the end – or the beginning of the end – of the Song Dynasty.

The citizens of Xiangyang had endured much, but these counterweight trebuchets were the last straw. Some began to panic, and want to escape the city. Nothing could stand up to these new weapons, which were called *hui-hui pao*, roughly translated as 'Muslim trebuchet'. Xiangyang commander Lu Wenhuan had done his best, but even he could see the battle was lost. When the Mongols threatened to put all the inhabitants to the sword, Lu Wenhuan surrendered. The Southern Song officially lasted until 1279, when the last heir to the dynasty died.

'The attacking soldiers were all blown to bits – not a trace was left behind.'

– Chinese chronicle on the use of thunder clap bombs

Feudal Japan and Modern Firearms

The Portuguese began Europe's age of exploration by sailing down the coast of Africa in the fifteenth century. Vasco Da Gama managed to reach India in 1499, after which all Asia beckoned. Portuguese ships were soon in China, landing in Canton (Guanghzou) in 1517 to tap the wealth of the Middle Kingdom. A permanent Asian base was founded in 1557 when the Chinese granted permission for the 'western barbarians' to settle and rule Macao.

Portuguese first landed in Japan in 1542, arriving aboard a Chinese junk. But soon they established a more permanent presence, beginning with Tanegashima island then spreading out over the country. The introduction of the Portuguese arquebus in 1543 caused nothing less than revolution.

Japan was in the midst of a civil war, with rival *daimyô* jockeying for power. The balance of power was tipped in favour of those who could appreciate the new military technology. At the battle of Nagashino in 1575, warlord Oda Nobunaga had literally blown away his enemies with 3000 arquebuses. It was a lesson that deeply ingrained itself on the Japanese psyche. Soon these matchlock weapons were all the rage and they were to profoundly influence Japanese history. The Japanese not only adopted European firearms but also made improvements and created heir own versions. Toyotomi Hideyoshi was a passionate advocate of firearms and it was he who united Japan by the end of the sixteenth century. Hideyoshi was ambitious, and sought to conquer Ming China. Once again, it was the arquebus that gave the Japanese an edge when they invaded Korea en route to China.

Though the details are disputed by scholars, it seems the Japanese used massed musket volleys in their battle tactics. Their matchlock was a cumbersome weapon, which took time to reload, but some have suggested a kind of 'fire and retire' system might have overcome that difficulty. After the first volley, musketeers could give way to the second loaded rank, each rank taking its turn.

For some reason, artillery did not produce as much enthusiasm as muskets. The superiority of European cannon was recognized by men like Ieyasu Tokugawa, and they were used on occasion, but only sporadically. This is curious, since the Japanese had been on the receiving end of cannon fire for centuries. When Japanese pirates raided Korean coastal areas, they were often driven off by artillery.

Stranger still, the Japanese brought few artillery pieces for their great invasion of Korea in 1592. They did not need siege guns in the opening stages, but not all Korean fortresses were badly designed or in disrepair. The Japanese learned a hard lesson at the siege of Chinju, where Korean cannon mounted on the walls forced a besieging army to withdraw in bloody ruin.

Their education continued when the brilliant Admiral Yi Sun-sin administered a series of crushing defeats on Japanese naval forces. The

Siege of Xiangyang

1267–72

The twin walled cities of Fancheng and Xiangyang straddled the River Han, guarding the main river routes to the south and stopping the Mongols from gaining access to the Song Chinese heartlands. Both cities were well-fortified with thick walls and extended moats to reduce the impact of the Chinese-built traction trebuchets. The garrison resisted numerous early assaults, and in 1272 Song Chinese supplies and reinforcements were able to break the blockade by river, although they were not powerful enough to lift the siege. Eventually, the Mongols sought to improve their fortunes by bringing Persian engineers from Mosul to build counterweight trebuchets. These larger, more powerful machines were able to hurl much heavier projectiles more than 500m (1640ft). Fancheng fell after a short bombardment, but Xiangyang lasted longer. It was only after the Mongols threatened to put all the inhabitants to the sword that the city finally surrendered, after a five-year long siege.

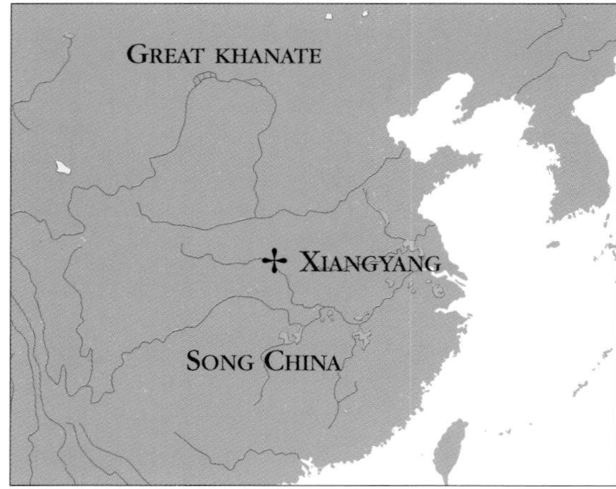

The twin cities of Xiangyang and Fancheng, located on the River Han, blocked Mongol progress towards the River Yangtze and the remnants of Song Dynasty China in the south.

FANCHENG

5 However, the Mongols introduce Middle-Eastern style counterweight trebuchets, which smash Fancheng's defences and the city falls.

3 From 1268 to 1273, the Mongols use of old-style traction trebuchets prove ineffective against Xiangyang's thick walls.

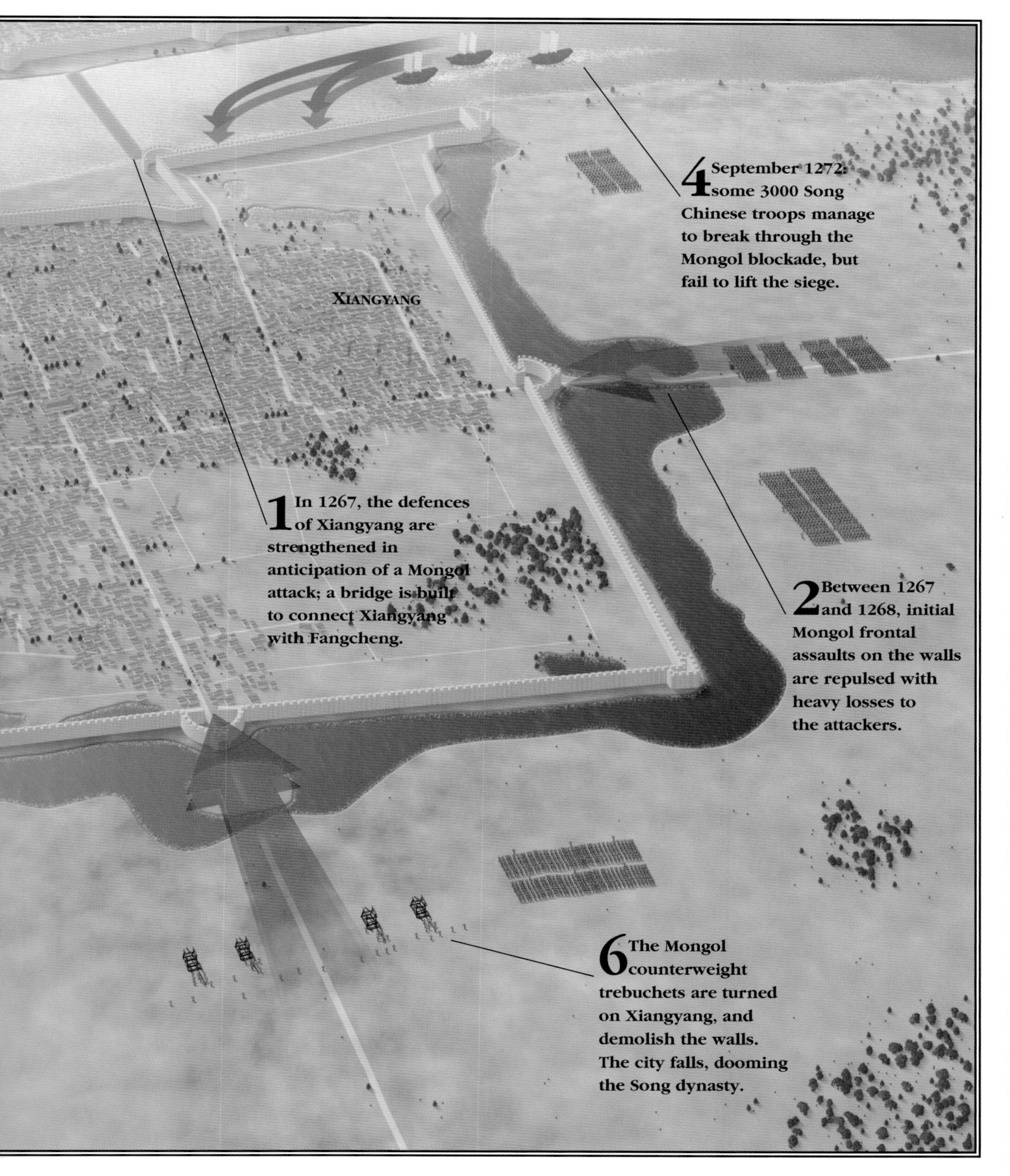
4 September 1272: some 3000 Song Chinese troops manage to break through the Mongol blockade, but fail to lift the siege.
XIANGYANG
1 In 1267, the defences of Xiangyang are strengthened in anticipation of a Mongol attack; a bridge is built to connect Xiangyang with Fangcheng.
2 Between 1267 and 1268, initial Mongol frontal assaults on the walls are repulsed with heavy losses to the attackers.
6 The Mongol counterweight trebuchets are turned on Xiangyang, and demolish the walls. The city falls, dooming the Song dynasty.

Japanese usually preferred hand-to-hand combat in naval engagements. Enemy ships would be grappled and boarded, the enemy bested with sword and lance. This was also very much in keeping with the samurai code of displaying individual bravery.

Admiral Yi's ships did not get close enough for boarding but instead blasted the Japanese with barrages of cannonballs and (artillery) arrows. The *kobukson* (turtle ships) were particularly effective because their upper deck was covered over with spikes and possibly iron plates. Boarders were likely to be impaled while seeking illusory glory. And that is assuming that a Japanese ship could even get close enough to try and board. Turtle ship cannons could, and did, pummel Japanese ships to destruction.

The Japanese did learn from these experiences, at least to some extent. Japanese armies began to use cannon during the Korean campaign, if only to respond in kind. Most of the pieces used were captured ordnance. Yet when the war was over the Japanese seemed to have lost interest in developing artillery. There were exceptions, of course, but those who still favoured cannon usually preferred European models.

Ieyasu Tokugawa never served in Korea but he was canny enough to recognize European superiority in artillery. A few Japanese guns were cast in bronze but Ieyasu preferred English and Dutch models. Mention is made of the European 'saker' and 'culverin' models being in his arsenal.

European cannon were used in Ieyasu's epic siege of Osaka Castle in 1615, but after that there was little need for powerful guns. There was a Japanese Christian revolt in 1638, and it is significant that Japanese authorities forced a European ship to add its firepower in crushing a rebel stronghold.

But when the last remaining embers of rebellion were stamped out the Japanese simultaneously lost interest in both artillery and the outside world. All trade was banned in 1639 and Japanese were forbidden to leave the country on pain of death. Shipbuilding languished, as did military technology. With no outside trade and exchange of ideas, Japan was cut off from western military advances that were changing the world.

As the years went by the arquebuses that the Japanese so highly prized became dated, then finally obsolete, but were still maintained as if they were state of the art. Even traditional samurai ways of war grew lax from the years of peace. Because Japan was an island nation it was protected from the outside world by surrounding seas, but this isolation was artificial and could not last forever.

Japanese Castles

The Japanese castle *(shiro)* played an important part in Japanese history, particularly during the sixteenth century. It was a time when internecine strife was so common the era is still known as *Sengoku Jidai,* or 'Age of the Warring States'. Nominally Japan was ruled by the emperor, a semi-divine being who was revered by all. In reality the Emperor was an honoured figurehead with little real power.

Instead the country was divided into a patchwork of domains ruled by *daimyô:* feudal lords who paid lip service to the emperor but jockeyed for power at every opportunity. Each lord felt the need to build a castle to defend and hold his lands, each according to his resources and the depth of his purse.

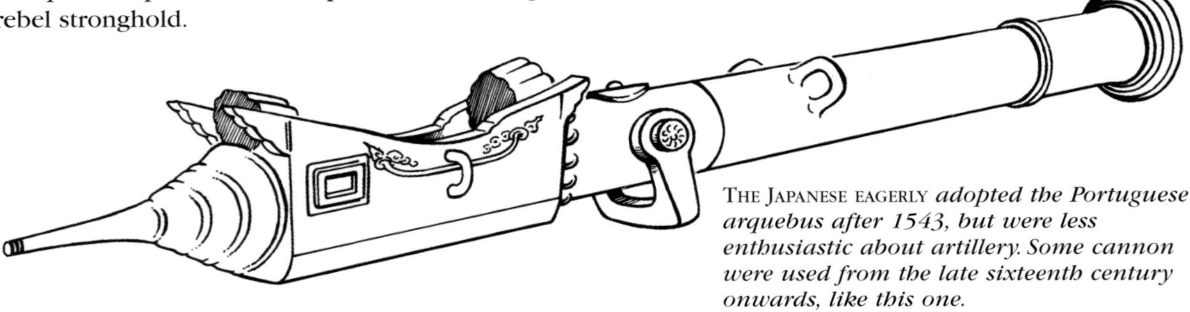

THE JAPANESE EAGERLY *adopted the Portuguese arquebus after 1543, but were less enthusiastic about artillery. Some cannon were used from the late sixteenth century onwards, like this one.*

Like its European counterpart the Japanese castle was a fortified building or series of buildings that had both defensive and offensive capabilities. The *shiro* was the seat of the *daimyô,* a place where he could dominate surrounding territory and rule his domain in reasonable safety. The castle's main function was defensive, but it was an admirable base from which to begin offensive operations as well.

When a castle was about to be built, a proper site had to be chosen. Once that was determined, the size and structure had to be planned, such as how much land the castle would encompass. In Japanese the word 'territory' is *nananari,* which means 'to lay down ropes', from when ropes were laid out to plan the size and dimensions of such a construction.

'When troops come (to Korea), have them bring as many guns as possible, for no other equipment is needed. Give strict orders that all men, even samurai, carry guns.'

— Asano Yukinaga, Japanese Samurai involved in the invasion of Korea

In theory a castle had three fortified compounds, or baileys. The most important was the *hommaru,* or inner bailey, the place where the lord had his residence and headquarters. The *hommaru* was noted for having one or more *tenshu* or towers. The *minomaru* was the second bailey, and the *sammaru* the third bailey. Both featured a labyrinth of passageways, storehouses, and living quarters. The chaotic design was purposeful. Should an attacker breach the outer defences he would get lost in the maze of passageways and buildings. While he searched for a route to the inner citadel he would be subjected to constant heavy fire.

Each bailey or *maru* was encircled by sturdy stone walls called *ishigaki,* basically stones embedded in an earthen embankment. But over time some Japanese castles went even further, carving away into hillsides, sculpting them into sturdy bases that could be faced with large stone blocks. Sturdy towers would be placed on top of these hill-mound bases, high points with plenty of machicolations for musketeers to fire down upon an advancing enemy. The casement stones on these mounds would be set without mortar and given a wide, flared base for further stability. These techniques could also help against earthquakes, which are common in Japan.

The castle buildings themselves were made of wood and plaster. The framework would be prefabricated in a master carpenter's workshop then assembled at the castle site. The wood construction of Japanese castles made them very vulnerable to fire. Buildings were plastered as a form of fireproofing, giving many castles a striking white appearance. Even so, lighting and interior fires destroyed many castles. Castles had killer whale talismans *(shachi)* perched tail-up on the edges of roofs as guardians against the danger of fire.

Dobei, small-roofed walls made from reed, bamboo and clay, often snaked on top of stone *ishikagi.* There were also *tamon yagura,* one-story galleries made of wood and plaster that were pierced by *tepposama,* funnel-shaped ports for defensive arrow or musket fire.

The main *tenshu,* or keep, was the heart of any castle, the place where the *daimyô* and his family lived, and the last bastion of defence. There were often secondary keeps, and important sites like Osaka Castle had these in abundance. In its heyday, Osaka Castle had 48 large *tenshu* and 76 smaller ones. Castle storehouses *(yagura)* had supplies of food, weapons and other items needed during times of siege. The word comes from *ya* (arrow) and *gura* (storehouse). They wee aptly named because they would be the repositories of thousands upon thousands of arrows as well as guns and other weapons.

Tenshu were also pierced with small protruding chutes called *ishi-otoshi,* translated as 'stone dropping'. If attackers tried to climb the sides of a wall or keep, defenders would drop stones, boiling water or hot oil. Though located in a slightly different position, the principle is much

MATSUMOTO-JO CASTLE, MATSUMOTO, HONSHU. *The* tenshu, *or keep, was the heart of any castle. This example is one the best preserved in Japan, dating from 1590.*

the same as the 'murder holes' found in gatehouse passageways in European castles and some Chinese city gates.

A castle may have been virtually impregnable but in the end it was only as good as its food and water supply. During the preparations for the siege of Osaka Castle in 1614, it was said that so much grain and other foodstuffs were stored the castle could have held out for several years. This is an impressive statement, considering the Osaka defenders numbered upwards of 90,000 men.

Osaka was also noted for its abundance of fresh water, which added to its overall invulnerability. There was one particular well within the compound that was particularly valued for its inexhaustible supply of good water. It achieved a kind of local fame as the place of *kim-meisui,* 'gold sparkling water', and an asset that was just as important as towers or strong walls.

Once the Togukawa shogunate established peace in the land, there was no need for such great fortresses. If anything, they were threats to the central government's authority. Some where dismantled, and some - like Osaka - were rebuilt on a much smaller scale. In fact, the Osaka Castle keep we see today is a twentieth-century concrete reproduction, although its stone-sheathed mound base is original. But there are still original castles left in Japan, treasured as relics of a distant and warlike past.

Siege of Osaka Castle, 1614–15

The siege of Osaka Castle marked an end to the civil wars that had plagued Japan for decades, and ushered in a period of peace and stability that was to last until the nineteenth century. Osaka was the stronghold of Toyotomi Hideyori (1593-1615), son and heir of former Japanese ruler Toyotomi Hideyoshi, the man who had tried and failed to conquer Korea.

In 1614 Hideyori was in his early twenties, really too young to have made a mark in the world. But he was maturing rapidly and his name was a powerful talisman to many who recalled his father.

That made him a threat to the continued domination of the Tokugawa family. Tokugawa Ieyasu had established himself as Shogun in 1603 after defeating all his enemies at the battle of Sekigahara in 1600. He stepped down in 1605, nominally handing the shogunate to his son Hidetada. It was a transparent move that fooled few people. Ieyasu would control Japan to the end of his life. But Ieyasu was still concerned about the Toyotomi threat. Ieyasu was 68, old for the period, and Hideyori was still a young man. Would the Togukawa family still be supreme in years to come? The best way was to deal with the problem now, once and for all.

Today's Osaka Castle is a modern tourist attraction and bears little resemblance to the original fortress. To begin with, the 1614 Osaka Castle was a multi-level defence system of dry and wet moats, towers, stone walls and galleries. The massive walls of the inner complex – still in existence today – rose 37m (120ft) and, if the outer defences are included, the castle perimeter stretched nearly 14.5km (9 miles). The heart of Osaka Castle was its *hommaru* or inner bailey, which featured a five-story *tenshu* of probably eight internal stories. In most castles the main *tenshu* was the primary residence of the lord and his family. At Osaka the main *tenshu* was mainly a storehouse; nearby was another tower which was a palatial home for Hideyori and his mother Yodogimi. Osaka Castle was not merely impregnable, it was dazzlingly opulent. It was said the roofs were gilded, and the interiors were famed for their unprecedented luxury.

Geography also played a role in Osaka Castle's defence. The Temma, Yoda, and Yamato rivers converged to the north, creating a labyrinth of muddy shoals, islands, and boggy rice paddies. The sea was much closer than it is today and formed a buffer on the castle's western flank. The River Huano and Nekoma stream flowed to the east, and the Ikutama canal hugged the outer defences to the west. By the summer of 1614 it was clear that war was going to break out between Ieyasu and Hideyori, though the younger man naively hoped for a peaceful resolution.

In the meantime, Ieyasu was gathering all the powder and European-style ordnance he could lay his hands on. He scored a major coup by purchasing five English cannons. Four of them were culverins, each weighing 1815kg (4000lb) and firing a shot of about 8kg (18lb). The other gun

TODAY'S OSAKA CASTLE *is but a pale reflection of its former glory. The stone work is original, but the tower-keep is a modern reconstruction.*

was a saker, which fired a 2.3kg (5lb) shot. There is some dispute over Osaka Castle's artillery. Some sources claim it was up-to-date, while other insist Hideyori had only inferior Chinese guns.

When war finally broke out Hideyori decided on a purely defensive strategy. He would sit in Osaka Castle, passively confident in its powerful defences. It was hoped that Ieyasu would waste time, men and precious resources trying to take an impregnable fortress. Within weeks Hideyori had an army of some 90,000 troops, most of then experienced and battle hardened. No major *daimyô* answered the call but Hideyori did have a major asset in a samurai named Senada Yukimura. Bold, unorthodox, and something of a swashbuckler, Senada was also a master of castle defence.

Though Osaka Castle was probably the most powerful stronghold in the country at the time, Senada was not satisfied. He had thousands of soldiers lay down their arms and pick up spades for more improvements. One major project was to link the Ikutama canal and the Nekoma stream with a large moat. The resulting excavation was an impressive 73m (240ft) wide and 11m (36ft) deep, and when the waters flowed they rose to a depth of 3.7–10.4m (12–24ft). Osaka's defences were upgraded just in time, because in November 1614 Ieyasu's Eastern army – so called to distinguish it from Hideoryi's Western political faction – approached the city. It was the start of what Japanese historians call the Winter Campaign.

The first objective was to take outpost forts that guarded the approaches to the main Osaka works. The outposts fell one by one, but the Eastern army suffered heavy casualties in the process. Ieyasu next ordered a major assault against the castle's southern flank, an extremely foolhardy manoeuvre. It was one of the most heavily fortified areas and commanded by Senada Yukimura himself.

The main southern assault was launched on 3 January 1615 and spearheaded by 10,000 crack troops. Senada purposely abandoned an outwork immediately fronting the main southern defences, lulling the enemy into a false sense of impending victory. The Eastern troops were jubilant, so overconfident that they blindly walked into a meat grinder. Ieyasu's men surged forward, only to be met by sheets of arquebus fire that shredded ranks without mercy. The attack was repulsed easily, but a little bit further down the wall one section was briefly taken by Eastern troops. Clambering up scaling ladders, Ieyasu's men gained the parapet then rushed into the outer bailey – and into the muskets of a mobile reserve waiting to receive them. Few, if any, of the attackers survived.

The Eastern army's grand southern attack was a bloody fiasco. There was nothing to do but settle in for a siege. Three hundred cannons were brought up, and once they were in place Osaka Castle was bombarded on a regular basis. Ieyasu ordered one artillery piece to be trained on a tower where he knew Hideyori stayed. One 6kg (13lb) shot smashed through the tower wall and killed two of Yodogimi's ladies.

The castle took some pounding but had been designed well. It was indeed almost impregnable, so Ieyasu tried trickery. He managed to persuade Yodogimi, the influential mother of Hideyori, that a truce could be worked out. Foolishly Hideyori agreed, and though most of the Eastern army withdrew, those that remained started to fill in Osaka's moats.

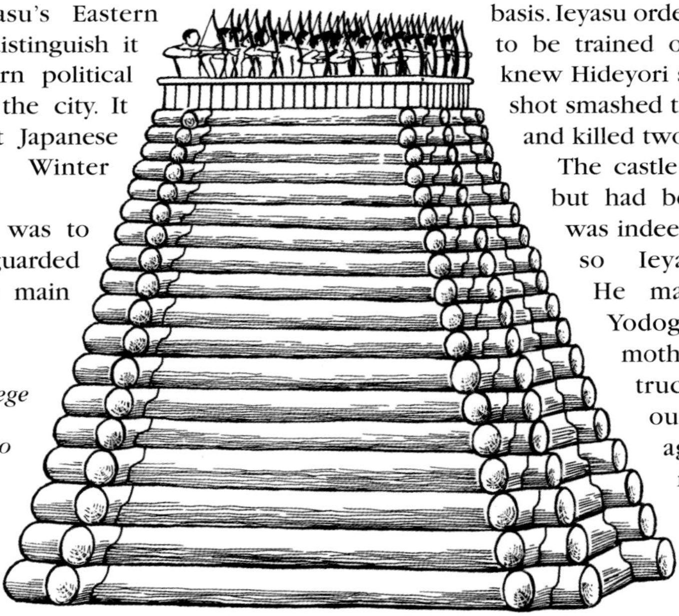

FIXED JAPANESE SIEGE TOWER. *Built of wood, a static siege tower provided an observation post when no high ground was available. In long sieges, the besieging force could also use the towers to rain missile fire on the defenders.*

Hideyori protested, but he could make no overt move lest he break the peace agreement. Ieyasu knew how to play the game, saying that since there was peace there was no need for moats! War came again some six months after the first siege. Hideyori was once again under siege at Osaka, but thanks to the earlier deceit its defences were not as formidable as they had once been.

Ieyasu's troops and cannon could get closer than before, and parts of the castle were blasted to rubble. Soon the main keep was aflame and all hope was lost. Hideyori committed suicide and his mother either killed herself or was dispatched by a retainer. Thousands of Toyotomi loyalists were also put to the sword.

The siege of Osaka marked the end of war and great castle building. Peace reigned for over 200 years, until the coming of the western powers in the mid-1800s began a whole new era for Japan.

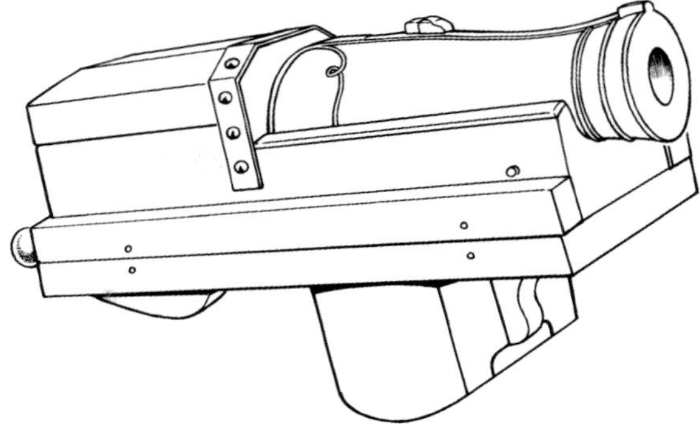

A JAPANESE CANNON USED AT THE SIEGE OF NAGASHINO. *This cannon is credited with destroying an enemy siege tower. In spite of such success the Japanese were slow to adopt artillery.*

Korea and the Imjin War

Korea's position between China and Japan – a Korean proverb says, 'a shrimp between two whales' – has been both a blessing and a curse. The curse part was certainly true in 1592, when Toyotomi Hideyoshi decided to conquer China by way of Korea. Toyotomi had recently unified Japan after a series of bloody civil wars and it seemed his ambitions knew no bounds

At first the Koreans were blind to the growing threat. Japanese pirates, who were called *waegu* by the Koreans, had raided on and off for centuries. In the Korean mind the Japanese were brigands, island barbarians who were beneath contempt. When Hideyoshi announced his intention to take over Ming China, it was treated as a macabre joke in Korean court circles. 'The Jap's attempt to attack the Celestial Empire [China] is like that of a snail which strikes a rock.'

But the laughter would soon stop, replaced by tears and cries of anguish. Hideyoshi assembled an expeditionary force of 158,000 men who were tough, well trained in the use of arms and hardened by years of internecine strife. They also had the arquebus, a matchlock musket of European origins that was to prove superior to anything yet seen in Asia. The Japanese have always

JAPANESE CANNON, PORTUGUESE STYLE, *early senenteenth century. In this illustration Japanese gunners are about to fire. Two crewmen hold ropes to haul the cannon back after its recoil. The crew are protected by bundles of green bamboo.*

Siege of Osaka

1615

In 1600 the Japanese emperor was a revered figurehead. Tokugawa Ieyasu was the real power in Japan, but the old samurai looked on Toyotomi Hideyori as a potential threat to his dynasty's continued rule. Politically unscrupulous and a master of intrigue, Ieyasu confronted Hideyori and forced him into a fight. Hideyori's Osaka castle was the strongest in Japan, but when Ieyasu's first siege bogged down into stalemate he resorted to trickery. Hideyori foolishly agreed to a truce, and during the 'peace' Ieyasu filled the moats so that they could be more easily crossed for assault. Six months later war broke out again: with the moats filled in, Osaka castle was less formidable, and was eventually taken. Hideyori committed ritual suicide and many Hideyori loyalists were put to the sword. With its last rival gone, the Tokugawa Shogunate ruled Japan until the nineteenth century.

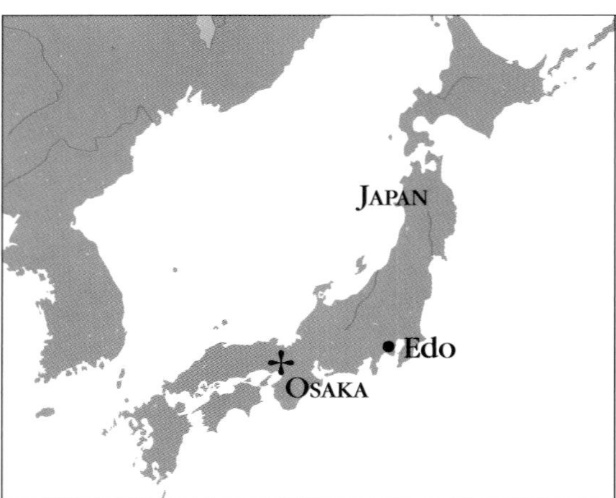

Osaka is located on the island of Honshu by the mouth of the River Yodo, which flows into Osaka Bay. Tokugawa Ieyasu's base was at Edo (Tokyo), 403km (251 miles) away.

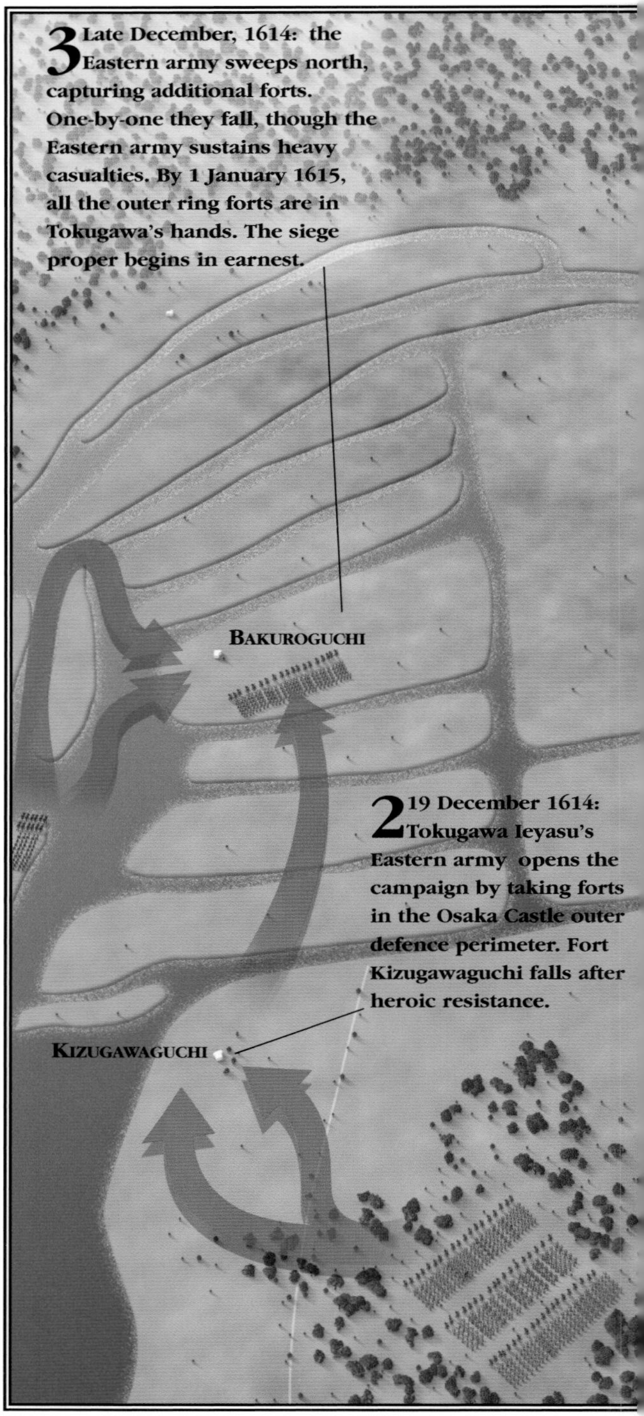

3 **Late December, 1614: the Eastern army sweeps north, capturing additional forts. One-by-one they fall, though the Eastern army sustains heavy casualties. By 1 January 1615, all the outer ring forts are in Tokugawa's hands. The siege proper begins in earnest.**

BAKUROGUCHI

2 **19 December 1614: Tokugawa Ieyasu's Eastern army opens the campaign by taking forts in the Osaka Castle outer defence perimeter. Fort Kizugawaguchi falls after heroic resistance.**

KIZUGAWAGUCHI

1 Mid-November 1614: the Osaka Castle defences, already the most powerful in Japan, are further strengthened. The Sanada Maru, a formidable barbican, is constructed and a dry moat excavated.
SHIGINO
OSAKA CASTLE
2 20 December 1614: Uesugi Kagekatsu captures the Shino fortress but is counter-attacked from Osaka castle and has to be reinforced.
SANADA MARU
4 3 January 1615: Tokugawa forces attack Sanada Maru, a defensive fortification named after its commander, veteran samurai, Sanada Yukimori. Sanada's defence is firm and the Eastern forces are repulsed with heavy losses.
5 3–21 January 1615: Tokugawa Ieyasu reluctantly settles for a long seige. Eastern army artillery bombard Osaka Castle, but the fortress proves impregnable. The siege ends with a negotiated peace – which Tokugawa Ieyasu secretly has no intention of following.

KOREAN CANNONS: *Korea designed its own artillery from the fifteenth century onwards, supplementing native models such as the 'Heaven cannon' (below) with Chinese designs such as the 'Crouching Tiger' cannon.*

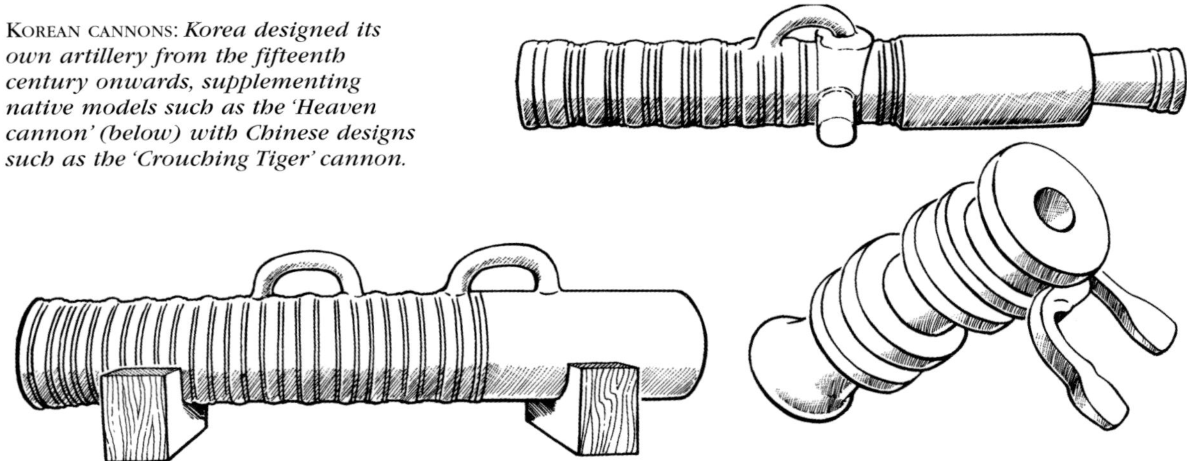

shown an interest in foreign technology and this new weapon had exciting possibilities. It had longer range than the Japanese bow and required less training to use. The Japanese had also developed massed volley techniques. Sixteenth century firearms were clumsy to reload, but by rotating weapons – bringing loaded musketeers forward while others retired to reload – they overcame this problem. Strangely, the Japanese made no effort to hide their new 'secret' weapon. In 1590, a Korean embassy came back home with Japanese gifts, including an arquebus, the first ever seen in that country. It was admired, treasured, then stored away and forgotten.

Korean Weaponry

Korea took pride in being a vassal of China, and the Middle Kingdom's culture and technology was copied with a kind of filial devotion. Chinese Emperor Taizu called the country *Zhaoxian,* (Choson), a name that stuck until modern times. It means 'morning freshness' or 'morning calm'.

But things were anything but calm in the late sixteenth century. The Korean army had 200,000 men on paper but the soldiers were ill-equipped and poorly trained. When the war started, the ranks were filled with raw recruits who were brave but who could not stand up to Japanese professionals.

Korean fortifications were also in a sad state. Walls were mainly vertical constructions made of small flat stones, all too often poorly maintained. Many were *sansongs,* mountain fortresses perched on high peaks and designed to shelter local populations in times of crisis. These mountain eyries were good for defence but were sometimes located too far from populated areas for easy travel.

When King Seonjo's (1552–1608) government finally woke up to the danger it sent out decrees ordering that city walls and other fortifications be repaired and refurbished. The work was done under a *corvee* system of forced labour, which was deeply resented by a people who still did not believe the Japanese were coming. Repairs were shoddy and made with 'quick fix' materials. Incredibly, one mountain fortress was relocated to a lower elevation for convenience. It was quickly taken by the Japanese invaders. Yu Songnyong was one government minister who believed war was rapidly approaching and he was dismayed by what he saw and heard. One friend flatly declared, 'Building fortresses is not a good idea... Why do you have to harass the people, making them work on the fortresses for no good reason?'

Korea did not have many cards to play but did possess a trump card in superior artillery. Many cannon of the period were heavy and hard to transport, especially in rough terrain. But they were perfect for other applications, such as walled fortifications and on board warships. And when placed in the hands of a military genius like Admiral Yi Sun-sin, they could alter the course of the war.

A minor government official by the name of Choe Mun-son is credited with introducing gunpowder to Korea in the late fourteenth century. At the time, the Chinese were trying to prevent the knowledge from spreading but Choe managed to get the essence of the formula by interviewing Korean merchants who had travelled in the Middle Kingdom.

As a reward, Choe was made head of the Superintendency for Gunpowder Weapons in 1377. During his tenure of office it was said many guns were developed. But it was during the reign of King Sejong (1379–1450) that cannon-making became a priority. In 1445 the king ordered a major project to upgrade Korean ordnance.

Advanced Chinese models were studied but Koreans made their own improvements. The project was such a success that a book detailing how to cast superior cannon and create prime gunpowder was secretly set down by the king's command. This secret handbook, the *Chongtungrock,* enabled the new techniques to be passed down though the generations.

Korean cannon were certainly superior to anything the Japanese possessed. There were four types of Korean ordnance in the late sixteenth century: *chonja* (heaven), *chija* (earth), *hyonja* (black) and *hwangja* (yellow). There is no particular significance to the names, they come from a book called *The Thousand Characters,* a primer to help learn Chinese writing. It seems as if *chija* and *hyonja* were the most common, the former being made of copper and the latter from iron. The *chonja,* also made of copper, was the largest of the quartet, with a barrel some 165cm (5ft 5in) long. Whatever the size all Korean cannon could fire stone or iron cannonballs, but the missile of choice was the iron-tipped, leather 'feathered' arrow.

Mention has to be made of an unusual weapon in the Korean arsenal, the *hwacha,* which consisted of a wheeled cart that could be moved by two to four men. A framework that was honeycombed with compartments could fire 100 small rockets, or 100 arrows that could be launched from small iron tubes.

THIS ARROW LAUNCHER *mounted on a cart was an effective anti-personnel weapon during the numerous sieges of the Imjin War (1592–1598). It could fire arrows or small rockets, causing devastation amongst infantry formations.*

The Japanese invasion of Korea finally began on 23 May 1592, when a huge armada of some 700 ships appeared off Pusan (now often called Busan). This was a 5000-man advance guard under a samurai general named So Yoshitomo (1568–1615). Pusan was a major port on the southern coast of Korea and well fortified. Unfortunately the invasion caught local authorities by complete surprise. So Yoshitomo led an attack on Pusan proper while a colleague named Konishi Yukinaga (1555–1600) simultaneously led an assault against a nearby harbour fort at Tadaejin. Curiously, the one Korean advantage – cannon – seems to have been entirely lacking. The Korean defenders sent clouds of arrows raining down on the Japanese, but were swept from the parapets by massed arquebus fire.

Musketry apart, the Japanese showed little finesse in their attacks, preferring brute force to sophisticated siege techniques. Perhaps it was an eagerness for battle, perhaps a contempt for their Korean foes. It seems their main tactic was to fill the moats with timber and rocks then use scaling ladders to mount the 4.3m (14ft) walls.

The Koreans fought bravely but they were no match for these seasoned soldiers. Pusan fell, its conquest accompanied by ruthless slaughter. Within a short time the main Japanese forces landed and three huge columns swept north. Fortresses fell one after the other and the Koreans were defeated in open battle time and again.

Yet in the autumn of 1592, the fortified town of Chinju (formerly known as Jinju) actually repulsed the invaders. It did so by giving the Japanese a taste of their own medicine by using both cannon and arquebus fire. Yet the Japanese took Seoul without a fight on 12 June 1592. The Korean king and his court, not to mention much of the populace, had already abandoned the city.

A KOREAN MOBILE MORTAR *armed with thundercrash bombs, 1592. The Japanese were taken by surprise by the Chinese-designed thundercrash bombs. When detonated they caused many casualties.*

But things were soon to change. Japanese atrocities produced a powerful popular resistance movement: the so-called 'righteous armies' of guerrilla warriors. But above all, Admiral Yi Sun-sin had won a series of naval engagements in the south, culminating in the battle of Hansando on 8 July 1592.

When the Ming Chinese came to Korea's aid, the war bogged down into stalemate. After a brief truce, hostilities flared again in 1597 but this time Korea was not the walkover it had been five years earlier. Hideyoshi died in 1598, and the Japanese finally withdrew their forces for good.

The Sieges of Chinju, 1592–93

The sieges at Chinju were among the bloodiest and most fiercely contested clashes of the entire Japanese invasion. The first siege was a Korean victory, the last a heroic defeat, but they are both remembered today as examples of national courage and patriotism in the face of overwhelming odds.

Chinju is a fortress that overlooks the Nam River in Southern Korea and guards the approaches to Cholla Province. By the autumn of 1592 the Japanese had occupied Seoul and most of the major portions of the county, yet their victory was far from complete. Samurai commander Hosokawa Tadaoki (1563–1646) decided to take the fortress for two reasons: it would open the road to Cholla province, and help suppress the activities of Korean guerrillas that swarmed in the south.

Admiral Yi was conducting a successful naval campaign against the Japanese from Cholla province, and it was imperative that he be deprived if his bases. Then, too, a literally colourful guerrilla leader (he wore red coats) by the name of Gwak Jae-u (1552–1617) was in the region, a perpetual thorn in the

Japanese side. One taken, Chinju would provide a base for anti-partisan activities.

Not all Korean castles were shoddy or badly planned. Chinju was a well-built fortress that was partly protected by nature. To the south it was shielded by towering cliffs and a bend of the Nam River, and on the other three sides there was a high sloping wall and four heavily fortified gates with towering gatehouses. This was all very much in the standard Chinese style. But it was the commander of Chinju, not its battlements, that was going to make a difference. Kim Si Min (1554–92) was a soldier of real brilliance, quite unlike the cowards and incompetents of the past. He recognized the need for firepower and acted accordingly. He acquired modern arquebus muskets and vigorously trained his men in their use. These were newly made Korean weapons, diligently copied from Japanese models.

> *'Both men, women and even dogs and cats were beheaded, and 30,000 heads were to be seen.'*
>
> *– Yoshino, describing the Japanese attack on Pusan in 1592*

The Japanese army arrived in front of Chinju in November 1592, fully expecting another easy victory. General Hosokawa had around 30,000 men, while the Chinju garrison numbered 3800. Once again, Japanese arrogance replaced common sense and caution was thrown to the winds. The samurai were more interested in being *ichiban nori,* the first person to enter an enemy fortification. The samurai went forward at a rapid pace, followed by the *ashigaru* close behind. As the Japanese neared the walls the Korean defenders sprung their surprise. Fingers of flame lanced through dirty blossoms of smoke as Chinju's *hyonja* cannon spat arrows and polished stone balls into the oncoming Japanese ranks. The cannonballs were complemented by a lethal storm of lead from the dozens of arquebus musketeers positioned on the walls.

The bloodied Japanese troops fell back and took some time to construct bamboo shields. When the shields were completed they renewed the attack, with their own musketeers providing coving fire. This time the Japanese got to the base of the walls, though at the cost of heavy casualties. The Koreans still hammered away with cannon and musket fire, but when the Japanese were close enough they hurled powder-filled exploding bombs – a kind of grenade – that did terrible execution. Hosokawa Sadaoki was one of those samurai who made it to the base of the wall and was determined to win the coveted title of *ichiban nori.* Before he climbed the wall he gave strict instructions to the soldiers around him. 'Until I have personally climbed into the castle' he declared, 'this ladder is for one person to climb. If anyone else climbs I will take his head.'

Sadaoki managed to get to the top, his efforts encouraged by cheering Japanese soldiers, but his triumph was short lived. He tried to get through an embrasure, but Korean defenders caught him and hurled him to the ground. His body joined the scores of Japanese corpses already carpeting the base of the wall. The Japanese seemed unwilling or unable to change tactics. For the most part unsophisticated 'human

A Statue of Kim Si-min. *Heroic commander of the first siege of Chinju, he provided inspirational leadership during the struggle.*

wave' tactics were employed, charges reminiscent of the *banzai* attacks of World War II.

While the Japanese soldiers struggled, labourers were building a primitive siege tower. It was meant not to push against the walls but to provide a high firing-platform for Japanese arquebusier. It seems to have little effect. The siege raged on for three days but Chinju showed no signs of surrender.

Reinforcements

Korean guerrilla leader Gwak arrived one night with a small band to reinforce the garrison. They were few in number so Gwak had each man carry five pine torches and shout loud war cries. Korean conch shell horns added to the cacophony, leading the Japanese to think a large relief army was arriving. The Japanese were desperate, but then an arquebus bullet struck Kim Si Min in the forehead while he was directing the defence of the north gate. Hoping that the north gate defenders would lose heart, the Japanese made a major effort there. But the Koreans proved to be of much sterner stuff and repulsed all attacks with a withering fire.

The Korean garrison was holding its own, but ammunition and other supplies were low. Just when all seemed lost, cargoes of fresh supplies, powder and shot came by boat up the Nam River. The timely arrival, one worthy of a Hollywood script, was greeted with jubilation by the weary defenders. Chinju was simply too hard a nut for the Japanese to crack. Hosokawa Tadaoki gathered up his battered army and retreated. It was a major Korean victory after a succession of disheartening defeats. When Japanese leader Hideyoshi heard of the reverse, he flew into paroxysms of rage. How could this happen to the warriors of *Dai Nippon,* Great Japan? The siege of Chinju became a symbol of the Korean people's determination to resist the invader. In time its symbolic importance loomed as large as its strategic value. The Japanese were determined to take it at all costs; it was a stain on samurai honour that had to be wiped away. Samurai general Ukita Hideie (1573–1655) was given 90,000 men, the largest number of Japanese troops ever committed to a single operation.

Korean general Kim Chon Il was placed in command of the Chinju fortress, a tough and resourceful soldier. The normal garrison of 4000 was augmented by many volunteers from outlining areas. In fact, Chinju contained some 60,000 souls, including women and children seeking safety from outlying villages.

The Japanese army arrived in July 1593 and lost no time in preparing for a siege. It was to be a three-pronged, all-out assault on Chinju's still formidable defences. This was no easy matter

AT THE SIEGE OF CHINJU *the Japanese attacks were unsophisticated, mainly using scaling ladders. They were met with missile fire and caldrons of boiling water.*

since the walls still bristled with cannon. General Konishi Yukinaga was stationed on the western side with 25,000 men; to the north Kato Kiyomasa had 25,000 men, and supreme commander Ukita Hideie anchored the eastern end. The rest of the army was fanned out in a semicircle to block any Korean reinforcements.

Earlier the Koreans had dammed some water from the Nam River thus creating a water-filled moat from what was usually just a dry ditch. The first Japanese attack was on 21 July, and its main object was to break the dam and drain the moat. Advance Japanese units managed to breech the dam, though undoubtedly they suffered casualties from Korean fire from the walls.

Once the moat was once again a dry ditch, they proceeded to fill it in with brush, earth and rock. Once that was accomplished, Japanese assault teams moved forward by sheltering behind large shields made of bamboo bundles, some of which may have been on moveable frameworks. The Korean defenders shot fire arrows and heavy cannonballs into these shields, which where not as effective as the Japanese had hoped. The shields were splintered, the impacts probably killing and wounding many Japanese soldiers as they hid behind them. Arquebus fire peppered the Japanese ranks, and Korean archers felled many with their arrows. This attack failed, as did others in the coming days.

Once again a Japanese 'arquebus tower' was raised, and once again it had little effect. At one point commander Kim spied some Japanese officers – the Korean accounts say generals – observing the siege from a mountain top to the east of the fortress. He ordered his cannon to open fire, 'knocking the generals to the ground'. Ukita Hideie sent Kim a message demanding surrender,

ASHIGARU ARQUEBUSIER, SIEGE OF CHINJU, 1592. *This common Japanese foot soldier is lightly armoured but carries an arquebus musket and wears a simple leather breast plate for protection.*

adding that if capitulation took place '10,000 [Korean] peasant lives would be spared'. The offer was rejected. The Japanese siege tightened, and all attempts at relief were foiled. Chinju was on its own.

The Japanese decided to try *kikkosha,* 'turtle shell wagons', to win the day. These were essentially movable roofed sheds that could be wheeled up against an enemy wall. Once in place, Japanese sappers could tunnel under the wall, hopefully to weaken it and cause it to collapse. While they dug they would be protected by the wagon's heavy roof from enemy missiles dropped from above. But the initial turtle-wagon effort ended in failure. The Koreans dropped bundles of combustible material soaked in oil or fat, which set the turtle sheds roofs ablaze. Flaming pine brands also set the surrounding grass alight.

The turtle shed wagons were rebuilt with ox hides as a form of fire protection then the assault was renewed. Progress was made on the fortress's northwest corner, and a heavy rain did the rest. A section of wall eventually collapsed and the Japanese rushed in. The Koreans fought to the last; battling with wooden sticks or clubs when ammunition ran out. It was said that when bow strings broke from constant use, Korean women supplied their own hair as replacement material.

When General Kim saw all was lost he calmly came down from a tower, bowed to the north (the direction of the king) and jumped into the river. The slaughter was truly horrific, since it was a custom for samurai to win prestige by collecting heads. At least 25,000 Koreans died in this holocaust, and perhaps many more. Thousands of headless corpses bobbed in the river, joined by those who had committed suicide or drowned while trying to escape.

Siege of Chinju

1593

By the autumn of 1592 the Japanese invaders had achieved great success on the Korean peninsula. However, Korean partisan fighters were making life increasingly difficult for the invasion force. Chinju was a fortified city that was on the fringe of Korean guerrilla territory. By capturing the city, the Japanese would deny the guerrillas a base of support and also open a new road to Cholla Province, which could then be conquered. During the first siege, the Korean garrison heroically beat off all attacks and the Japanese withdrew. The Japanese returned again in 1593. The Japanese used covered 'turtle shed' wagons to approach and undermine the walls. Eventually, a section of the wall collapsed, allowing the Japanese assault troops to storm in. Fighting was fierce with the Koreans fighting to the last, and the whole garrison was put to the sword.

1 November 1592: Japanese troops try to take the fortress city. Their methods are crude and Korean resistance stalwart. They are repulsed and forced to lift the siege.

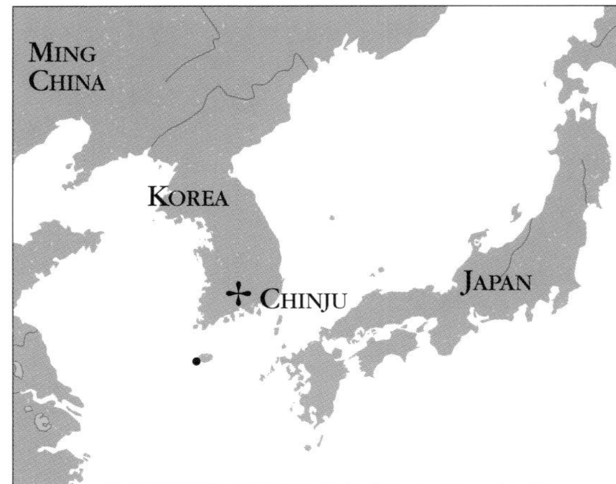

Chinju was an important fortress town situated along the River Nam, a tributary of the River Naktong, west of Busan. Its capture would open another road to the south and Cholla Province.

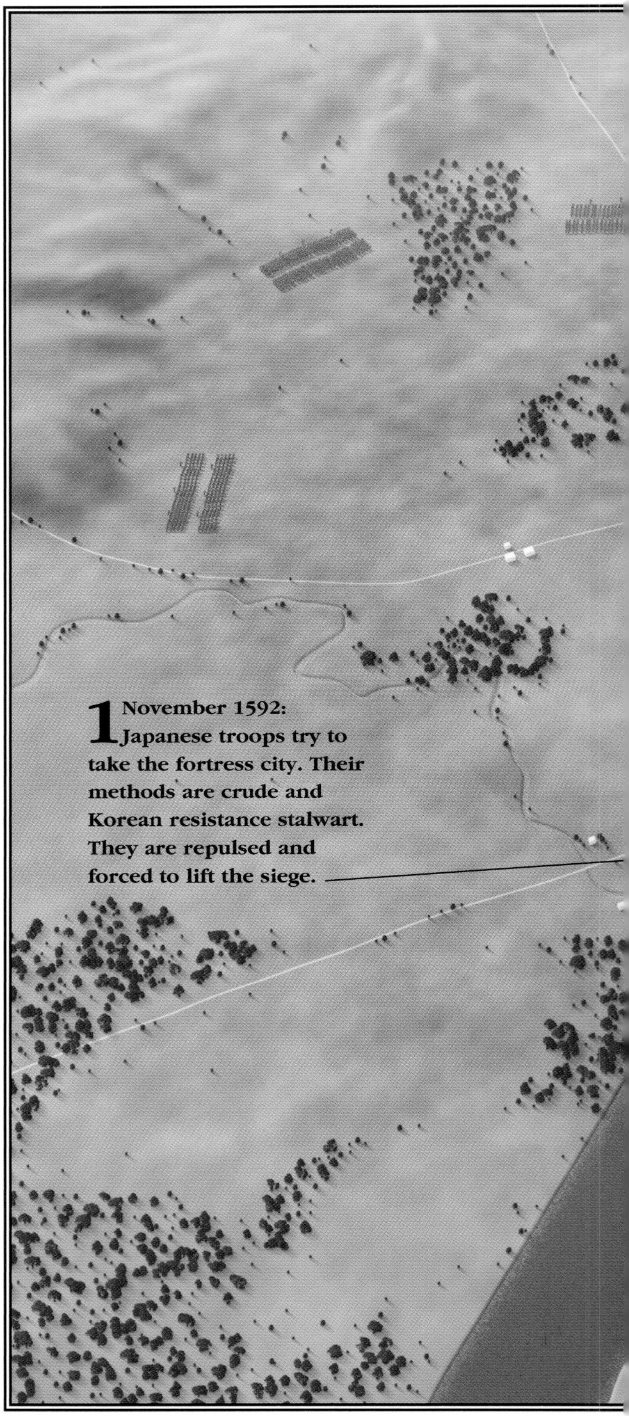

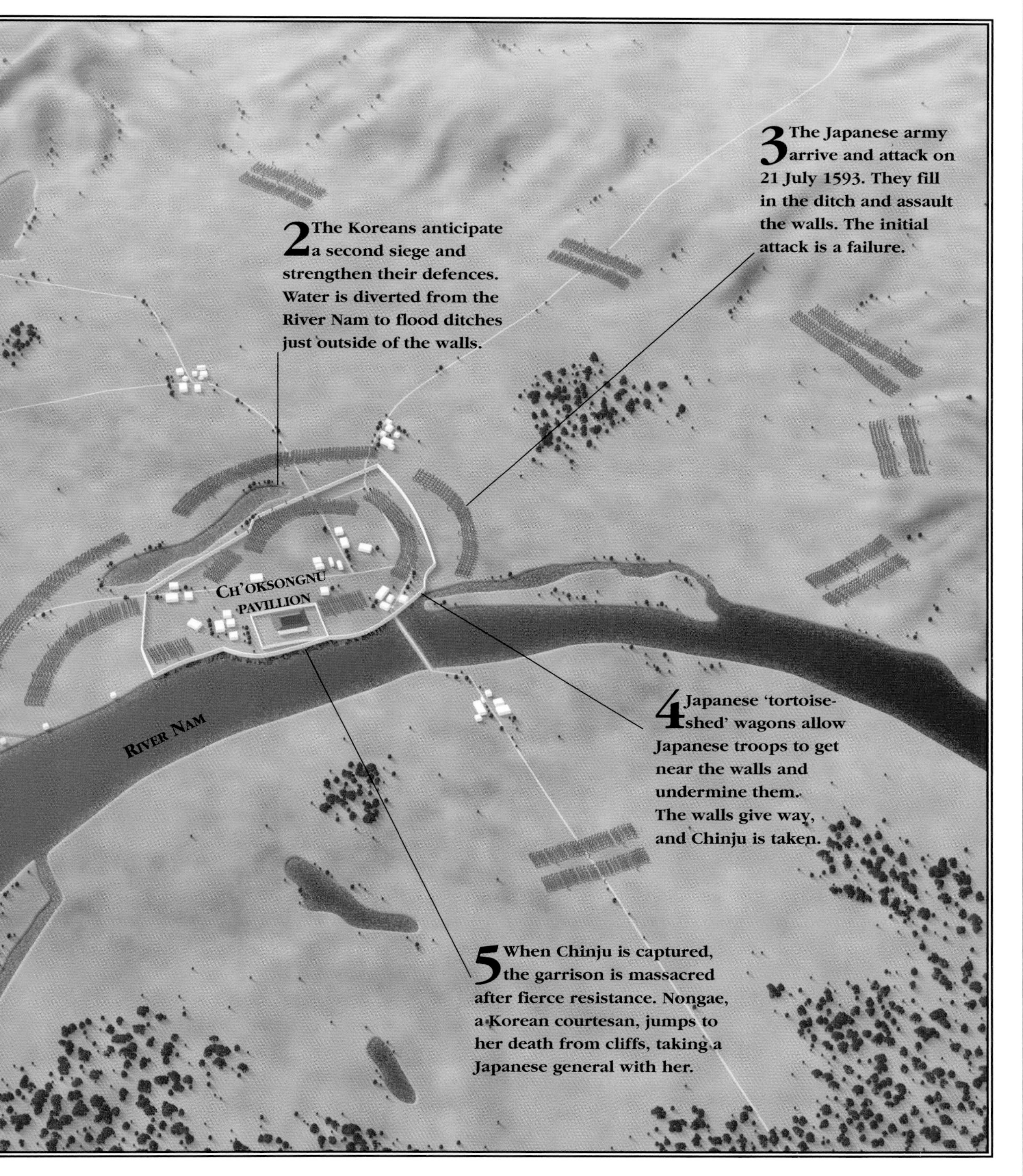
3 The Japanese army arrive and attack on 21 July 1593. They fill in the ditch and assault the walls. The initial attack is a failure.
2 The Koreans anticipate a second siege and strengthen their defences. Water is diverted from the River Nam to flood ditches just outside of the walls.
CH'OKSONGNU PAVILLION
RIVER NAM
4 Japanese 'tortoise-shed' wagons allow Japanese troops to get near the walls and undermine them. The walls give way, and Chinju is taken.
5 When Chinju is captured, the garrison is massacred after fierce resistance. Nongae, a Korean courtesan, jumps to her death from cliffs, taking a Japanese general with her.

The Japanese had won, but it seems to have been a pyrrhic victory. The sieges at Chinju are a tantalizing glimpse of how the Imjin War might have had an entirely different outcome. If every Korean city and fortress had been defended by cannon and muskets, the Japanese might have been stopped in their tracks much earlier. Suppose Pusan, the great port of entry, had been bristling with cannon and musket-wielding soldiers? Perhaps the great invasion might never even have come ashore.

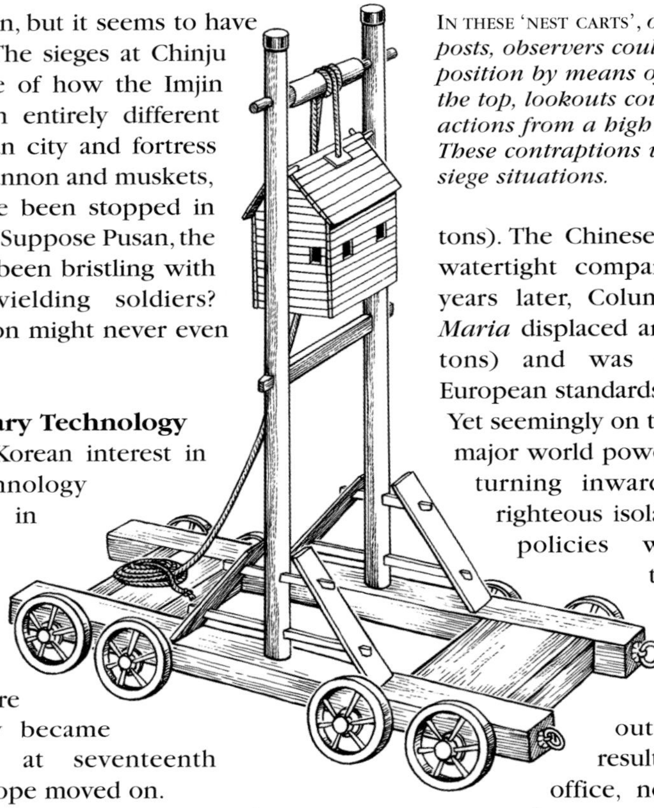

Decline of Asian Military Technology

Chinese, Japanese and Korean interest in western military technology seems to have peaked in the early seventeenth century. Thereafter, all three societies turned inward, keeping outside trade and cultural contacts to a bare minimum. The military became 'fossilized', preserved at seventeenth century levels while Europe moved on.

China's military arts, once so innovative that they taught the world, atrophied as time went on. Inventiveness was replaced by complacency; open-mindedness with a smug ethnocentrism. In the early 1600s the Mings showed some interest in western science and technology by welcoming Jesuit missionary-scholars to their Peking court. On occasion they even allowed 'red haired barbarian' cannon to be forged, but the interest proved ephemeral.

Even so, in the early seventeenth century China was still leading the world in science and technology, and could have been a world-shaping, even conquering, power. A famous series of voyages by Admiral Zheng He (1371–1433) from 1404 to 1433 illustrate the point. There were seven voyages in all, vast undertakings that reached as far as the Arabian peninsula, and aptly demonstrated the power and might of China. But it was the ships that were the most impressive. Zheng He's 'treasure ships' were 113m to 143m (370 to 440ft) long and probably displaced 3050 tonnes (2950 tons). The Chinese knew the concept of watertight compartments as well. Fifty years later, Columbus's flagship *Santa Maria* displaced around 102 tonnes (100 tons) and was considered large by European standards.

IN THESE 'NEST CARTS', *or mobile observation posts, observers could be raised into position by means of pully ropes. Once on the top, lookouts could observe enemy actions from a high place of relative safety. These contraptions were especially useful in siege situations.*

Yet seemingly on the brink of becoming a major world power, China stepped back, turning inward in a kind of self-righteous isolation. These isolationist policies were reinforced by the traditional Chinese mindset, which placed China as the only island of true civilization in a sea of outer barbarians. As a result China had no foreign office, no diplomatic network and certainly no foreign representatives in Beijing. There was a tribute system instead, where China graciously allowed inferior nations to acknowledge its superiority by sending gifts. This tribute system was the only form of trade allowed for centuries.

While China deluded itself with notions of its own perpetual superiority, European military technology steadily advanced. By the late seventeen and early eighteenth centuries the flintlock, a far superior method of ignition, had replaced the matchlock. Cannon also improved, and by the 1770s the British used flintlock mechanisms to fire their massive 12-, 24-, and 32-pounder shipboard cannon. European warships became great floating batteries. First-rate ships like HMS *Victory* had three decks and carried 104 cannon. A 32-pounder could hurl a cannonball of 14.5kg (32lb) more than a mile, and at 915m (3000ft) could smash through 0.8m (2ft 6in) of solid oak. By the eighteenth century China had no navy and only a few anachronistic 'war junks' armed with dated cannon types. Yet China refused to awaken from its complacent slumber. The Europeans also improved their weapons on land. The flintlock was soon replaced by rifled muskets that used percussion caps to ignite the charge. Artillery also became more powerful, with rifled barrels to ensure greater accuracy.

The Chinese did permit a trickle of foreign trade, but only at Canton (Guangzhou), and only though authorized Chinese merchants known as *hongs*. By the mid-eighteenth century Britain was developing a taste for Chinese tea, silk and porcelain, and started to chafe under Chinese restrictions. George III dispatched an embassy under Lord Macartney to try and establish regular diplomatic relations with the Middle Kingdom.

CHINESE TROOPS ATTACK BRITISH FORCES *as they land in a creek off the River Canton in the Opium War of 1856–58. Though highly stylized, the illustration shows the vast technological gap that existed by the 1850s. Chinese soldiers with pole arms advance against British troops armed with modern percussion rifles and bayonets. Offshore, British warships wait to provide covering fire with cannon.*

The mission was a total failure, though cordial and friendly. By coincidence the British came during the reign of Qianglong, one of the greatest emperors of the Qing dynasty. It was also a time when Chinese power and wealth was at its height.

But there were cultural misunderstandings from the beginning. Most of the problems stemmed from the fact that the Chinese would accept no other nation as their equal. To the Chinese the British were another nation of 'red-haired barbarians' seeking to curry favour with the mighty Celestial Empire. When the embassy travelled to Beijing, they were accompanied by Chinese banners with characters that identified them as 'tribute embassy of red barbarians'. Qianlong was the perfect host, but made it clear he wanted no part of the British or their goods. The restrictive Canton system would be maintained, nothing more. He wrote a condescending letter to King George III, accepting the British monarch not as an equal but as a faithful vassal. The emperor explained that 'we do not have the slightest need for your country's manufactures' and advised King George to

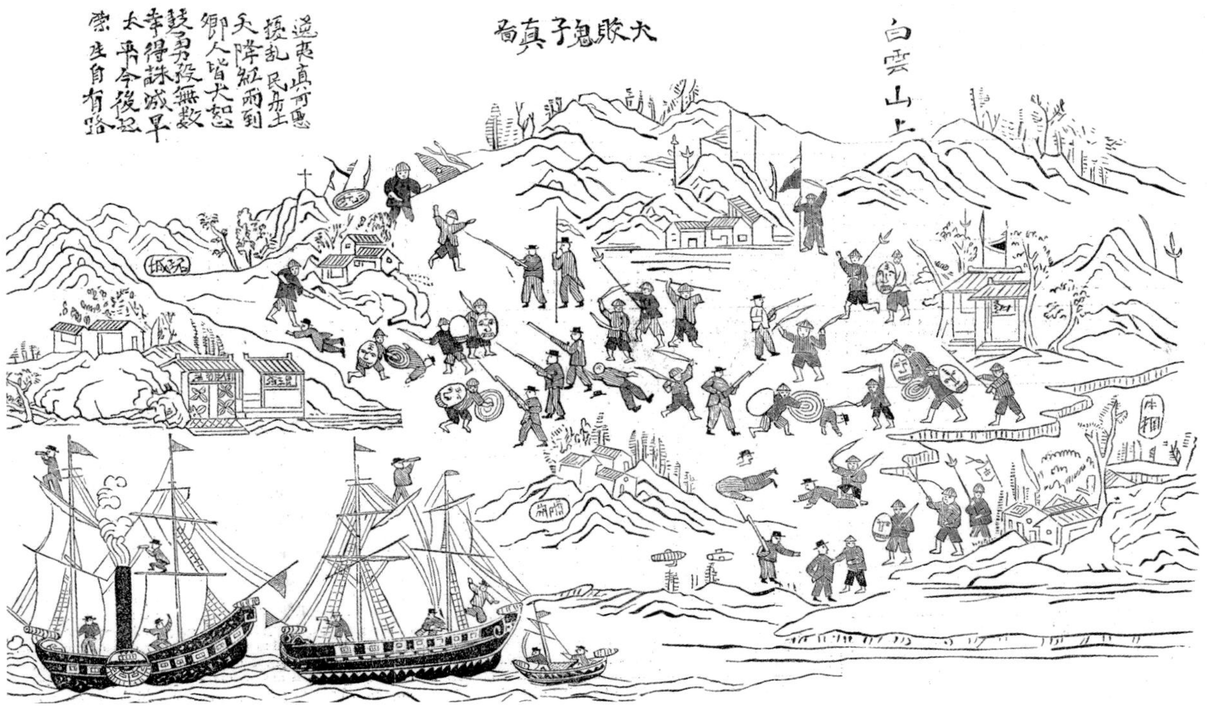

'strengthen your loyalty and swearing perpetual obedience'. Such an arrogant attitude was understandable in 1600, dangerous in 1800, and positively suicidal in 1850. The situation was made even worse after Qianlong's death, when a series of weak, youthful or ineffectual rulers occupied the Dragon Throne. Corruption increased, decay spread and the dynasty was in obvious decline.

The reckoning came in the mid-nineteenth century when the Opium War of 1839–42 finally forced China to open its doors to outside commerce. Chinese war junks were sunk almost at will and Chinese armies were routed by superior European weapons. Yet the dead hand of tradition was so prevalent China still did not seriously modernize. Chinese armies still used cavalry that wasn't too far removed from Genghis Khan, while Chinese infantry wielded swords, spears of various kinds and matchlock muskets.

In the mid-nineteenth century, *the Chinese navy used converted trading junks for miltary purposes. This junk has a number of cannon mounted on its main deck. These ships were no match for the leading European ships of the line, which were larger and carried far greater firepower.*

In a series of clashes from 1857 to 1860 China was once again bested by European firepower and forced into more concessions. It was then, and only then, did the Chinese start on the road to military modernization.

Korea was even more isolated than China. It was a tributary state of China, and had no other ambitions. It became the 'Hermit Kingdom,' where even shipwrecked western sailors faced possible enslavement. Korea simply did not want to have anything to do with the outside world.

The Korean military were also frozen in time, virtually unchanged for almost 300 years. As late as the 1870s, Korean troops carried bows, spears or matchlocks unaltered since the sixteenth century. In 1871, the United States landed a punitive force of marines on the Korean coast after a series of incidents. The marines easily overcame the opposition, and captured several cannon as prizes. Later, when these guns were examined, the Americans were amazed. Some of the cannon bore the Korean equivalents of 1607, 1665, and 1680. Two pieces dated from 1313. The Americans had been attacked by some of the oldest functioning artillery in the world.

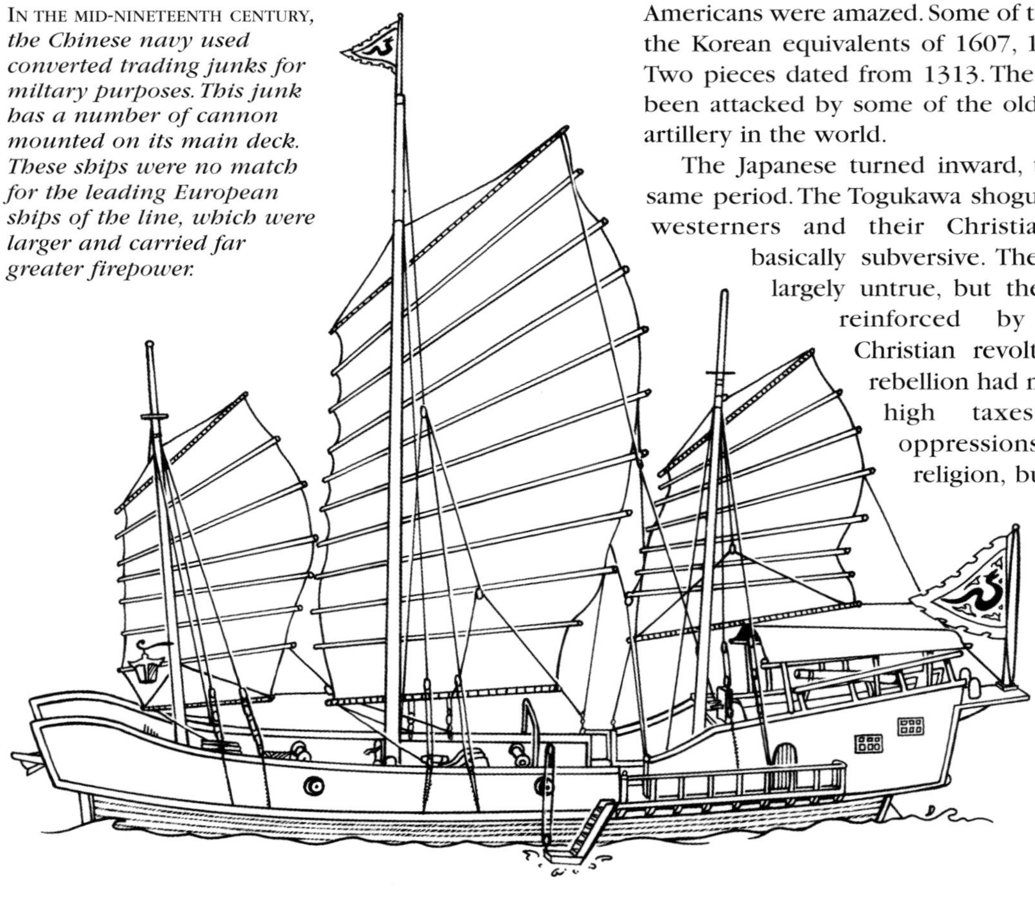

The Japanese turned inward, too, and in the same period. The Togukawa shoguns came to see westerners and their Christian religion as basically subversive. The charges were largely untrue, but the paranoia was reinforced by a Japanese Christian revolt in 1638. The rebellion had more to do with high taxes and other oppressions than with religion, but the Japanese government did not see it that way. After the rebellion was crushed Japan was sealed off from the rest of the world for the next 200 years. Taking a leaf from the Chinese book, the Japanese did allow a small Dutch trading mission at Nagasaki, but only on an island with very limited access. When Ieyesu Togukawa established peace throughout the home islands there was no need for European guns or artillery. Even the samurai lost their martial edge.

Battles at the Dagu Forts, 1859–60

China's military establishment received its first shock in 1839-42, when it engaged Great Britain during the Opium War. The Chinese tried to use sixteenth-century technology in the nineteenth century, with predictable results. The Middle Kingdom's armies were routed and its war junks sent to the bottom with impunity. The war ended with a humiliating defeat, and China was forced to open five of its ports to foreign trade. Much of the blame could be laid at the feet of the Qing (Manchu) government. Early nineteenth century emperors were either young, incompetent, or both. A definite rot had set in, giving rise to incompetence and corruption. The government took refuge in a false sense of superiority, of scorning outside nations as worthless barbarians.

In 1858 James Bruce, the Eighth Earl of Elgin negotiated a new treaty that required the Chinese to open 10 new ports to British trade. In addition, the British were allowed, in principle, to have diplomatic representation in Beijing. The Chinese considered this point as a serious loss of face, but did agree to have the treaty ratified in their capital.

In June of 1859, a British mission of one battleship, two frigates, and a number of gunboats hove to off the mouth of the River Hai in northern China. Frederick Bruce, brother of Lord Elgin and new British envoy to the Dragon Throne, was aboard HMS *Magicienne,* ready to sail up the river to Beijing. The River Hai was guarded by a series of forts that clustered near its northern and southern shores where its mouth emptied into the Gulf of Bo Hai. These were the Dagu Forts, though some old spellings say Tagu. In addition, there were a series of booms that stretched across the river to bar the passage of any unwanted vessels.

Rear Admiral James Hope was in charge of the military aspects of the mission. When the Chinese refused to remove the booms and let the 'foreign devils' move up the river, Hope lost his patience and decided to force a passage. To the Admiral British prestige demanded action - there were French and American ships in the area, watching the proceedings.

The river booms were made of wood cross-lashed together to form a solid mass 37m (120ft) wide and 1m (3ft) deep Each boom was studded with iron spikes that threatened to puncture the hull of an unwary vessel. Some accounts maintain that some of the booms were wooden while others were made of chain links.

In any event, Admiral Hope and his eleven gunboats went forward at about 2 p.m. Each vessel had four to six guns, with complements of 40 to 50 men. The first boom was cut without incident, but when the tiny British gunboat fleet approached the second boom, all hell broke loose.

The gunboats were subjected to a heavy fire from around 40 Chinese cannon, and the ranges were so close that even antiquated artillery was effective. Dozens of British sailors were cut down in this iron hurricane and even the Admiral received a painful wound in 'the fleshy part of his thigh'. HMS *Plover,* the Admiral's lilliputian flagship, was particularly hard hit, with nine dead and 22 wounded.

Failed Attack

The attack was an embarrassing failure. Casualties had been high and several gunboats damaged and even sunk. Hope relinquished command to Royal Navy Captain Shadwell. Shadwell now foolishly decided on a direct land assault on the south fort. The landing party consisted of about 600 Royal Marines and 60 French sailors, accompanied by engineers and sappers with scaling ladders.

But the tide had fallen, exposing mud flats that girded the south fort like a glutinous moat. While the bogged-down allied force struggled, the Chinese garrison opened up with everything it had - grapeshot, iron roundshot, matchlock musket balls, rockets and arrows. It was a lesson review of 2000 years of Chinese military technology, delivered all at the same time.

The mud was bad enough but the allied attackers had to also cross a series of muddy ditches. Somehow about 50 attackers managed to reach the last ditch, only to be pinned down at the base of the fort's main wall by heavy musket fire. The British and French were forced to withdraw. Accounts differ on British losses, but at least 89 were killed and 434 wounded.

For perhaps the last time in history, outdated military weapons, some of them dating back to the Middle Ages, had won a battle. But it was the actions of Admiral Hope and Captain Shadwell that had contributed to the defeat. They neutralized the advantage of modern weapons by just pushing ahead with no real plan and inadequate intelligence.

'Never did the interior of any place testify … to the noble manner in which it had been defended.'

– Lt. Colonel Garnet Wolseley, surveying Dagu Fort after its capture

Second Expedition

The defeat caused uproar in Britain and before long an Anglo-French expeditionary force was dispatched to China. The British force consisted of 13,000 British and Indian troops under the command of Lieutenant General Sir James Hope Grant. The French contingent of 7500 men was led by General Charles Montauban, a touchy man always taking offence at real or imagined insults.

The allied forces landed at Beidang on 1 August 1860. Beidang was about 13km (8 miles) miles from the Dagu Forts. As the British and French moved towards their objective, they were met by Chinese cavalry, probably bannermen. The British unveiled a new weapon, the powerful Armstrong breechloading gun, which was very effective as it blasted bloody holes in the 'Tartar' ranks. Many of the horsemen were riding too fast to be hit by British artillery, so the cavalry advanced. British Indian lancers went forward along with the 1st King's Dragoon Guards and soon scattered the Chinese cavalry.

The advance continued and soon the allies were approaching the Dagu Forts area. It was decided that the small north fort was the key to the whole area. Once taken, the other forts would be untenable. The post was a compact structure measuring some 40m (130ft) in diameter. Its walls were of mud, though some accounts insist there was also unburned brick. The fort also had a cavalier, or raised, platform: a man made 'mountain' that hovered over the battlements and allowed the Chinese to fire in all directions.

The 9m (30ft) high parapets were crenellated, the row of embrasures looking like a line of jagged teeth. Guns in each embrasure, some 47 in all, poked out like stubby black fingers. The Chinese had a wide variety of ordnance, dated for the most part, but there were also two 32-pounders taken from the British attack the year before.

The attack began at 5 a.m. on 21 August with an artillery barrage. Royal Artillerymen and engineers had performed wonders, manhandling 16 heavy guns though muddy terrain to within a few hundred yards of the fort. Besides the Armstrongs, the British also brought up 203mm (8in) howitzers.

The barrage began as the sun was lightening the eastern sky, the roar of cannons so great that large flocks of birds burst into the sky in flights of panic. The Chinese guns flamed in response, but it was obvious modern technology was going to win the day. The muddy walls were breached in several places, until the fort seemed practicable for assault. But then a lucky howitzer shell pierced the roof of the fort magazine, igniting its contents with an ear-splitting roar.

The Chinese garrison still refused to surrender so the assault forces went forward. Most of the defenders' antiquated cannon were by now out of action, but they still had muskets, spears and jingals. A jingal was a large, cumbersome matchlock that had a 38mm (1.5in) calibre and fired balls from 110 to 450g (4oz to 1lb). It was so long and heavy the musketeer had to prop it up on a wall or even on another man's shoulder. The British assault force was comprised of 2500 men under Brigadier General Reeves. The British infantryman was armed with the 1853 percussion

BODIES LITTER THE RAMPARTS *of a fort at Dagu following its capture by Anglo-French forces during the Second Opium War, 1860. This early photograph shows the interior of the small north Fort shortly after its capture. French scaling ladders still mount the walls.*

Enfield rifle-musket, sighted to 1100m (3610ft). The French also had their own sector and contributed about 1000 men.

The fort was ringed by two water-filled moats 13.7m (45ft) wide and 4.6m (15ft) deep, obstacles the British called 'wet ditches.' Between these waterways the ground was studded by sharpened bamboo sticks, further barring easy progress.

The Royal Engineers had constructed a pontoon bridge to span the ditches, although it proved cumbersome and vulnerable to Chinese fire. It was smashed by Chinese cannonballs so some soldiers simply took the plunge into the muddy, brackish waters. After much effort, including 'tightroping' over an old bridge beam from a dismantled span, the assault forces reached the base of the fort walls.

The mud-spattered redcoats climbed the walls under heavy Chinese matchlock fire and Ensign Chaplin of the 67th Foot planted the Queen's Colour on the top of the parapet. The Chinese fought bravely but were betrayed by their misguided faith in their own superiority. Spears, trident-like pole arms and matchlocks were no match for bayonets and rifled muskets in a hand-to hand fight.

Most of the garrison was slaughtered, though a few slipped over the walls and escaped. Some 1500 Chinese soldiers were casualties, while the British tally was 17 dead and 184 wounded. The French also had 17 killed, but 141 wounded. The neighbouring forts surrendered as predicted.

The battle of the Dagu Forts was one of great irony. The Chinese 'teachers,' the ones who had developed gunpowder, firearms, and cannon centuries before, had been bested by their European 'students.' The westerners had taken Chinese inventions, improved them, then turned them on their original creators.

Battles at the Dagu Forts

1859-60

In June 1859, Admiral Hope was charged with the responsibility of forcing an opening into the River Hai following failed diplomatic negotiations. The operation ended in humiliating failure after a naval assault was repulsed by Chinese missile fire and an amphibious attack became bogged down in the mud flats around the estuary. British prestige was hurt and the following year, on 1 August 1860, a 21,000-man Anglo-French expeditionary force landed at Pei Tang (Beidang) and occupied the city. On 12 August, the Allied forces marched down to the Dagu forts, about 13km (8 miles) away. European artillery and infantry proved to be too much for the Chinese defenders, many of whom engaged in hand-to-hand fighting with spears and matchlocks.
After the fall of the small north fort, the other fortifications surrendered and the river was opened to British naval craft.

Beijing
Pei Tang
DAGU FORTS
CHINA
Shanghai

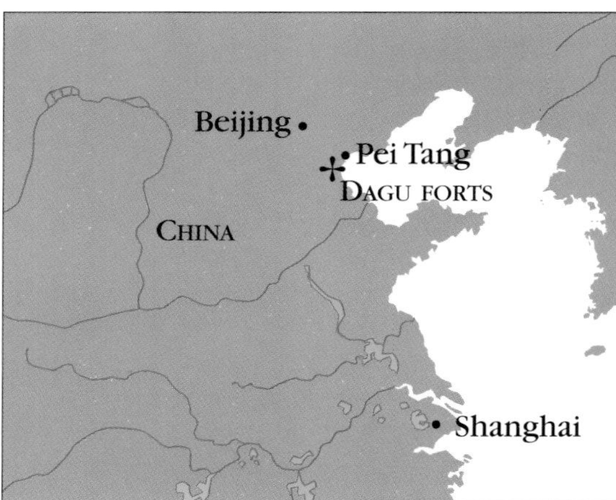

The Dagu Forts, situated on the estuary of the River Hai, near the Bohai Gulf, guarded the approaches of Tianjin. Ultimately the forts protected the Chinese capital at Beijing, some 180km (111 miles) away.

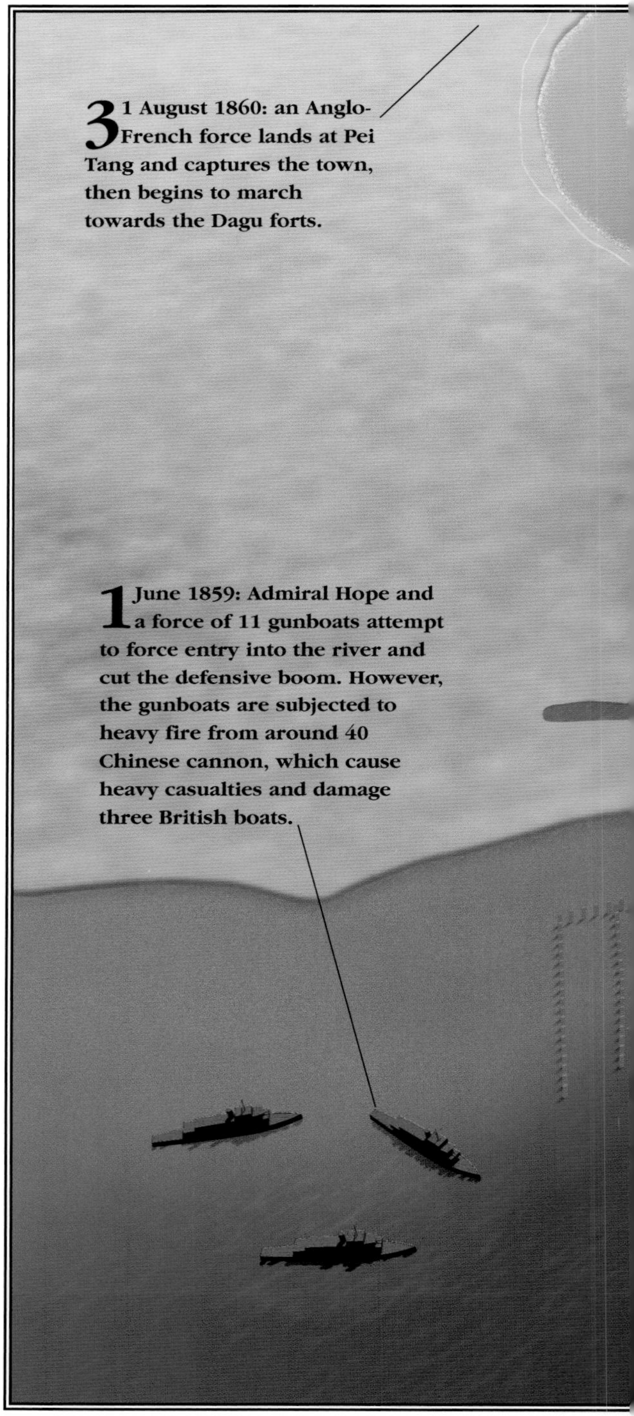

3 1 August 1860: an Anglo-French force lands at Pei Tang and captures the town, then begins to march towards the Dagu forts.
1 June 1859: Admiral Hope and a force of 11 gunboats attempt to force entry into the river and cut the defensive boom. However, the gunboats are subjected to heavy fire from around 40 Chinese cannon, which cause heavy casualties and damage three British boats.

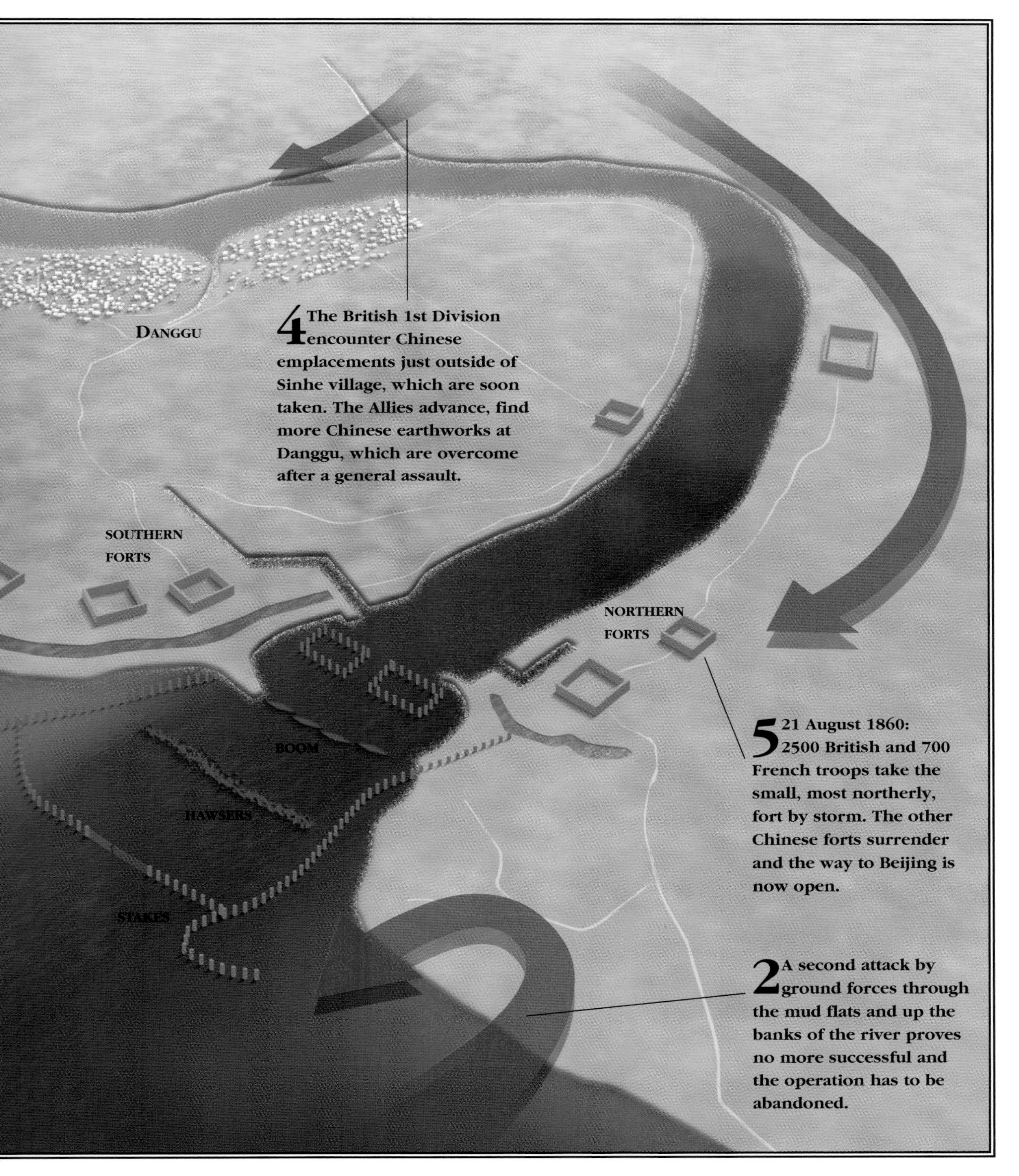
DANGGU
4 The British 1st Division encounter Chinese emplacements just outside of Sinhe village, which are soon taken. The Allies advance, find more Chinese earthworks at Danggu, which are overcome after a general assault.
SOUTHERN FORTS
NORTHERN FORTS
BOOM
HAWSERS
STAKES
5 21 August 1860: 2500 British and 700 French troops take the small, most northerly, fort by storm. The other Chinese forts surrender and the way to Beijing is now open.
2 A second attack by ground forces through the mud flats and up the banks of the river proves no more successful and the operation has to be abandoned.

CHAPTER 5

NAVAL WARFARE

The study of naval warfare in the Far East contains much that is different, much that is fascinating and much that is strange to the new arrival. Three universal constants allow understanding in the midst of unfamiliar, transcripted names, dates and battlefields. Asian objectives and attitudes towards combat differ in many ways from the wars of the west – but upon closer examination, the same factors that controlled the Greeks at Salamis and the Byzantines at Constantinople appear in the waters of the Lake Poyang or off the blood-spattered coast of Korea.

In every epoch and every geographic location, tactics, technology and terrain determine the nature and direction of conflict at sea, in the Far East as elsewhere. In our analysis of four battles ranging chronologically from 1274 to 1842, the effects of all three determinants are clear. Asian

SEA AND SAMURAI CONTEND *as human conflict finds itself at the mercy of the shifting waters of the Inland Sea. The feuding Taira and Minamoto clans fought it out to the finish at the battle of Dannoura, 1185, depicted here in this painted panel.*

admirals found themselves using the ships and weapons the artificers of the east provided them in seas, lakes and rivers where the wind could determine the weapon. The combatants in each of our battles had differing objectives, but they all worked within the same three universal constraints: tactics, technology and terrain.

Naval Tactics

Sun Tzu (c.544 BC–496 BC) and von Clausewitz (1780–1831) would agree with the least-stated and most obvious objective of all warfare from the earliest records to the cloudy mists of the future: victory. One group makes war with the objective of overcoming another group's resistance to their objectives, which can be as varied as the different tribes, nationalities, clans and sects who have resorted to violence in the course of human history. China, to the Chinese, has long been 'the Middle Kingdom,' directly under Heaven, in the centre of all the world, and as such, drawn more inward than outward for most of its long history. Once the idea of Chinese unity had rooted firmly in the minds of the Chinese, warfare also turned inward. Warfare – and there has been a great deal of it in China – has largely taken place in defence of or rebellion against the established order. The re-establishment of domestic tranquillity in China has been a shared goal of those who wished to protect or supplant and profit from the status quo.

Accordingly, naval warfare in China has been a matter of transferring armies across bodies of water or of interdicting the movement of forces that could: seize control of the cities, support an organized central regime, prevent the incursions of hostile external ethnic groups, or resist a warlord's will. That the centuries to come would render invaders indistinguishable from the Chinese themselves mattered little to contemporary rulers with the energy to resist external influence, very little of which came from over the sea.

Japan's geographic isolation also drew that nation's attention inward, creating a national narcissism from which there would be rude awakening with catastrophic results. Such was not the case for Korea, whose people inhabited a peninsula and whom centuries of invasion and piracy had forced into an inescapable awareness of their coasts as long areas of potential vulnerability in need of organized professional defence. The tactical upshot of both China and Japan's isolation from overseas threats meant that such battles as did occur on the water – and there were many – would be fought in a manner and style as close to a land conflict as both sides could arrange.

Perhaps the most decisive battle in Japanese history was fought at Dannoura, in 1185, but in that battle, as in the Chinese battle of Lake Poyang in 1363, the movements of the ships involved were primarily to bring the warriors on board into range for missile or hand-to-hand combat.

Battle on rivers, lakes or coasts did have much to make them attractive to even the most land-centred commander. Under ideal conditions the flatness of a lake or calm ocean allowed a decisive commander a superb view of his own forces, and of his enemy's – providing tactical opportunity or final proof of disaster. Wind and water could carry large amounts of men and materiel far more quickly than wagons and pack animals, allowing the possibility of tactical surprise. And when the Koreans and British equipped their warships with cannon this could have devastating consequences for both naval and land-born defenders.

The means of bringing a waterborne enemy to battle varied enough for considerable tactical elaboration, affected both by terrain and the application of military technology. Both the Japanese and Chinese had pre-determined tactical formations, described in military manuals with poetic names such as 'swan's wing'. The accounts of such groupings dutifully mention them as ideals, although their actual utility and the various ships' complements success in executing their complicated evolutions are matters for some speculation.

Only the Korean navy, in the four battles of our study, had professional combat sailors. In both China and Japan the warrior castes ordered the civilian crews of their ships to do their bidding.

How well fisherman and river watermen could suddenly fight ships in combat varied considerably.

From the very earliest Chinese recorded history there is an emphasis on the use of incendiary warfare not to be found in the west until much later. Writers from Sun Tzu on mention the dangers and possibilities of fire in warfare – and in no case more devastatingly than in burning an enemy's fleet and the warriors on board.

The danger of fire attack was the counterweight to the desire in China and Japan to convert sea battles into land battles. Archers and warriors could chain ships together for more stable platforms, while the Japanese and Chinese both built 'tower ships' that were literally floating castles. Thick bulwarks and high missile platforms were all very well – unless they made the ship possessing them so unwieldy that the vessel could not avoid an enemy's fire attack.

Naval Technology

Naval warfare in Asia contains the same mix of the familiar and the inexplicable that other aspects of Asian culture lay before the western observer. Blanket statements claiming that either east or west had superior or more advanced technology at a given time tend to provide more shadow than light. The effects of the technologies peculiar to warfare in the oriental world at a given time are manifest, however, upon any examination.

In terms of the ships themselves, over the centuries the river *sampan* and the deep-sea junk evolved into two supremely capable basic designs that could carry men, cargo, weapons and warriors. The keel-less 'three plank' design of the *sampan* made the most efficient use of shallow water and human labour, while its wedge-shaped high-sterned design expanded easily and naturally into the seagoing junk. Both types were easy to beach and unload quickly – a capability useful to

ARCHERS AND OARSMEN *fought standing when the Japanese linked* sampans *into a mobile platform for archery. Twin hulls reduced rolling, each boat the other's outrigger, at a significant cost in manoeuverability and speed. Boatmen working their oars in a 'screwing' motion propelled and steered the combined unit.*

the trader as well as the commander. The 2735 navigable kilometres (1700 miles) of the Yangtze river and its larger tributaries offer the ship builder and navigator almost any imaginable challenge. It is not hard to see how the reefs, depths, and shallows of the South China Sea led to some of the features of Asian marine architecture.

Asian vessels sported nimble, balanced sternpost rudders that could function as centreboards and be raised in shallow water, with their helmsmen high above the sea for safe and continued operation. Even the smaller junks sported as many as five masts with strong bamboo-brailed sails to shift them quickly with astonishing manoeuvrability, while the larger Chinese vessels sported damage-resistant double partitioned hulls which it took the west centuries to copy despite pleas from such luminaries as Benjamin Franklin. One of the first British ships so equipped would end up confronting the Qing Dynasty's navy at River Wusung.

The *sampan,* and the junk, especially, could be stretched into vessels far larger than their Neolithic progenitors would have seemed to promise. Accounts from the Han Dynasty (206 BC–AD 220) speak of monster seagoing junks 30m (100ft) in beam growing to an astonishing length of 90m (300ft) with the transverse bulkheads of their partitions adding strength and rigidity to their construction. Chinese metallurgy fitted these vessels with strong iron bolts, nails and other fittings as well as iron anchors, and, in time, cannon of native or derivative European design.

The Chinese were well acquainted with both the windlass and capstan to convert human strength into mechanical work. Windlasses could raise anchors or sails, while the Han Chinese and their successors actually put into practice an idea proposed by a fourth century Roman writer – using the capstan to power a paddle-wheel to move a ship. There are accounts of Han ships with as many as seven such devices thrusting them along in a calm.

Oars, too, could move ships, but here western technology and Asian ingenuity took markedly differing paths. Long parallel benches, straight, layered or angled, characteristic of western galleys from Salamis to Lepanto do not appear in Asian nautical architecture. Instead, long sculls similar to those of the modern gondolier served to push *sampans* and even war galleys along. Multiple standing oarsmen thrust their counter-weighted paddles down from an overhang in a screwing motion to force their vessels forward. No ship with the speed, manoeuvrability and ramming characteristics of the Greek trireme appears in Asian naval history, but oarsmen on their feet could fight as needed more readily than their seated western counterparts, and were somewhat less vulnerable to missile fire. The Koreans would send the most deadly ships constructed by Asian hands into battle employing this 'standing' system.

The wind, of course, could move more mass with greater velocity than human muscle, although not in every direction. Military naval architects in all cultures were faced with the traditional dilemma of building a powerful ship that could be left helpless by a lack of wind, or, as was the case in the battle of Lake Poyang in 1363, unable to fight in the shallows against smaller antagonists. Multiple ships linked by platforms offered a partial solution to those problems and made convenient transports for siege weapons or other special equipment.

Fire and Gunpowder

In terms of weaponry, Asian designers were second to none in applying human intellect to human destruction. In keeping with the tactical emphasis on incendiary warfare, Chinese military manuals mention fast and slow burning incendiary balls or grenades, and 'fire lances' similar to the *bomba* the Spanish sought to employ in the Armada battles. Tubes of explosive on the ends of poles would evolve into the 'gonnes' of Medieval Europe as well as China.

Ships tied together with platforms could make use of weapons such as long booms equipped with heavy weights to smash down upon an enemy foolish enough to approach too closely, or hooks like the old Roman boarding bridges to pin and hold an enemy for boarding. Broad-beamed conventional vessels could also employ such devastating armament – very slowly.

Fire ships and rafts driven by wind or currents were among the most feared of all naval weapons, while catapults could throw clay containers filled with molten iron – an idea the British would emulate in the age of Queen Victoria. There are no accounts of the British pelting an enemy's deck with bird eggs, a tactic recommended in the Chinese manuals with an eye towards rendering uncertain the enemy's footing. Ships would sport fire-resistant coatings of specialized ceramic washes. Rockets could carry explosive and shrapnel from even the smallest craft, although accuracy from such weapons was more to be hoped for than expected.

Gunpowder functioned at first as much as an incendiary as an explosive, but the samurai charging down to the beaches at Hakata Bay found themselves facing exploding bombs discharged by the Mongol fleet's batteries of ship-borne trebuchets. The inventors of gunpowder would have to wait until that technology had filtered through the Arab world into the western world and returned to the east. It was the fifteenth century before the Chinese, Japanese and Koreans encountered, purchased, emulated and employed European cannon and small arms extensively in the naval wars of the East. The Koreans would, however, come to have a unique appreciation of cannon – much to the eventual consternation of the Japanese.

Terrain

The vastness of the Pacific explains the lack of formidable overseas enemies for most of the states of the Far East. The first recorded deep-water battle in Chinese naval history did not take place until 485 BC, and was fought between forces of two bordering states endeavouring to effect or forestall a flanking movement around terrestrial defences. China most feared invaders across the vast land

FIREPOWER, NOT FIREWORKS, *was the objective of the gunpowder in this European-style cannon of the late eighteenth century mounted on a nineteenth-century* sampan. *The Chinese had enclosed 'the fire drug' in a tube for rockets and 'Roman' candles, but the Arabs and Europeans refined the concept into carriage-mounted weapons such as these – and sold them back to the Chinese at a handsome profit.*

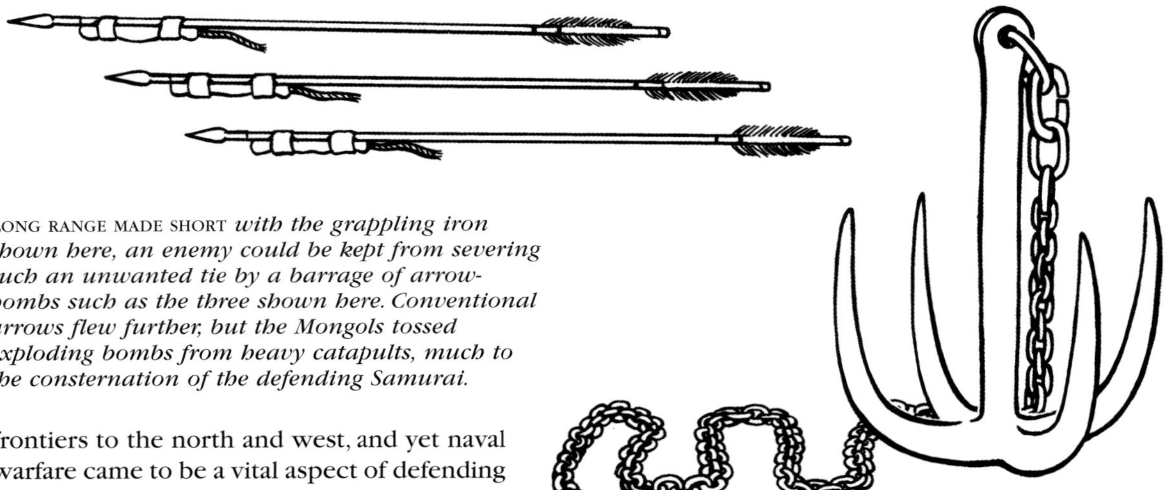

LONG RANGE MADE SHORT *with the grappling iron shown here, an enemy could be kept from severing such an unwanted tie by a barrage of arrow-bombs such as the three shown here. Conventional arrows flew further, but the Mongols tossed exploding bombs from heavy catapults, much to the consternation of the defending Samurai.*

frontiers to the north and west, and yet naval warfare came to be a vital aspect of defending the empire.

The reason, of course, was the rivers that could physically halt the onrush of barbarians from the steppes, especially if powerful warships waited to prevent any efforts at a crossing. In such a way did the navy of the Southern Song Dynasty forestall the Mongols until 1279, making the Yangtze a moat more formidable – and effective – than the defences of the Great Wall itself ever proved. As corridors for transportation, as means of supply, as limiters or facilitators of movement, the rivers of China have directed, permitted, or necessitated naval combat up to the present day.

The geography and position of the Japanese archipelago also explains much of the naval history of the islands. The relatively calm and sheltered waters of the Inland Sea allow easy travel between Honshu and Kyushu, as well as provide ideal conditions for land warriors to fight at sea, such as at Dannoura in 1185. Travel to Hokkaido was only slightly more difficult so the Japanese were free to trade and fight each other as their distinctive culture developed into its historical form.

The geography of another stretch of water exercised its own dramatic effect upon the course of Asian history. In the winter, the Sea of Japan is stormy and difficult to navigate, while in the summer, prevailing currents and winds sweep nearly continuously from northeast to southwest past the Tsushima Strait with its two conveniently placed islands.

Consequently, Japanese pirates could, over the centuries, descend upon the coasts of Korea for swift and profitable depredations, in time convincing the Japanese of their military superiority over the Koreans while persuading the Koreans of their own need for a powerful and effective naval arm. Moreover, such conditions allowed at least the more visionary Japanese of our period to dream of an easy large-scale invasion of both Korea and China while rendering counter-invasion correspondingly difficult. There had been one historically recorded effort to reverse that geographically determined flow of seaborne aggression, and that will be the first of the four battles in our analysis.

Tactics drove the direction of technology, technology allowed or forestalled choices or alterations of conventional tactics. Terrain determined the efficacy of a given tactic or weapon, or the simple feasibility of either. How they impacted on individual battles can be seen in the stories of four great naval battles in Oriental history.

Mongol Invasions of Japan: 1274 and 1281

China's unenviable status as the objective of wave after wave of non-Chinese invaders throughout its existence has an explanation besides the traditional allure of silks, cities, tea and treasures.

The wealth of China was all very well, but the Mongols had more than the softer and softening aspects of life in mind when in 1211 they began a relentless pressure upon the Chinese culminating in their conquest of the Song Dynasty in 1279.

Technology explains not only the success of the Mongol conquest, but also offers another explanation for its motives, as well as explaining why Kublai, the grandson of Ghenghis Khan, would hurl his two monster fleets against the distant samurai of Japan.

Consider the Mongols themselves - their shaggy little ponies and the grassy steppes of Eurasia had existed for millennia before the life of Ghenghis, as had the forerunners, ancestors and rivals of the Mongols, such as the Huns, Seljuks, Jin and Qidan. What swept the Mongols into control of a world empire was a new technological element the steppe barbarians had previously lacked. The proto-cannon of the twelfth century were slow to load and prone to slaughtering their own gunners. The counterweight trebuchet, however, proved capable of ripping open the defences of the walled cities that earlier generations of invaders from the steppes had menaced largely in vain.

SEEKING HEADS AND GLORY, *Samurai in small boats take advantage of the calmer waters between Korea and Japan and sally forth to board the ships of the second Mongol Invasion fleet. In the morning the Mongols would find entire ships drifting with decapitated crews, forcing them to link their ships together into a defensive formation.*

Exactly how the Mongols got the weapon is somewhat unclear - the Chinese had a 'traction trebuchet' that used simple force on the short side of the weapon's arm to lever a slung missile into flight. Adding a heavy weight to that arm and applying the tractive force to the long arm of the weapon produced a siege engine that could throw a 140kg (300lb) projectile for as far as 300m (1000ft). Against this onslaught, fortifications from

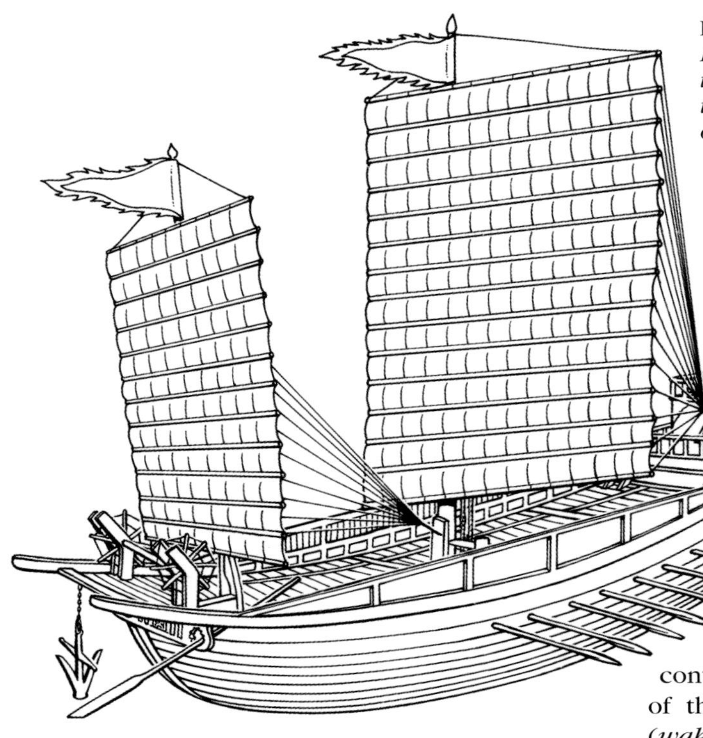

FREIGHTED WITH A SECRET WEAPON, *the Chinese and Korean junks of the Mongol Empire carried the trebuchet that served the Mongols as their key to Empire. The trebuchet threatened Samurai on the beach with bombs, fortifications with heavy stones, and Japan with Mongol conquest.*

Baghdad to Canton collapsed into rubble and became absorbed into the Mongol Empire.

It was under Kublai Khan (1215–94) that China fell and the Mongol Empire reached its final, breathtaking extent – from eastern Russia to the very coast of the China Sea, an area of some 33 million square kilometres (12 million square miles), almost a quarter of Earth's total land area. Kublai's actions show that he was aware of the hidden secret of the nature of Mongol domination – the fierce warriors and aggressive Khans of the lesser hordes needed to expand outwards, or they would turn inwards, warring upon each other and causing the collapse of the empire's unity, as indeed did happen within a century of Kublai's death. Accordingly, Kublai's warriors would constantly need new objectives and the means to subdue them.

China itself could provide the apparatus for future conquests. Not all the Song navy had burned in the battles along the Yangtze by which the Mongols had finally conquered the last defenders of China. In addition to those vessels were the heavy ships of Korea, which had resisted the Mongols for 30 years but became the property of the Khan along with the fortresses and foundries of that peninsula. The Japanese were slow to realize their peril. After Dannoura, the victorious Minamoto family established the shogunate and by means of puppet emperors engaged in expanding their control throughout the archipelago. One function of this control was the suppression of pirates (*wako*) by the conquest of their havens on the Japanese shore as a gesture of support for trade and good relations with Korea.

In 1268, while Kublai was still mopping up the last of the Song resistance in China, emissaries from the court of the Great Khan arrived at Kyushu with a letter from the Emperor of China directed to the 'King' of Japan. Neither the deliberate insult nor the letter's references to the 'liberation' of Korea were lost upon the rulers of Japan at Kamakura, who were familiar with threats in flowery language from their own centuries of internecine warfare. The messages received no official reply.

In the emperor's name, the Shogun urged the samurai to defend home and heartland, but there was little the Japanese could do besides mobilize their warriors and await the Mongol onslaught. Their own activities against the *wako* had stripped them of an irregular force that might have alerted them to the invasion fleet's movements and at the least raided the Mongol supply vessels and transports.

Long lines of Chinese ships moved down the coast, gathering in the southern Korean ports.

Prior decades of resistance had so impoverished the peninsula that this initial force collecting for the invasion was forced to retreat back into China until the Mongols could establish supply depots to support their numbers. The Mongols had some final mopping of Korean resistance to complete as their troops poured over the Yalu and moved southwards.

By November 1274 the Mongols were ready for the invasion. Some 200 remnant Song ships joined 700 commandeered Korean warships and transports to deliver some 15,000 Mongols and 8000 Korean levies and crews across the 200km (120 miles) of the Korean Strait, the first leg reaching the two linked islands of Tsushima. So Sukekuni, one of the last surviving scions of the Taira clan vanquished at Dannoura, defended his isolated fiefdom well enough to provoke a Mongol response in victory that the Japanese considered savage. The population of the next island closer to Japan, Iki, fared no better.

The defenders of the two islands had at least purchased time with their lives and severed heads, allowing the first large Japanese army collected in 50 years to move to the coast opposite to face the Mongols. The details of the land combat can be found elsewhere in this volume, but samurai who proclaimed their pedigrees in offer of single combat on the beach found themselves and their horses stunned by trebuchet-discharged bombs hurled over their heads from the fleet as the Mongols moved inshore for a massed landing of organized troops in close formation. Modern scholars have faulted Kublai for the relatively small size of the force he sent to sea, but the Mongols were out to secure a beachhead, not to effect a conquest of all Kyushu, and were well on their way to doing so when nature intervened.

Japanese resistance had been enough for the Mongol commanders to keep the landing force close to the fleet, partly because their archers had discharged a dangerously large proportion of the available arrows. The fleet's experienced navigators knew a horrific storm was bearing down upon the invasion fleet and, faced with the prospect of being marooned on a hostile shore, the invaders re-embarked.

The typhoon caught the fleet as it made for the open sea and lasted for two days. According to the Korean estimates, most of the older Song vessels foundered, while the total casualties numbered some 13,500 men. The surviving ships returned homewards, with the great Khan undaunted. Again, it was in Kublai Khan's interests to direct his warriors outwards at an enemy, so to succeed where others had failed would be a major inducement to his commanders to try again.

> *'Cublay, the Grand Kaan who now reigneth, having heard much of the immense wealth that was in this Island, formed a plan to get possession of it. For this purpose he sent … a great navy, and a great force of horse and foot … and there they landed, and occupied the open country and the villages, but did not succeed in getting possession of any city or castle. And so a disaster befell them…'*
>
> – MARCO POLO

Kublai Tries Again

After the final suppression of native Chinese rule concluded in 1279, Kublai began reallocating resources for a second try at Japan. The Japanese had spent the interval executing Mongol envoys, fortifying the heights above Hakata bay and constructing smaller versions of their own design of junk for close-in combat. Their intelligence, apprehensions, and the returning envoys demanding submission, convinced them that the invaders would return.

Kublai's plans seven years later called for Japan to be ground between two millstones – a second

Mongol invasion of Japan

1281

The Mongol Empire would either expand outward or turn against itself and disintegrate, a fact well understood by Kublai, the grandson of Ghengis. The Japanese archipelago was an obvious target for the Khan's troops and generals, eager for possessions and glory. Determined to succeed after the disaster of seven years before, Kublai Khan prepares an invasion force five times the size of his previous effort. Chinese river and naval craft set out independently of the heavier warships of the subject Korean kingdom in May of 1281, and descended upon the fortified islands in the Korean Strait, where they encountered and overcame heavy resistance by the Japanese garrisons.As in 1274, the Khan's massive invasion collapsed from a failure to secure a harbour proof against the typhoons that again devastate the invaders' fleets.The Japanese acquired a faith in their divine destiny and military superiority that would endure until 1945.

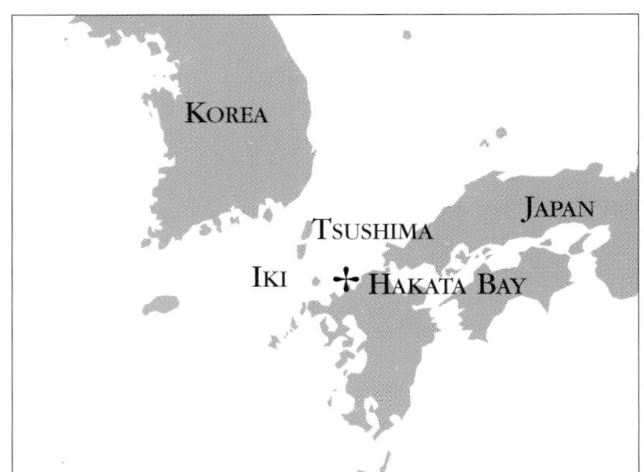

Islands such as Tsushima and Iki made the voyage from Korea to Japan, or the reverse, one conceivable for even the most heavily laden of supply or troopships. The Tsushima Strait proper ranges from 64km (40 miles) to 97km (60 miles) in width - not an easy voyage for a riverboat, but one conceivable in good weather.

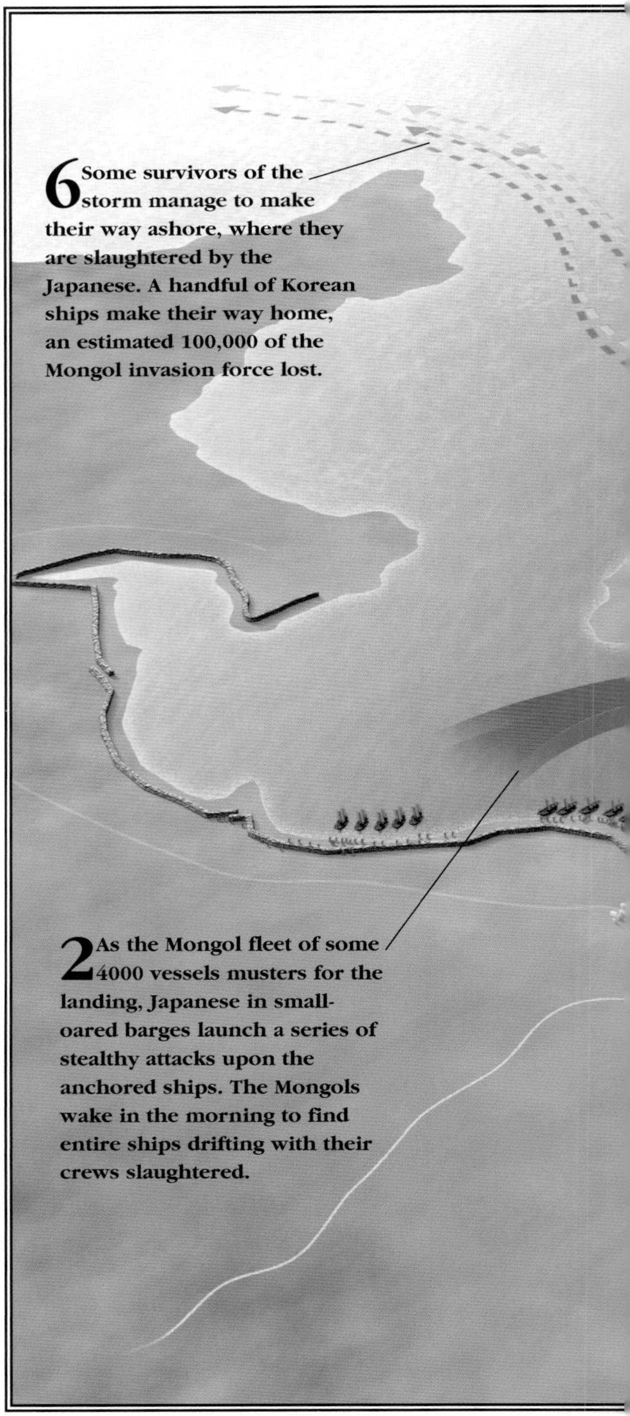

6 Some survivors of the storm manage to make their way ashore, where they are slaughtered by the Japanese. A handful of Korean ships make their way home, an estimated 100,000 of the Mongol invasion force lost.

2 As the Mongol fleet of some 4000 vessels musters for the landing, Japanese in small-oared barges launch a series of stealthy attacks upon the anchored ships. The Mongols wake in the morning to find entire ships drifting with their crews slaughtered.

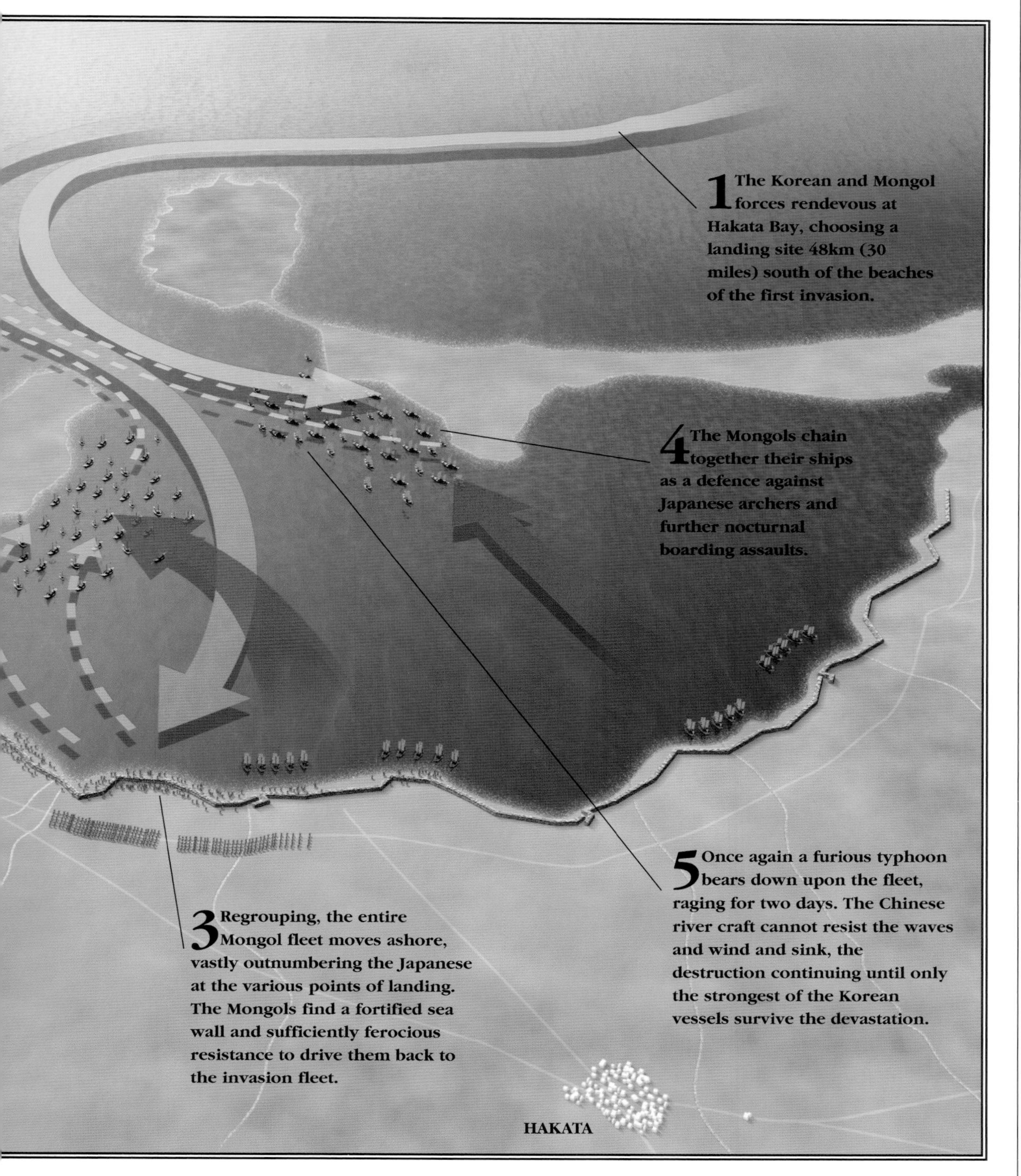
1 The Korean and Mongol forces rendevous at Hakata Bay, choosing a landing site 48km (30 miles) south of the beaches of the first invasion.
4 The Mongols chain together their ships as a defence against Japanese archers and further nocturnal boarding assaults.
5 Once again a furious typhoon bears down upon the fleet, raging for two days. The Chinese river craft cannot resist the waves and wind and sink, the destruction continuing until only the strongest of the Korean vessels survive the devastation.
3 Regrouping, the entire Mongol fleet moves ashore, vastly outnumbering the Japanese at the various points of landing. The Mongols find a fortified sea wall and sufficiently ferocious resistance to drive them back to the invasion fleet.
HAKATA

levied fleet from the Koreans would combine with what could be forced from the merchant fleets along the Yangtze and coastal traders and freighted with some five times as many Mongols as had participated in the earlier assault. The fleet of 900 Korean vessels put to sea in May of 1281 with 17,000 sailors transporting 10,000 Korean levies, 30,000 Mongols and their supplies.

Fighting at Iki and Tsushima was much more severe than during the previous invasion as the Japanese had been given time to prepare the islands in the strait, but eventually the Korean arm of the Mongol attack arrived at Kyushu on 21 June.

The Japanese were more experienced and more numerous than they had been before so the Korean fleet pulled out and rendezvoused with the Chinese portion of the second invasion force at the offshore island of Hirado.

This 'Chiang-nan' division consisted of whatever remained of the Song navy and the River Yangtze fleet: some 3500 warships and transports carrying 100,000 Chinese and Mongol soldiers. The united fleet bore down on Takashima, some 48km (30 miles) to the south of the previous landing site at Hakata Bay.

The time of year was different from before, the numbers were different, and this time the invaders found an island fortress with a coastal defence force waiting for them. By design or good fortune the Japanese had come up with a way to effectively employ their tradition of individual combat against the Mongol hordes. Samurai in their small, oared vessels sailed out to individual invasion ships, by day or night, then boarded and engaged their crews in an environment where the superlatively deadly *katana* and heavy armour of the individual Japanese could be best applied.

Entire crews of the assault fleet fell under the gleaming swords of the samurai, the ships found drifting and abandoned in the morning off the coast of Takashima.

The Mongols countered in a manner typically Chinese but also in a fashion that presaged a measure taken some 1200 years later during the Normandy invasion. Instead of isolated individual ships, the marauding samurai found themselves facing a massive floating harbour-fortress of linked ships in the island's Imari Bay, from which Mongol trebuchets and archers could bring fire upon the smaller craft of the Japanese as men shifted on plank bridges between the vessels of the fleet. The struggle raged on for two weeks, with neither side prepared to withdraw.

> *'And it came to pass that there arose a north wind which blew with great fury, and caused great damage along the coasts of that Island, for its harbours were few. It blew so hard that the Great Khan's fleet could not stand against it.'*
>
> — *Marco Polo*

For a second time, what would thereafter be called the *kamikaze,* the 'divine wind,' blew down upon the Mongol fleet. On 30 July 1281, a typhoon bore down upon the Japanese coast. The samurai retreated inland but the typhoon's power ripped into the linked ships of the Mongol fleet, jostling and destroying the clustered ships in a density marine archaeologists are still exploiting profitably. Some ships managed to escape from the bay's narrow entrance, but the power of the wind rendered what might have been shelter in a lesser tempest into a death trap for an estimated nearly 4000 ships.

From the bones of the vessels submerged by the fury of the storm, it is possible to learn more than the historical sources of the battle communicate on their own. Around 1170 of the surveyed wrecks were large war junks of about 73m (240ft) in length, again Song remnants or new Korean construction. These vessels would have required a crew of 60 and each towed a small flat-bottomed craft called a *batoru* and intended to land 20 infantrymen at a time. By far the bulk of the sunken ships were the River Yangtze *sampans,* whose mere arrival in Japanese waters had already tested their sea-keeping qualities to the breaking point. Scattered among those vessels were around 300 two-masted Korean war junks. Historical accounts put the numbers of Mongols and subject allies drowned or slaughtered onshore at about 100,000 men.

Kublai Khan was making plans for a third assault upon Japan at the time of his death in 1291.

Lake Poyang, 1363

The foundries and the fleets of China remained under control of Mongolian dynasts, now calling themselves Yuan, for another two centuries after the disasters off the Japanese coast. The successors of Kublai ruled efficiently with a relatively easy hand, but still considered themselves separate from their Chinese subjects. The ethnic Chinese reciprocated these feelings and the seeds of rebellion grew.

One of the spurs to Chinese discontent had its origins in the invasions of Japan. The central government in Beijing suddenly and arbitrarily revoked the opportunities granted under Kublai for the transport of grain along the rivers and coasts. They accused the ship owners of treason and piracy and violently put an end to their extremely profitable operations. The people suffered along with the commerce, and suddenly a large quantity of unemployed and disgruntled sailors could be numbered among the enemies of the Mongol regime.

There are two recovered wrecks from the Yuan Dynasty. One was excavated at Quanzhou, the port Marco Polo visited and left his description of the construction of a Chinese sea-going ship. Underwater archaeologists found the other at Shinan, off the Korean coast. The Quanzhou ship was 34m (112ft) long with a beam of 11m (36ft), substantiating Polo's account of 13 compartments separated by bulkheads and layered wooden hulls held together by myriad iron nails. Like the Shinan wreck, the Quanzhou ship has the deep, V-shaped hull of the deep-sea vessel.

Kublai and his immediate successors understood that overseas trade brought wealth even from areas beyond the reach of practical conquest. Kublai's later successors, however, did not value such trade enough to protect it, particularly as the Japanese *wako* made national revenge an excuse for profitable assaults along the southern Chinese coast. Coastal trade collapsed under that and the government reversal, but along the internal rivers of China the Mongols set about constructing fleets such as the Song had used to keep them at bay for centuries.

The objectives were different. Rather than internal defence against an external attacker, the new and powerful riverine squadrons were meant to maintain the control of the central power at Beijing. Such efforts at internal policing collapsed in a rush of floodwater in 1351, when massive overflows of the Huang and Huai river basins created waterways and disorder beyond the Yuan Dynasty's

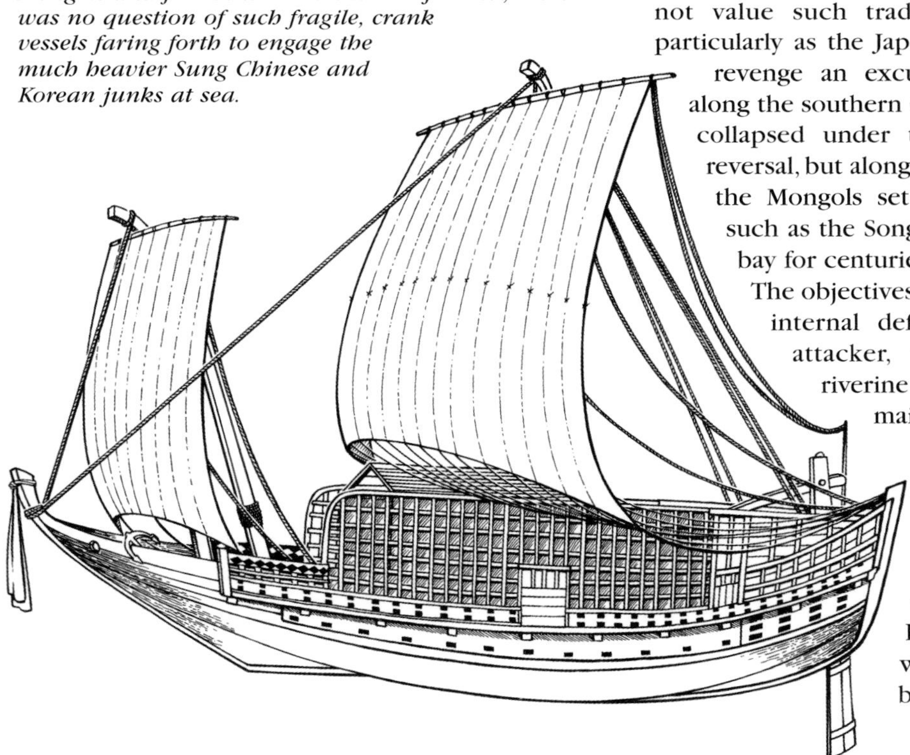

JAPANESE JUNKS *such as this one were built for commerce and warfare on the Inland Sea and among the islands of the archipelago. Although used as transports both to muster samurai to repel the Mongols and for the later invasions of Korea, there was no question of such fragile, crank vessels faring forth to engage the much heavier Sung Chinese and Korean junks at sea.*

ability to control. Central authority disintegrated as Mongol generals turned against each other and the central government and as the pre-conquest states of ethnic China began to reassert themselves.

Two such were the rival powers of Han and Wu, while an army under Zhu Yuanzhang supported the renewed Song Dynasty, now with a capital at the industrial city of Nanchang.

Tower Ships

The ruler of Han, Chen Youliang, after some initial setbacks, chose to move against Zhu while the bulk of his forces were outside of Nanchang, held in combat against the Wu, the nominal rulers of Nanchang and the lower Yangtze. To do that, it seems that Chen recalled Song precedent by constructing a fleet with a core of monstrous 'tower ships,' (*lou chuan*) capable of reducing a city or crushing the smaller Ming fleet.

We are fortunate to have a description of these behemoths, giving their characteristics as having three decks, presumably above the water line in almost a 'pagoda' style, pierced with loopholes for missile fire. Around each of these and the 'spar' deck were crenellated battlements to protect crossbowmen and spearmen while the size of the vessels allowed the mounting of such equipment as trebuchets, weighted booms for defence against close approach, and possibly even horses in covered galleries to exploit an opening shoreward.

PUGNACIOUS PAGODA, *the tower ships (*lou chuan*) offered their commanders and crews every advantage but mobility. Height mean extra range for archers and gunners to direct against hostile ships or towns, while heavy booms on trip-lines stood ready to smash down upon any smaller vessel rash enough to approach closely.*

From more specific details of this particular battle, scholars have estimated that the crews of the Han tower ships could have numbered as many as 2000–3000 sailors and marines. The accounts speak of them as 'iron plated,' a term which very much requires clarification in the light of marked incomprehension from later study.

The term 'ironclad' calls to memory the prototype vessels of the Crimean and American Civil War: vessels such as *Monitor* with 180mm (7in) of laminated plating on her turret. Such vessels were meant to resist shot from recently invented 300mm (11in) or larger cannon, which was very much not the case for ships from earlier centuries. Other ships in the American Civil War which were armoured similarly to their Chinese and Korean counterparts (only against fire from infantry and field guns) bore the sobriquet of 'tinclads,' a concept of far more utility to the modern reader than the other misleading expression.

An admiral commanding from such a flagship would have safety and a field of view not to be found in smaller, albeit more handy craft. Missiles fired from the lofty upper decks of such a leviathan would travel further than their counterparts from smaller ships. However, even the original manual mentions the vulnerability of such vessels to the sudden onset of a squall. Even so, it was not wind-resistance that eventually proved to be the greatest liability of the Han monsters.

By the year 1363 Nanchang, once on the lakeshore, was 24km (15 miles) as the crow flies and 80km (50 miles) by way of the River Kan away from the largest freshwater lake in China, Lake Poyang. The surrounding area was one of bogs and other wetlands left behind by the receding water level of the lake. Down from the upper Yangtze through the lake descended the Han fleet, comprising an estimated 300,000 sailors and marines. Chen Youliang was making the wise military investment of throwing the maximum force and the greatest investment of resources he could muster at his first target, with the idea of taking his enemy's centre of production and control and then 'steamrolling' on to a further series of victories.

It should have worked. The first check came at the very walls of Nanchang soon after the Han arrived there on 15 April. One very basic truth in amphibious warfare is that fire from the stability of firm emplacements is necessarily more accurate than the discharge of missiles from the shifting decks of even the largest of warships. Movement and larger ordnance can overcome that advantage, but swift movement was entirely beyond the capacity of the tower ships. The Han monsters could not reduce the city by their missile fire.

The garrison commander of Nanchang, Zhu Wenchang, the Ming commander's nephew, held out against bombardment with palisade walls across the breaches produced by the Han cannon and trebuchets. Marines trying to land along the riverside were met with red-hot pikes, while 2000 picked troops in a central reserve preserved the city's perimeter intact.

'When two vessels are closing, and conquest and overthrow loom in the offing, fire-weapons of some variety should be used, but due to the difficulty of their employment, at sudden need they may not be discharged in the correct manner …'

– CHAN-CH'UAN CH'I-YUNG SHEH

Zhu Leads the Counterattack

In the sixth month of the siege, word reached the Han that the Ming army and navy were on their way to the city's relief down from Lake Poyang, in between the Han and their own territory on the upper Yangtze.

Chen Youliang's decision to remove his ships from the siege on 16 July and engage the arriving Ming conceded a tactical defeat in the hope of preventing a strategic one. Zhu Yuanzhang himself was in command of the relief force sailing up the Kan, and the death of such a commander could quite possibly bring about the collapse of the entire army. Zhu himself had much the same idea in mind. He had left enough of his own fleet behind to keep the Wu from advancing up the Yangtze, but the forces that confronted the Han on Lake Poyang on 29 August numbered some 200,000 men in an uncertain number of smaller ships – but in a quantity that grew larger as the days went by and additional forces arrived in the lake from the waterways controlled by the Ming. The stage was set for the largest naval battle, in terms of human beings employed, in the whole of world history.

In the face of the Han's superior numbers and the great size of their ships, Zhu had only two advantages, both of which he utilized to the fullest – time and position. When the spring rains of May had filled the vast basin of the lake, the Han behemoths enjoyed a great deal of water under their massive keels, giving them plenty of room to manoeuvre. With the August sun boiling the lake's waters rapidly away, not only were the large ships of the Han constrained in their movements, but Zhu and his smaller squadron lay directly across the only viable line of retreat – the deep entrance leading back up to the Yangtze. The Han situation would only get worse if the battle could be delayed, and Zhu had no intention of rushing things. He divided his own fleet into 11 smaller flotillas and positioned those in the shallower water where they could fire on the Han without coming to close and disastrous quarters.

Zhu himself initiated action the following morning, sending his ships directly towards the looming towers of the Han line, firing incendiary missiles as they came into range. The result was not to the Ming's advantage. Chen Youliang kept his fleet in the widest and deepest part of the lake in close formation. With his all-out assault Zhu's fleet was able to kindle flames in 20 of the Han monsters, but it was as a defence against incendiary attacks that the thin iron plates on the tower ships were most effective, and Zhu found his own large flagship in flames as the Han behemoths bore down.

Zhu's greatest moment of peril came when his own vessel, rather than that of the larger Han armada, went aground on a sandbar as the long-range fire of the advancing tower ships continued unabated. His lesser commanders rushed to Zhu's defence in a demonstration of loyalty that would be utterly unreciprocated, and the wash of their oars eventually allowed the battered flagship and other grounded Ming vessels to escape up the nearest river mouths for emergency repairs.

Chain Gang

Dawn of the next day found the Han tower ships still in the middle of the lake, their formation made impenetrable by a network of chains in the classic Asian style of converting mobile ships into a single unified fortress. Zhu's second assault on this collection of floating towers was such an utter disaster that he executed the commanders of his 11 flotillas. The Han held the field and the advantage of the second day.

The traditional counter to the chained formation was the deployment of fire ships, and these Zhu and his new, presumably enthusiastic, commanders launched the following day. The accounts speak of both fire-rafts and seven more complex vessels, these equipped with straw dummies outfitted in weapons and armour to deceive the Han. As the same narratives mention that fisherman volunteers sailed these weapons into the lake, these were probably what the Chinese military manuals describe as 'mother-son' vessels, their cargoes of straw, oil, and gunpowder left unfired until such time as their contact with an enemy ship was inescapable. The concealed, smaller 'son' vessel would then carry the crew to safety.

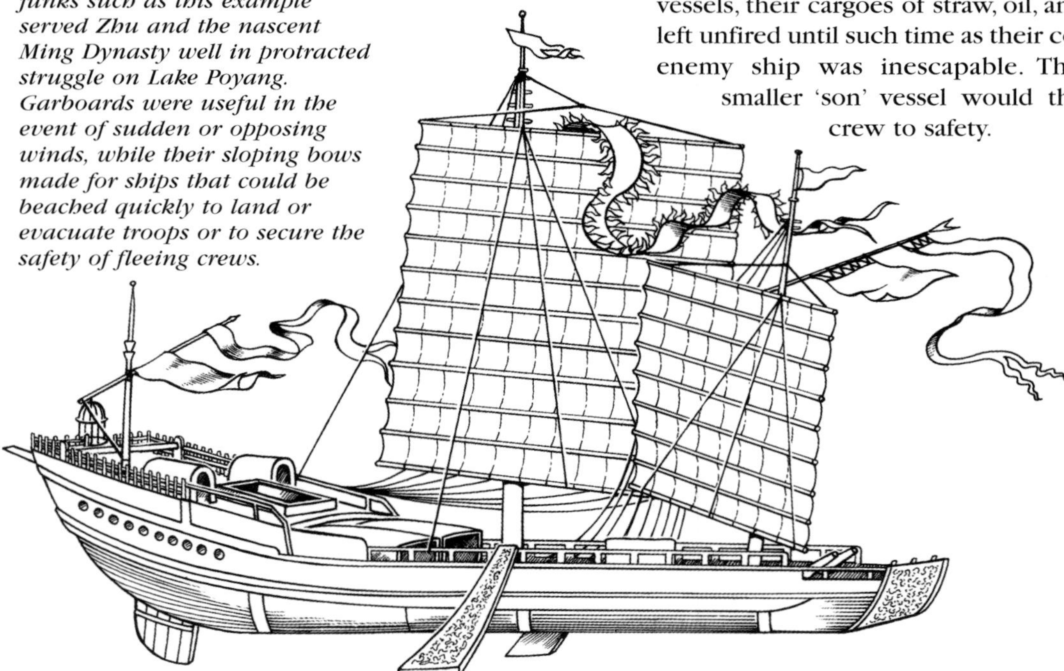

LIGHT, LITHE AND LETHAL, *smaller junks such as this example served Zhu and the nascent Ming Dynasty well in protracted struggle on Lake Poyang. Garboards were useful in the event of sudden or opposing winds, while their sloping bows made for ships that could be beached quickly to land or evacuate troops or to secure the safety of fleeing crews.*

Wind and ruse combined to create a conflagration that destroyed half of the Han fleet.

Combat did not resume until 2 September, with the repaired Ming fleet facing the remaining Han vessels with their connecting chains removed. Chen Youliang determined to use the weight and power of his still formidable remnant to crush Zhu's distinctive white flagship and settle the issue definitively in what modern planners call a decapitation strike. At least as old a tradition in Chinese warfare as incendiary attack was defection to what appeared to be the winning side, and a traitor's warning produced the spectacle that morning of the entire Ming fleet in gleaming whitewash moving in to the more open Han formation with grapping hooks and missile fire. The befuddled Han took sufficiently severe damage for the slow process of utter disintegration to set in.

Chen Youliang had left troops with supplies to continue the siege of Nanchang and provide for the Han fleet. Over the course of the ensuing month Zhu's land forces arrived to complete the relief of the city and sever that line of relief and resupply. Ming batteries and barriers appeared around the lake's five outlets, while Zhu withheld open battle over the balance of September, perfectly willing to stand back and let his wounded quarry bleed to death as the waters of the lake receded still further.

On 2 October, Chen Youliang took his remaining 80 to 100 largest ships and made a final direct effort to break out and back into the Yangtze outlet to the north. The Han did manage to break into the river itself, but still more fire ships coming down the current met the monsters struggling back up against the current. Combat ended when Chen Youliang took an arrow or a crossbow bolt to the head. His death ended all fighting, with 50,000 Han and the remaining tower ships surrendering and becoming part of what was soon to become the Ming Dynasty.

Sacheon, 1592

The Ming Dynasty's appreciation of sea power – perhaps 'water power' would be more accurate a term – led to their control over nearly the whole of mainland China and fleets of war and merchant ships larger and of more advanced construction than any in use elsewhere in the world. At its

FLOATING FORTRESSES *such as this tower ship held the fate of the Han Dynasty. Linked together, such vessels formed a virtual walled city, within which defenders could transfer marines and munitions with the advantage of interior lines. So large an object became a target for repeated and successful Ming onslaughts.*

Battle of Lake Poyang

1363

From the ruins of a fading dynasty grew two nascent powers and the largest naval battle, in terms of human numbers, in recorded history. Chen Youliang and the fleets and armies of Han challenged the Ming armies of Wu and the remaining Song Dynasty with a huge fleet of fortress-warships laying siege to the industrial centre of Nanchang. The city held firm while Zhu Yuanzhang gathered a relieving force of smaller vessels with shallower draft. Superiority in the resulting weeks of combat on the vast, swampy Lake Poyang varied back and forth, but the summer sun and the dwindling waters of the great lake finally forced the Han behemoths into a finish fight with the Ming warships across their only channel of escape into the River Yangtze. The resulting victory of Zhu's forces would pave the way for the artistic and technological glories of the three-century Ming Dynasty.

YUAN CHINA
LAKE POYANG
Nanchang
SONG CHINA

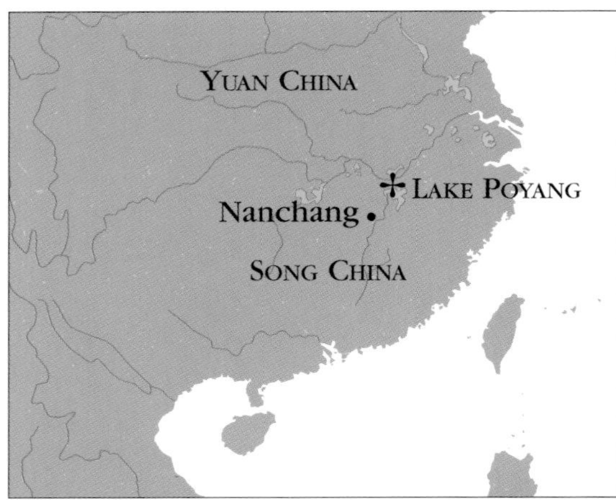

Lake Poyang was a crucial strategic channel leading to the industrial heartland of Nanchang and into the River Yangtze. Victory for Zhu Yuanzhang proved decisive in establishing the power of the Ming Dynasty.

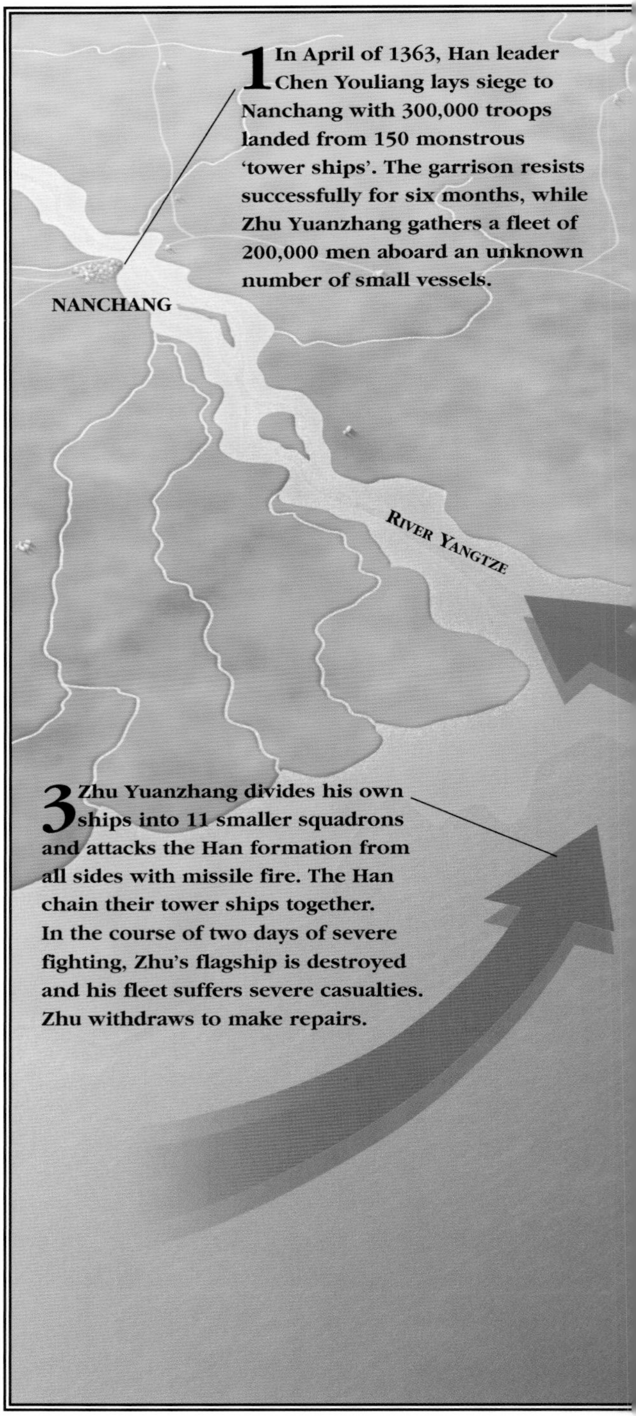

1 In April of 1363, Han leader Chen Youliang lays siege to Nanchang with 300,000 troops landed from 150 monstrous 'tower ships'. The garrison resists successfully for six months, while Zhu Yuanzhang gathers a fleet of 200,000 men aboard an unknown number of small vessels.
NANCHANG
RIVER YANGTZE
3 Zhu Yuanzhang divides his own ships into 11 smaller squadrons and attacks the Han formation from all sides with missile fire. The Han chain their tower ships together. In the course of two days of severe fighting, Zhu's flagship is destroyed and his fleet suffers severe casualties. Zhu withdraws to make repairs.

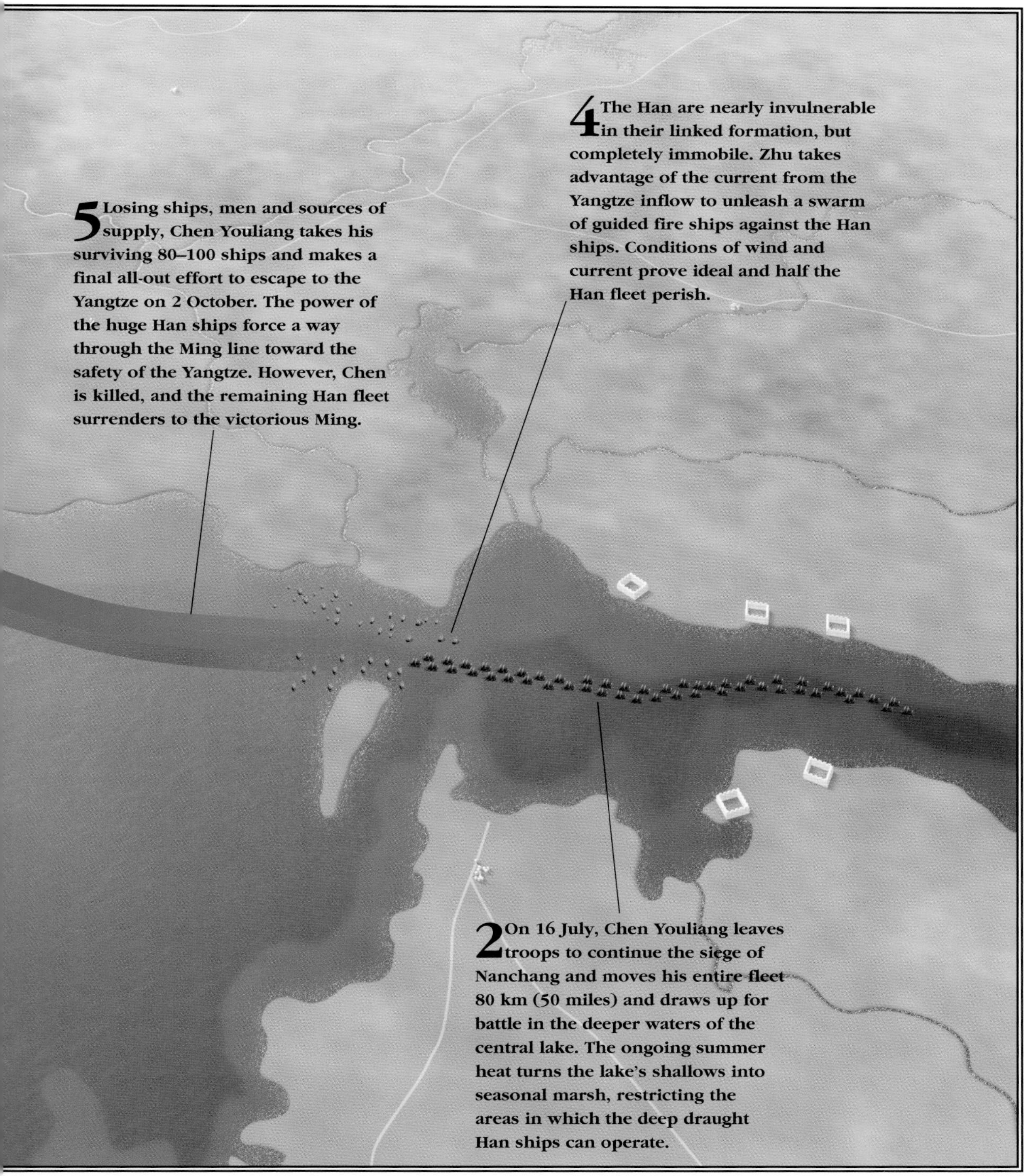

4 The Han are nearly invulnerable in their linked formation, but completely immobile. Zhu takes advantage of the current from the Yangtze inflow to unleash a swarm of guided fire ships against the Han ships. Conditions of wind and current prove ideal and half the Han fleet perish.

5 Losing ships, men and sources of supply, Chen Youliang takes his surviving 80–100 ships and makes a final all-out effort to escape to the Yangtze on 2 October. The power of the huge Han ships force a way through the Ming line toward the safety of the Yangtze. However, Chen is killed, and the remaining Han fleet surrenders to the victorious Ming.

2 On 16 July, Chen Youliang leaves troops to continue the siege of Nanchang and moves his entire fleet 80 km (50 miles) and draws up for battle in the deeper waters of the central lake. The ongoing summer heat turns the lake's shallows into seasonal marsh, restricting the areas in which the deep draught Han ships can operate.

height in the 1420s, the Ming Navy contained 400 war junks as large as 60m (200ft) based in Nanjing, 1350 riverine warships and other patrol craft scattered throughout the interior, and 250 'treasure ships' whose recorded great size, armament and nature as transports of precious cargo pose a fascinating parallel to the great East Indiamen the Dutch, French and British would have sailing the same waters off India some 400 years later.

The emphasis on considerable river forces echoed the collapsing Yuan Dynasty's efforts to stay in power, as did an effort to encourage coastal grain transport with the construction of 400 large grain ships, on the naval rolls as available for conversion to troop transports at need. Just as the iron foundries of northern China had poised at the very brink of an industrial revolution in the eleventh century, in the early fifteenth the Chinese stood in a position where their great ships and naval policies could have anticipated the Europeans in voyages around Africa and in overseas trade and empire.

Instead, policy made a sudden shift in the opposite direction – back inwards. Perhaps the echoes of the Mongol invasions of Japan still resonated, perhaps, as was often the case in Chinese history, those in power saw nothing outside of China worth seeking, and turned to defending what they had rather than adding to the scope of their knowledge and power. By 1433, the emperors had turned away from foreign contact, repeated the Mongol persecution of the ship owners and found it easier to accuse their own navigators of being pirates than to suppress the Malay, Vietnamese and Japanese pirates that swarmed to prey upon them. Japanese pirates there continued to be. As the power of the shogunate became the objective of warring families instead of the domain of a dominant one, what the Japanese called *gekokujô*, 'the oppressed oppressing', erupted throughout Japan. The archipelago had been at something of a military ebb at the time of the Mongol invasions, but the eruption of domestic civil war re-militarized Japanese society and politics with far-ranging results.

In 1542 or 1543 yet another fateful typhoon drove a Chinese merchant junk with three Portuguese traders on board to an island off Kyushu. Soon the Japanese had encountered their first Europeans and the two matchlock smoothbore arquebuses that they carried. Paying a vast sum for the weapons and for lessons in their use, the local *daimyô* soon had his own craftsmen producing duplicates and ammunition. Small arms spread like wildfire throughout Japan.

There was little that could go wrong with an arquebus: the Japanese version was usable by a single individual, could be mastered to the limits of its accuracy quickly and had a faster rate of fire with more power than all but the heaviest crossbows. The battles of the samurai and lower-caste *ashigaru* became more sanguinary and simultaneously more decisive.

The *wako* reappeared when the central government of the declining shogunate no longer

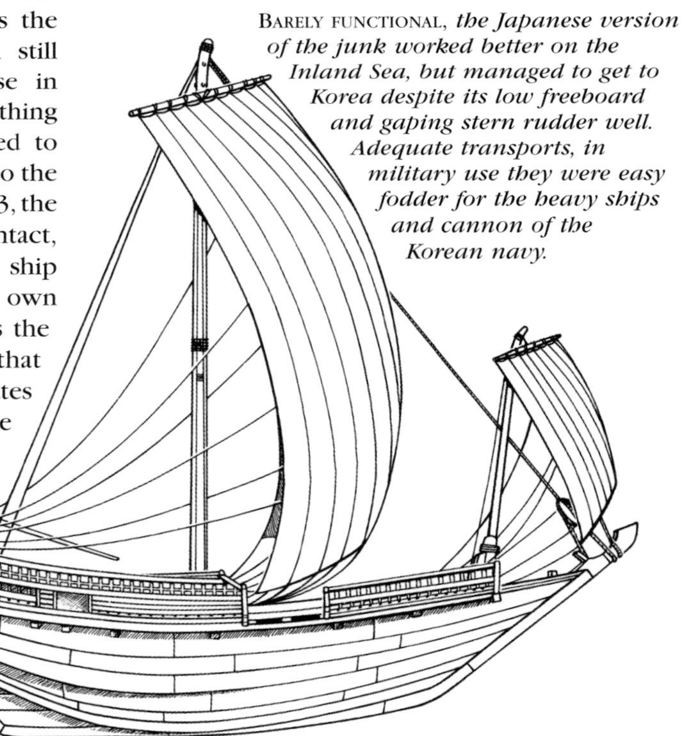

BARELY FUNCTIONAL, *the Japanese version of the junk worked better on the Inland Sea, but managed to get to Korea despite its low freeboard and gaping stern rudder well. Adequate transports, in military use they were easy fodder for the heavy ships and cannon of the Korean navy.*

had the ability to enforce trade agreements made with the Ming and as freebooters found it more profitable and safer to fight and plunder foreigners than their fellow countrymen. The resurgent brigands raided both Korea and China with considerable success, giving ambitious leaders, such as Toyotomi Hideyoshi, ideas for the future.

Hideyoshi Takes the Offensive

At the age of 55, Hideyoshi had united Japan and, with all the opposing armies surrendered, disbanded, or else enlisted beneath his banner he had the most formidable military Japan had ever seen and no further domestic use for it. The success of the *wako* against the Chinese and Japanese provided immediate evidence of the superiority of Japanese tactics and equipment, while the two Mongol invasions had demonstrated, conclusively, that large armies could be transported across the Korea Strait. Korea had a large population subordinated to a ruling caste by bonds that had fractured when the Mongols had conquered Korea just before the attempted invasion of Japan. Moreover, Hideyoshi's fondness for ornate tea services undoubtedly acquainted him with Korean trade goods – including iron ones that were superior to much of native Japanese manufacture. The plan to use Korea as a foothold for the conquest of Ming China was not a particularly extravagant flight of a megalomaniac's fanciful imagination.

In 1555, 70 ships' worth of *wako* had gone ashore, scattered the Korean defenders, and returned to Japan with their booty. By 1585, Toyotomi Hideyoshi was asking Portuguese Jesuits in his country about the possibility of securing two European galleons to assist in his planned invasion of Korea, without success. Ships and crews that had to sail around Africa to reach Asia were far too valuable as cargo carriers, whatever the potential gain in Japanese goodwill offset against the cost of Korean hostility. In 1587 Hideyoshi gave orders for an ultimatum to go to the Chosun king in Hanyang (now Seoul) offering a choice of either submission or conquest.

It was not first time that the Korean peninsula had found itself trapped in the middle of a conflict between China and Japan and it most certainly would not be the last. Hideyoshi was not mistaken in his assessment of the inherent weaknesses of late sixteenth-century Korea. A two-tiered upper caste ruled the peninsula directly under the royal family, the military establishment officially subordinate to the civilian bureaucracy of well-educated Confucian mandarins. The lower castes ranged from outright slaves to serfs and landed peasants with the bonds of loyalty in both directions often tenuous at best. But one constant in Korean life had been an emphasis on technical and theoretical education, and the practice of crafts that would prove to be the

MING VASES AND MORE *travelled overseas in large and technologically advanced sea-going junks such as this example from about the time of the Japanese invasion of Korea. The Ming sent soldiers and supplies to the aid of the beleaguered Koreans, but the cost and expense of such monster vessels made China's decision to abandon overseas trade that much easier.*

Battle of Sacheon

1592

Despite the warnings and even limited preparations by the more foresighted of Korean officials, the Japanese invasion by the orders of Toyotomi Hideyoshi quickly overpowered Korean garrisons and walled towns throughout the peninsula. Some Korean admirals deserted or scuttled their squadrons, but the fleet under Admiral Yi Sun-sin moved to trap the invaders by the systematic destruction of their ships of transport and supply. With the help of Yi's shrewd use of tides and currents and the advanced technology of his 'turtle ships', the Japanese had no grounds for the overconfidence they showed at Sacheon and subsequent disasters, when their lightly-armed *atake bune* sailed forth into the guns and rams of the apparently retreating Korean squadron. At Sacheon, given conditions of tide and overlooking terrain, Admiral Yi lured the over-confident Japanese out into a straightforward and brutally effective ambush.

MANCHU
CHINA
KOREA
SACHEON
JAPAN

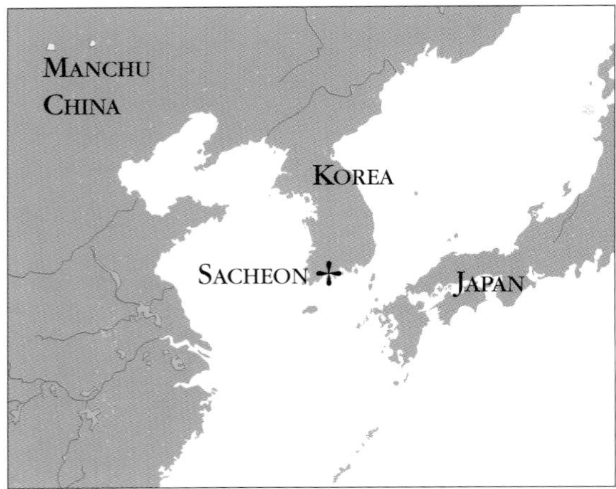

Fatally contemptuous of the Korean military, the Japanese neglected to include the Korean navy's vital bases and shore installations in their initial run of conquest.

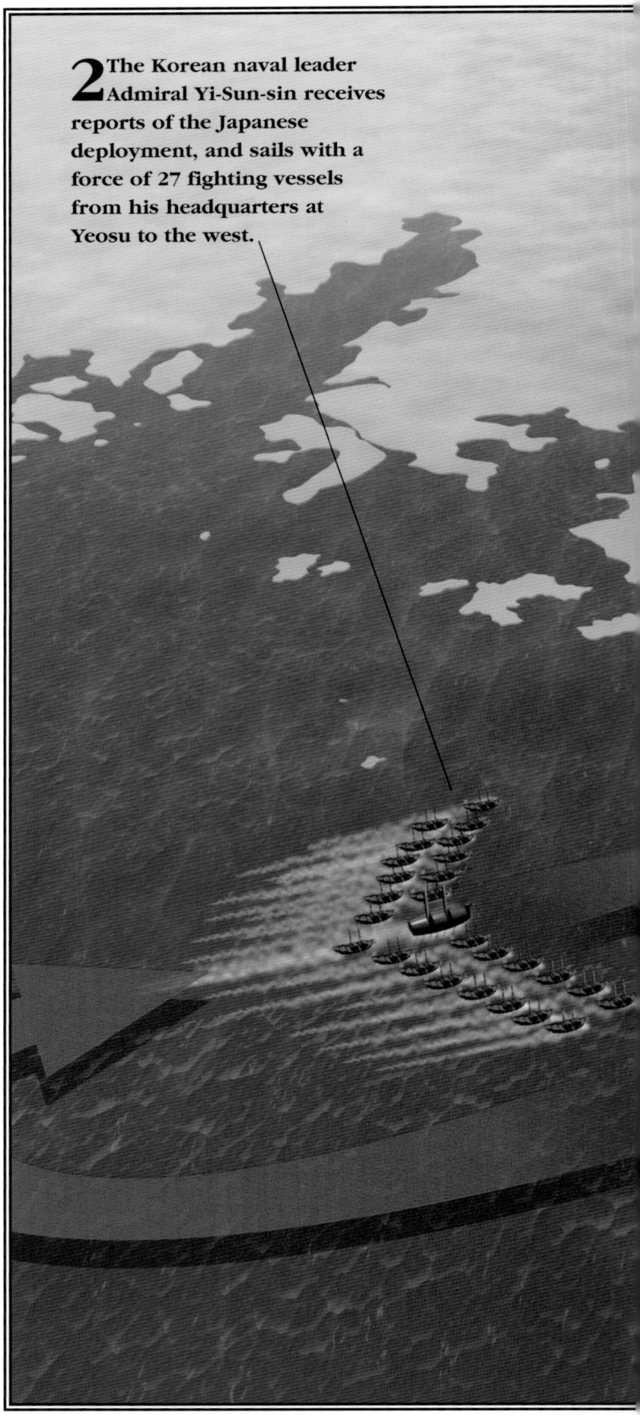
2 The Korean naval leader Admiral Yi-Sun-sin receives reports of the Japanese deployment, and sails with a force of 27 fighting vessels from his headquarters at Yeosu to the west.

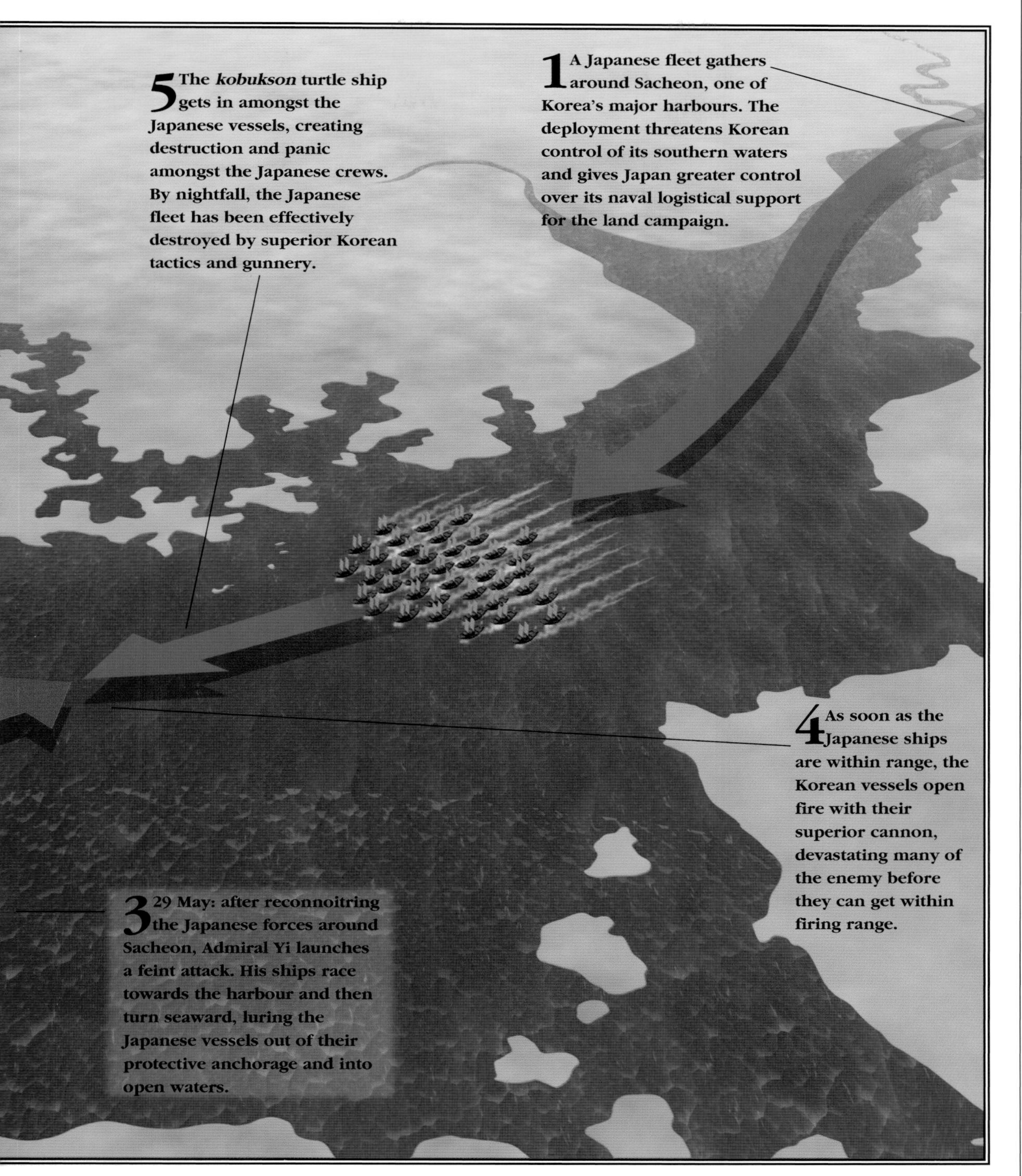
5 The *kobukson* turtle ship gets in amongst the Japanese vessels, creating destruction and panic amongst the Japanese crews. By nightfall, the Japanese fleet has been effectively destroyed by superior Korean tactics and gunnery.
1 A Japanese fleet gathers around Sacheon, one of Korea's major harbours. The deployment threatens Korean control of its southern waters and gives Japan greater control over its naval logistical support for the land campaign.
4 As soon as the Japanese ships are within range, the Korean vessels open fire with their superior cannon, devastating many of the enemy before they can get within firing range.
3 29 May: after reconnoitring the Japanese forces around Sacheon, Admiral Yi launches a feint attack. His ships race towards the harbour and then turn seaward, luring the Japanese vessels out of their protective anchorage and into open waters.

peninsula's salvation. When the Korean king and his ministers received Hideyoshi's final communication the response was the same as it had been to previous Japanese requests for safe passage through their country into China – a polite refusal. From the recent Mongolian suzerainty, the Koreans had become acutely aware of the strength of the Chinese, whose vassals the Chosun monarchs still ostensibly were. As much of an annoyance as the *wako* had been – there had even been an uprising by the inhabitants of Japanese trading settlements along the Korean coast in 1510 – Korean defences had had so far proved adequate. There was nearly complete incomprehension of what Japan had become over the last half century, a lack of understanding illuminated by a single incident.

Along with the ultimatum in 1591, Toyotomi Hideyoshi had sent along a single arquebus. The reasons for the gift are debatable – the weapon was as much a proof of Japanese technical ability as it was an ominous threat. But in an era when two arquebuses and European instruction had served to arm all Japan with gunpowder small arms, in Korea there is no record of any effort to employ the superior weapon in the place of the clumsy Chinese-style hand-cannon to be found scattered throughout the Korean levies. One provincial minister, Yu Seong-ryong, argued in vain that the weapon should be copied on a large scale in the face of the impending invasion. He undoubtedly expressed his frustrations to a boyhood friend of his in the navy.

The Attack Begins

On 23 May 1592, Japanese ships began landing what would come to be a total of 158,850 men Hideyoshi had allocated for the invasion. Operating in two main army groups under two of Hideyoshi's best generals, the Japanese invaders surged over city walls and militia levies with almost incomprehensible speed behind sheeting curtains of musket fire. The courage of individual Korean commanders or soldiers was the one vital element not lacking in the defence, but without armour, organization, centralized intelligent command, or even consistent supply, the Koreans could die in front of, but not halt an advance that by that August stood poised to cross over the Yalu and into Manchuria.

'The enemy vessels kept pursuing ours until they came out to open sea... Our ships dashed forward with the roar of cannons ... breaking two or three of the enemy vessels into pieces. The other enemy vessels, stricken with terror, scattered and fled in all directions in great confusion...'

— Yi Sun-Sin

They got no further. The samurai of Japan had put the king and most of the generals of Korea to flight, slaughtering in the process literally thousands upon thousands of Korean civilians and soldiers as farms and cities burned. It was to the supreme good fortune of the very idea of Korea as a nation state that a verifiable naval genius, patriot and hero rose to the defence of the peninsula at a moment when every other means of preservation had proved incapable of halting the Japanese.

The Korean Saviour

It is hard to avoid hyperbole when discussing the career of Admiral Yi Sun-sin (1545–98). He was born in the same year as Sir Francis Drake, but utterly lacked Drake's unpleasant habits of executing subordinates and putting personal profit and glory before military obligation. Like Nelson, he was of humble origins, was completely undefeated in his naval campaigns and was slain at the end of his greatest victory – but Nelson was not the technological innovator Yi was. Where Nelson received decoration and reward from his sovereign, Yi was twice broken and sent to the ranks by his, and put to severe torture on grounds later confirmed to be specious. He would be

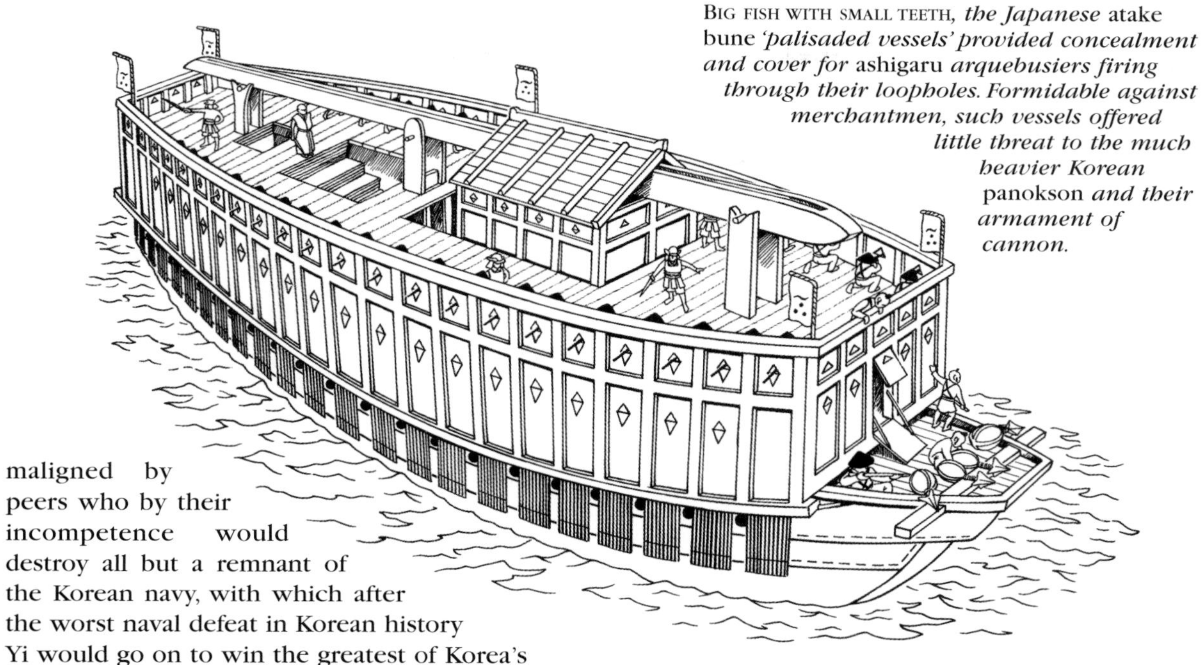

BIG FISH WITH SMALL TEETH, *the Japanese* atake bune *'palisaded vessels' provided concealment and cover for* ashigaru *arquebusiers firing through their loopholes. Formidable against merchantmen, such vessels offered little threat to the much heavier Korean* panokson *and their armament of cannon.*

maligned by peers who by their incompetence would destroy all but a remnant of the Korean navy, with which after the worst naval defeat in Korean history Yi would go on to win the greatest of Korea's naval victories and complete his rout of the invaders in 1598.

Yi began his career with less than no family influence - his own father had been convicted on what later proved to be false charges and sent away from the court in disgrace. As a boy he met Yu Seong-ryong, whose political support at court would twice be the difference between life or death for Yi - and, incidentally, Korea. Eligibility for either civil or military rank depended upon performance in Confucian exams and Yi did well enough to begin his rise through the ranks of the Korean military.

Like the Athenians before them, the Koreans did not differentiate between land and naval service, in both cases ending up with both incompetent and superior admirals. An excellent example of the former was Won Kyun, who had benefited from family connections to receive command of one of the four squadrons of the Korean navy, the Western Gyeongsang fleet based at Geoje Island south of Pusan. The Eastern Gyeongsang fleet was lost with the rest of Pusan itself when its commander, Park Hong, from a safe distance ordered his 75 ships scuttled and his base destroyed. Having saved the Japanese the trouble, he then fled inland, spreading panic as he went. The Western Gyeonsang kept four ships - the four that Won Kyun, who had thought the Japanese fleet some kind of huge trading mission - kept to protect him as he made his own escape from danger.

The Koreans had built their navy with some other sort of action in mind. Yi, by then a senior admiral, commanded the Eastern Jeolla fleet based at Yeosu in that province. He had shared Yu Seong-ryong's mixture of apprehension and frustration at the state of Korea's defences and preparedness, and had done what he was allowed to do to repair things. While Yu bustled about frantically attempting to repair city walls and fortresses, Yi had constructed his own version of a type of ship he thought might prove useful against the Japanese.

History rejoices in the preservation of the multiple volumes of Yi Sun-sin's war diary, which he kept from his promotion to admiral until two days before his own death. In it, the admiral recounts his ongoing horror at the savagery with which the Japanese were treating his people, his frustration with the failure of the Korean armies and generals to counter the invaders, and his desperate pleas to the king to be allowed to take his ships into action.

The collapsing bureaucracy of Korean government was powerless to halt the Japanese, but somehow it found time in the ongoing collapse to declare one Kwak Chae-u, a patriotic landowner who had raised a guerrilla force behind the Japanese lines, to be a rebel and to dispatch troops against him! It took some time for the valiant Kwak to receive official authorization to fight back against the Japanese. Yi risked court martial or worse if he left his own assigned area to counterattack the Japanese, and it was not until Won Kyun and his four surviving ships entered his defensive area that he secured authorization to counterattack the invader.

Yi's intent and intelligent efforts to make ready both before the invasion had brought the most solid of all Korean military institutions to its highest degree of readiness at the time it was most needed. As has been noted previously, Korean ships tended to be of the most seaworthy construction, the surviving vessels of both the Mongol invasion attempts being predominantly Korean. With upper-caste commander and lower class sailors literally in the same boat, artisans and Confucian scholars united to forge a tremendously formidable naval arm.

A bit of clarification can provide an idea of how seriously the Koreans took their navy. The Western reader might draw the wrong conclusions from the names of four standardized units of ordnance in Korean naval artillery - although the very concept of having but four standardized types of cannon would have been a tremendous assistance to the Spanish Armada! But the poetic images in the translated names of the 'yellow', 'black', 'heaven', and 'earth', cannon are best cast aside. Those characters were the first four in a standard reading text and as grimly utilitarian as the progressively larger heavy iron tubes on wheeled carriages they denominated.

When Yi assumed command of his fleet, the primary fighting unit of the Korean navy was a powerful and effective line of battle ship called the *panokseon,* a heavily built 'board-roofed' ship, in nature and design very similar to the heavy war galleys the Hellenistic naval architects labelled 'cataphract' with exactly the same meaning. In their long years of fighting the *wako* pirates, the Korean sailors had come to understand the Japanese penchant for hand-to-hand boarding combat and taken preliminary steps to prevent it. The cannon, high bulwarks and 'board covering' of these vessels made them almost impossible to board.

Panokseon ranged in size from 15m (50ft) at the waterline to the largest at 35m (110ft), with oarsmen on the lower deck and marines working the guns above numbering some 125. Admiral Yi and his captains occupied a combination watchtower and command deck at the highest level of the ship. Slow as their heavy construction and heavier cannon armament made them, properly handled *panokseon* seldom had to run away from an enemy that wished to fight - at least, with a crew and a commander that wished to fight back.

Turtle Ships

As formidable as the *panokseon* proved to be, Yi himself thought that there was yet room for improvement. Chinese military manuals from centuries previous referred to a sort of ramming rowed junk called the 'covered swooper' (*meng chong*) with a covered upper deck sheathed in fire-resistant hide and apparently meant for ramming, and not, the manuals say, 'conventional fighting ships.' The Korean naval architects had toyed with a similar vessel in the fourteenth century, but Yi Sun-sin took the old idea and created what was perhaps the most powerful naval weapon of his day.

There is a great deal of confusion about the details of what Yi and his fellows called the 'turtle ship' (*kobukson*), which should not exist, given two fairly extensive descriptions, one in Yi's own diary and another in an account composed by his nephew. It is possible in this treatment to end some four centuries of confusion about the appearance of the vessels by the simple expedient of actually believing their employer's own account in place of poor-quality illustrations produced some two centuries after Yi's version of the turtle ship first saw combat.

That the 'turtle ships' were turtle-backed is the easy part of our understanding of the design, which was lower, lighter, and consequently faster

SECRET SUPER WEAPON, *the* kobukson *turtle ship's rectangular lines allowed a power-to-weight ratio that gave the new design strength, stability and manoeuverability. Teams of oarsmen standing at their sculls could row or fight as the tactical situation demanded, cannon and the ram alternately lethal as opportunity offered. Iron plating on the arched back of the vessel shrugged off fire arrows or arquebus shot, although the Japanese preserved a story of a valiant team of boarders hacking their way into one vessel's interior. This illustration (right) shows the accepted reconstruction, the dragon's head depicting how the eighteenth-century illustrators took 'dragon's head' to mean a figurehead, such as seen on the 'dragon boats' the Chinese still race today.*

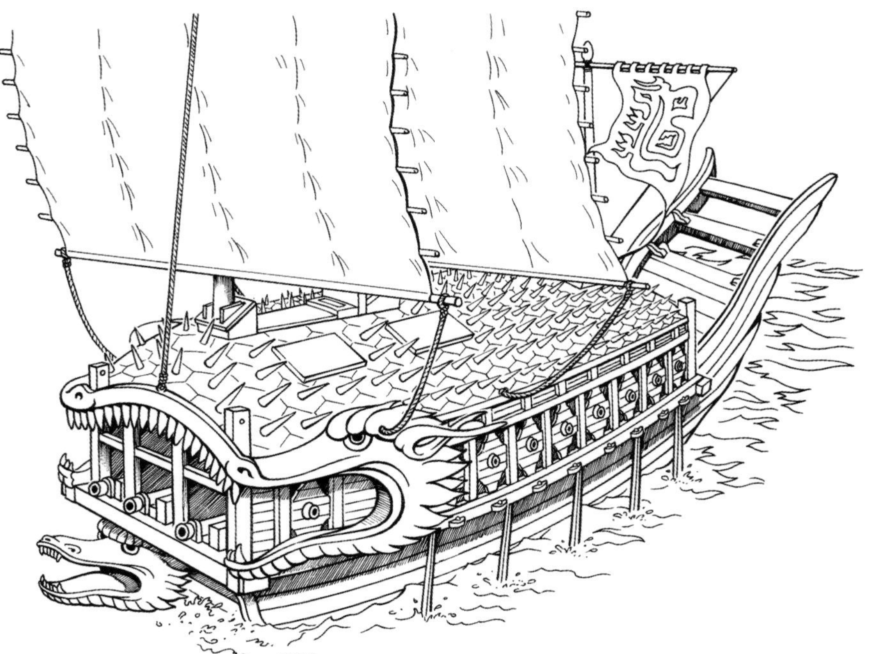

'DRAGONS' MOUTHS *through which we shot our cannon...' The words of Yi Sun-sin himself prompt this new and accurate reconstruction of the* kobukson, *a design Yi himself based upon earlier Chinese and Korean concepts. Note also the correction of the design showing the vessel's ram as something that would strike an enemy vessel at the waterline, with the entire momentum of the attacking 'tinclad' focused on a single shattering wedge.*

than the lumbering *panokseon* they were meant to supplement. Standing oarsmen and gunners would work side by side from the single rectangular upper deck, which, like *Monitor's,* overhung the lower deck in which the crew had their quarters.

A great deal of needless ink has been spilled in argument over the 'ironclad' nature of the turtle back, another manifestation of the confusion between 'iron' and 'tin' clad ships noted above. Any iron armour on the back of a turtle ship would have been in thin anti-incendiary plates, not cannon-proof heavier material that would have and did cause one putative replica to collapse under the immense and utterly unnecessary weight. The other form of protection on the armoured backs of the vessels consisted of spikes, which Yi's nephews mentions as being concealed in the hope of impaling a boarder's foot.

Yi's and his nephew's accounts agree on something upon which modern scholars can find little agreement in the light of the flawed illustrations, that cannon fired through the dragon's heads on the bows of the ships. The eighteenth-century illustrators' reading of the term was to put the dragon figureheads of the traditional 'dragon boat' racing sculls on the top of the turtle back, in a position from which not so much as a musket could have been fired. Yi and his nephew obviously meant that the *entire front end* of the ship took the form of a dragon's head, as the accompanying illustration shows, with the forward two or three bow guns firing from ports among the 'teeth' in the front of sides.

Modern reconstructions also find themselves at a loss to explain how the turtle ships, in the fashion attributed to their earlier antecedents, rammed. Eighteenth-century illustrations are correct in that detail, in depicting a *second* dragon's head on the cutwater, a ram not at all unlike the proven model of the Ancient Greeks – and like the one dredged up from the sea floor at Athlit in 1980. The modern reconstruction's flat mask on the reverse angled bows of the reconstruction means that the actual ram of the turtle ship would never strike the enemy at all – the impact would instead be upon the forward end beam, which would have the effect of diluting the momentum of a ramming attack into uselessness against the side of the target's hull. By a combination ramming and cannon attack, a turtle ship would severely maul the *Nihon Maru,* a tower ship originally built as a floating castle for Hideyoshi, at the battle of Angolpo in 1592.

Yi's single prototype turtle ship missed the admiral's first engagement with the Japanese off the island of Okpo, in which Yi paused to listen to the fishermen who had tried to warn Won Kyun about the Japanese. Won himself and some other surviving ships managed to rendezvous with Yi's fleet of 24 *panokseon* and descend upon a completely unprepared Japanese naval force primarily engaged in loading loot. The Japanese ships, *atake bune,* bore a superficial resemblance to the *panokseon,* but the difference was telling.

Unarmoured Korean infantry had fallen like wheat before the scythe in the face of rows of arquebuses fired through the bulwarks of Japanese vessels, but the Korean ships stayed well beyond musket shot, their crews behind thick boards as their cannon did their work. One of the most feared projectiles thrown by the 'earth' cannon of the *panokseon* was a large iron arrow wrapped with incendiary material. Whether beached or trying to escape, a great many of the *atake bune* burned.

Over the course of the next two days, the Korean fleet sank some 37 of the Japanese by long-range bombardment without the loss of a single ship. Won Kyun spent the aftermath of the battle driving Korean sailors from the prizes they had taken so that he himself could claim them, while Yi prepared his prototype turtle ship to intercept a Japanese squadron making its way along the Western coast. The Japanese never realized it, but Yi undoubtedly did – given the superiority of the Korean fleet made manifest at Okpo, the best hope the Japanese had of defeating it was the conquest of its remaining bases on shore.

The Turtle Ships Attack

On 8 July 1592, 23 of Yi's squadron and three of Won Kyun's bore down upon the Japanese fleet anchored below the heights at Sacheon. As usual, Yi had a clear understanding of the enemy's numbers and position gained from fishermen and refugees. Showing the awareness of a truly great

HAN SAILING JUNKS *brought Chinese conquerors and supplies to Korea, sailing as far as India and Africa before China made its own decision to turn inward and let the world come to them. One such vessel changed the course of world history by delivering three Portuguese traders and two matchlock arquebuses to the emulous Japanese.*

commander he was also aware that when his fleet made contact with the twelve large *atake bune* tied up on the shore, the tide was on the ebb. Yi had to preserve his own vital fleet while luring the enemy into deep water where they could be utterly destroyed.

To help him, he had one great asset - Japanese contempt for the Koreans as warriors. Japanese savagery upon the peninsula had not been tremendously more pronounced than it had been towards hostile clans and cities back in Japan, but the disaster at Okpo was still unknown to the samurai. Yi's ships feigned a retreat as if put into panic by the mere appearance of the Japanese vessels, the Japanese pursued in poor order - and the new turtle ship did deadly work by destroying every Japanese ship to leave the bay. As darkness fell, the turtle ship under Yi's chosen commander, shot flame, iron and smoke into the Japanese from bow, stern and broadside ports as it shattered the first of a series of Japanese formations.

Yi would go on to victory after victory behind his turtle ships - and without them. The Japanese would eventually retreat from Korea. In his final battle, when giving chase to the remnants of the last Japanese fleet, Yi Sun-sin would be mortally wounded. His last order was to a sailor to cover his body with a shield so that his death would not discourage his men.

Wusung-Shanghai, 1842

It is possible to understand, as well as to criticize, the decision of the Ming and their succeeding dynasty, the Qing, to turn inwards after the incredible outlay and technical success of the Ming navy in the 1420s. Why sail around Africa when the Europeans were perfectly willing to save the Chinese the trouble and when the trade goods they offered were about the same as the Ancient Romans had offered their ancestors over the silk road, primarily spices and specie? Trade and threats with the Japanese proved an acceptable solution to piracy - acceptable, at least, to an inland government suspicious of seaborne trade. China had so much the world wanted, it seemed to make as much sense to let the world beat the proverbial path to their door and pay for it. That the 'barbarians' of the outside world might decide to do something besides pay for the desired trade goods was a realization too long in coming.

In 1557, the Portuguese, considered mere pirates by the Ming due to their connections and trade with those who were indeed pirates, seized and occupied the fortified peninsula of Macau. The Dutch East India Company dominated Taiwan's coast from 1624 to 1661 and there were incursions from the British and the Russians. It is curious that the Qing Dynasty, which had expanded into rule from Manchuria assisted by a fleet, followed the Ming into turning its back on the sea and into eventual eclipse.

With no desire for a navy, despite the growing need for one, the Chinese allowed technological development to proceed without them. Confucian

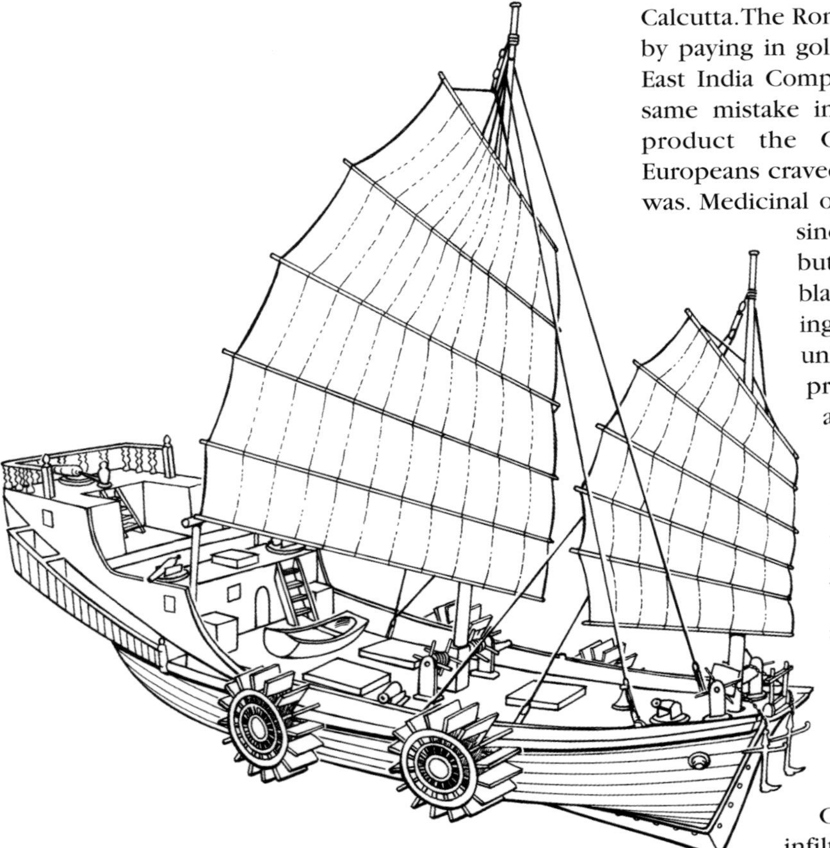

FUNCTIONAL FUSION OF *Chinese and Western technology, this nineteenth century war junk combines European-style cannon with the classic battened sails that served sailing junks and* sampans *superbly throughout the centuries. Capstan-powered paddle wheels go back at least to the treasure ships of the Song era, but muscle and wood could not overcome the advantages iron and steam bestowed upon the attacking British vessels.*

reverence for the past recalled the indeed admirable achievements of the Ming and earlier generations in ship construction and armament, and enough of the ancient military manuals survived so that the great ships of the past could be constructed again. There was also a tendency to dismiss a device as 'not invented here'.

Disaster for the Chinese came from an economic triumph. The British East India Company had evicted the Portuguese from most of India by 1612, but British consumers soon came to prefer Chinese tea to cotton cloth from Calcutta. The Roman Empire had bankrupted itself by paying in gold and silver for Chinese silk; the East India Company was not about to make the same mistake in purchasing Chinese tea. Some product the Chinese desired as much as Europeans craved tea had to be found, and it duly was. Medicinal opium had been known in China since the sixth or seventh century, but the idea of putting the sticky black poppy sap into a pipe and ingesting the narcotic fumes was an unhappy result of combining the practice of native Americans with an old world drug. Addiction and demand for opium swept over China. At long last the Europeans had found something to trade with the Chinese instead of precious metals.

The costs of any narcotic addiction quickly become unacceptable to an organized society and the Chinese government banned the sale of non-medicinal opium in 1729. 'Country traders' with Company opium from India infiltrated the Chinese coasts with no fleet to stop them and by 1838 the volume of the opium trade had grown to 40,000 chests per year. In 1729, the figure had been one-two-hundredth of that.

The Chinese were faced with the disturbingly modern choice of legalizing the popular narcotic or taking extreme measures to suppress it. In 1838, a determined official named Lin Tse-Hu received the Emperor's commission to enter the foreign warehouses in Canton and destroy the opium stockpiled there, which he did with decisive efficiency. Rioting erupted between British sailors and the Chinese.

Private Enterprise

In 1840, the British East India Company responded by diverting its profits into something besides stock dividends. The company's purchase was a ship named *Nemesis,* privately built in three months by the Birkenhead foundries and manned

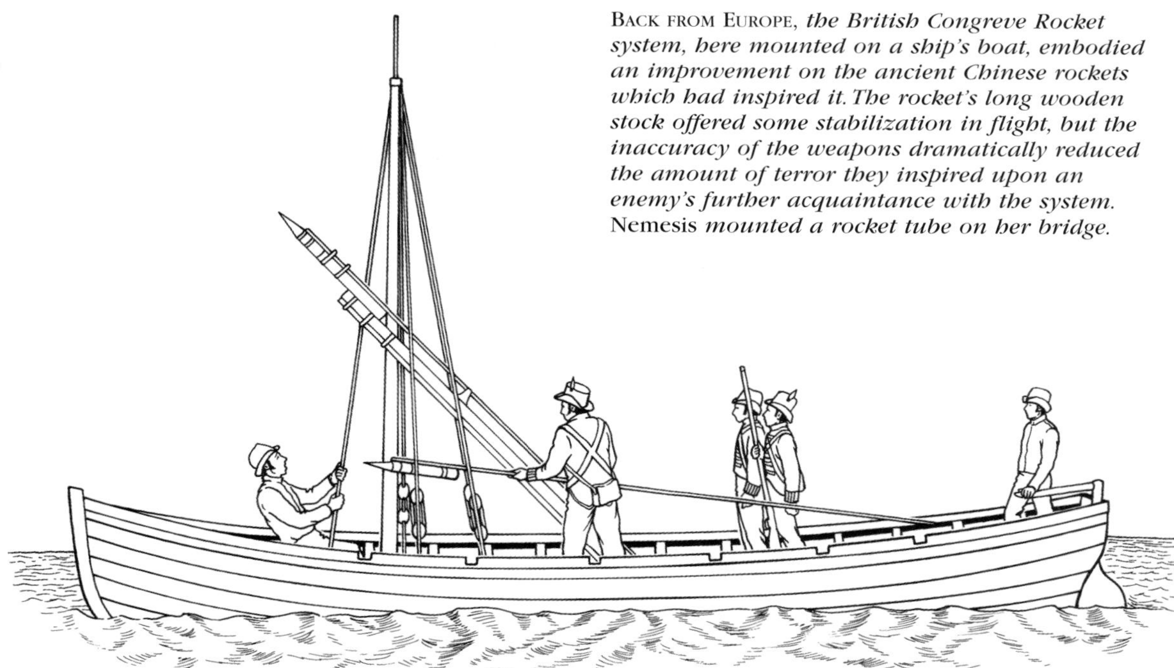

BACK FROM EUROPE, *the British Congreve Rocket system, here mounted on a ship's boat, embodied an improvement on the ancient Chinese rockets which had inspired it. The rocket's long wooden stock offered some stabilization in flight, but the inaccuracy of the weapons dramatically reduced the amount of terror they inspired upon an enemy's further acquaintance with the system.* Nemesis *mounted a rocket tube on her bridge.*

by Company sailors to counter Lin Tse-Hu's efforts to suppress the opium trade. The 630-tonne ship included many unusual features, such as an iron hull and interior fittings, very rare for the age; two cannon firing 14.5kg (32lb) shot, with smaller guns added later; and a very shallow draught of 1.5m (5ft), built with an eye to the navigation of China's river system. Her engine produced 120 horsepower driving twin paddlewheels – a system vulnerable to cannon fire or obstruction, but one that would give her tremendous manoeuvrability in shallow water or tight quarters.

In addition to her iron hull, from the Chinese viewpoint, *Nemesis* freighted a full cargo of irony. Four of *Nemesis's* features, and among her most vital, would have been familiar to a Ming shipbuilder, starting with the two centreboards fore and aft of her engine room that could be winched up or down by means of her deck windlasses and chains into recesses in the flat bottomed hull, from fully recessed to as far down as 1.5m (5ft).

Chinese ships lifted the entire rudder up and down on tackles, while *Nemesis's* rudder terminated in an iron plate which another winch and chain could lift or lower for greater or shallower draught as desired. Such features would allow *Nemesis* easy passage across the sandbars of the Yangtze shallows and enough steerage way for her to sail and steam around the Cape of Good Hope on her way to China.

A Chinese invention saved *Nemesis* when she struck a reef at the entrance to St Ives Bay in Cornwall – a compartmented hull such as those found in the Quanzhou and Shinan wrecks, which allowed her to make port and effect repairs. The collision was quite possibly caused when yet another Chinese invention – her compass – was thrown off by the incorrect positioning of magnets meant to counter the effect of *Nemesis's* iron hull upon its readings.

The final echo of China's past manifest upon her intended destroyer was a tube for the launching of rockets mounted upon the bridge, although Sir William Congreve's rockets proved to be no more accurate than their Chinese counterparts of some 1000 years before. The Chinese had used them as incendiaries – and so did the British, although Captain Hall, *Nemesis's* commander, had his hand severely burned in pushing one that had failed to leave the tube away from his bridge at Canton.

By July 1840, a British blockade fleet including three line of battle ships, four armed steamers, 13 other armed ships and a troop transport had assembled at Canton, the shore of which the Chinese had fortified under Lin Tse-Hu's direction with the Imperial legate offering considerable sums for the capture of the British or their vessels, graduated by the size of the ship or the rank of the individual captured.

Cannon-armed war junks arrived at the inner harbour of Canton, with the Chinese forts firing at isolated British ships and their shore parties landed to spike the guns of isolated Chinese batteries. By August *Nemesis* was at Canton and by January of 1841 the British squadron had destroyed a Chinese fleet armed with Portuguese cannon and reduced the Cantonese forts. The fleet then proceeded north towards Shanghai, with the intention of forcing the mouth of the Yangtze and in time-honoured fashion dictating a peace at the Qing capital at Nanjing.

Nemesis's role over the months had been to reconnoitre and to tow prizes, transports and landing craft, her shallow draught enabling her to sail close to the shore and disperse Chinese troop concentrations with blasts of grapeshot from her batteries. The Chinese, for their part, were consulting the military manuals and employing what had worked before. When the British fleet paused at the Chusan Archipelago the Chinese sent nearly 100 fire rafts such as had proved so devastating at Lake Poyang down the ebb tide towards *Nemesis* at anchor, forcing her crew to destroy them with cannon fire or send boats out to tow them away from the ship. The British sent parties ashore to forestall confiscation of *sampans* for subsequent fire attacks before proceeding towards the main naval base of the Qing at Shanghai.

The British Move Upstream

At the mouth of the Yangtze estuary lay the forts and smaller town of Wusung, the site of the last stand of the Chinese navy. The British stood off the mouth on 5 June 1842 with launches and smaller craft stopping the trading junks bound for Shanghai and examining the shore for batteries. They did not have it all their own way, however, losing the steamer *Ariadne* to a rock in the channel. *Nemesis* herself had grounded twice, requiring on one occasion the use of captured junks as pontoons to lift her off. The Chinese had mounted some 175 largely foreign-made guns in earthworks where the smaller, deeper channel of the River Wusung led up into the navigable parts of the Yangtze.

The battle of Wusung began on 16 June, with *Nemesis, Phlegethon* and *Pluto,* all steamers, lashed to *Modeste, Columbine* and *Clio,* respectively, a tactic the Union Navy would employ extensively two decades later in the American Civil War. The larger *Tenasserim* pulled *Blonde* and *Sesostris* towed flagship *Cornwallis.* As the British moved up against the river current into the estuary, their squadron came under the most intense fire experienced in the course of the war.

Simultaneously, down the river came the final product of the traditional Chinese military manuals. At the core of a fleet of 14 conventional war junks, five new constructions moved under power towards the British as the Chinese shore battery did their best to bar the invaders from the Yangtze and the inner heart of China. Accounts of the Song 'treasure ships' recorded vessels with as many as twelve paddle wheels but this quintet of 'wheel boats,' as the British called them, carried two on either beam.

The means of their propulsion was simple human effort on the bars of a standard capstan, linked to the paddle wheel by a large wooden cog. Whatever these vessels lacked in firepower or mechanical power, they had a clean run. Breaking through the wall of fire but severely damaged, *Nemesis* engaged and found the Chinese retreating against the current at a speed of three and a half knots. Human muscle could neither equal nor escape *Nemesis's* untiring 120 horsepower engines in an iron hull.

The steamers moved the larger British ships to where their crushing firepower could suppress the more obstreperous of the batteries and the smaller ships to where they could send landing parties ashore to complete the work of silencing the guns. *Nemesis* ran aground pursuing two of the 'wheel boats', with *Phlegethon, Nemesis's* sister ship, overtaking and boarding others still attempting to escape down the river to Shanghai. The British were met by fire from arquebuses and a shower of grenades including what they called 'stink pots', sulphur bombs more in the line of a chemical gas grenade. The British towed off such junks as were run ashore or left drifting by their abandoning crews, and sailed into Shanghai with no further resistance on 19 June.

STEAM AND CANNON *versus muscle-bound tradition, the mobility and firepower produced by British technological expertise was more than even the most determined Chinese could counter as their forts and harbour fleets distintegrated. China's early lead in technology had utterly vanished by 1842, a legacy to its foes.*

The British would effect repairs at Shanghai, and proceed in strength up the River Yangtze to where they would blockade the mouth of the Grand Canal and secure a treaty promising no further interference with the opium trade by 29 August.

A second round of hostilities over the trade would erupt from 1856–60, but there would be one major change, just as there had been in Japan after Commodore Matthew Perry's squadron of American steamers and sailing ships of war had forced trade concessions in 1854. Both the Chinese and the Japanese would begin and continue to demonstrate an acute interest in acquiring the most modern – and lethal – products of Western technology.

Wusung-Shanghai

1842

China's efforts to expel imported opium from their cities ran afoul of the British East India Company's economic drive to sell the one product they found the Chinese willing to exchange for their own trade goods, especially tea. The iron-clad *Nemesis* and other products of the Industrial Revolution steamed at will off China's coast, towing the older warships of the Royal Navy with their overwhelming firepower. When the British fleet made for their largest naval base at Shanghai, the Chinese mustered their heaviest foreign-made artillery and capstan-paddled war junks in an effort to block the mouth of the River Yangtze. British steam and gunnery were more than the Chinese could stand and the victorious British took Shanghai without further setbacks. A dictated peace would soon follow, on British terms.

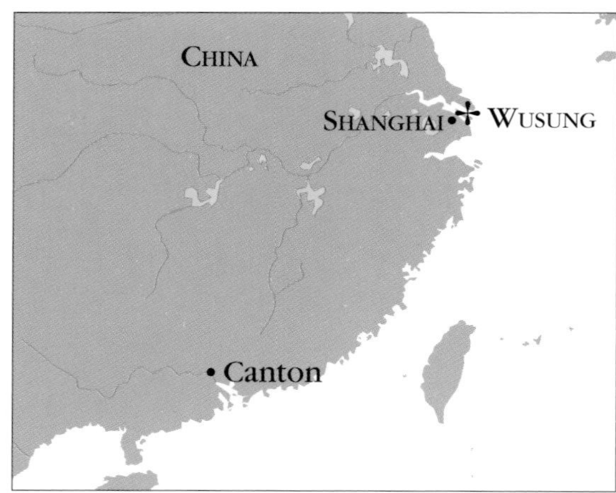

The high tides and converging currents made seafaring difficult along the Chinese coast. Ports were rare and required good designs to allow traders and fishermen to venture out. China's ports were her overseas lifeline, but also very vulnerable to attack.

5 The British have additional trouble in the river's bars and shallows but continue, landing marines and other infantry to approach their objective by land. Having passed batteries and captured or destroyed the defending fleet, the British sail up the Wusung and occupy Shanghai on 19 June 1842.

SHANGHAI

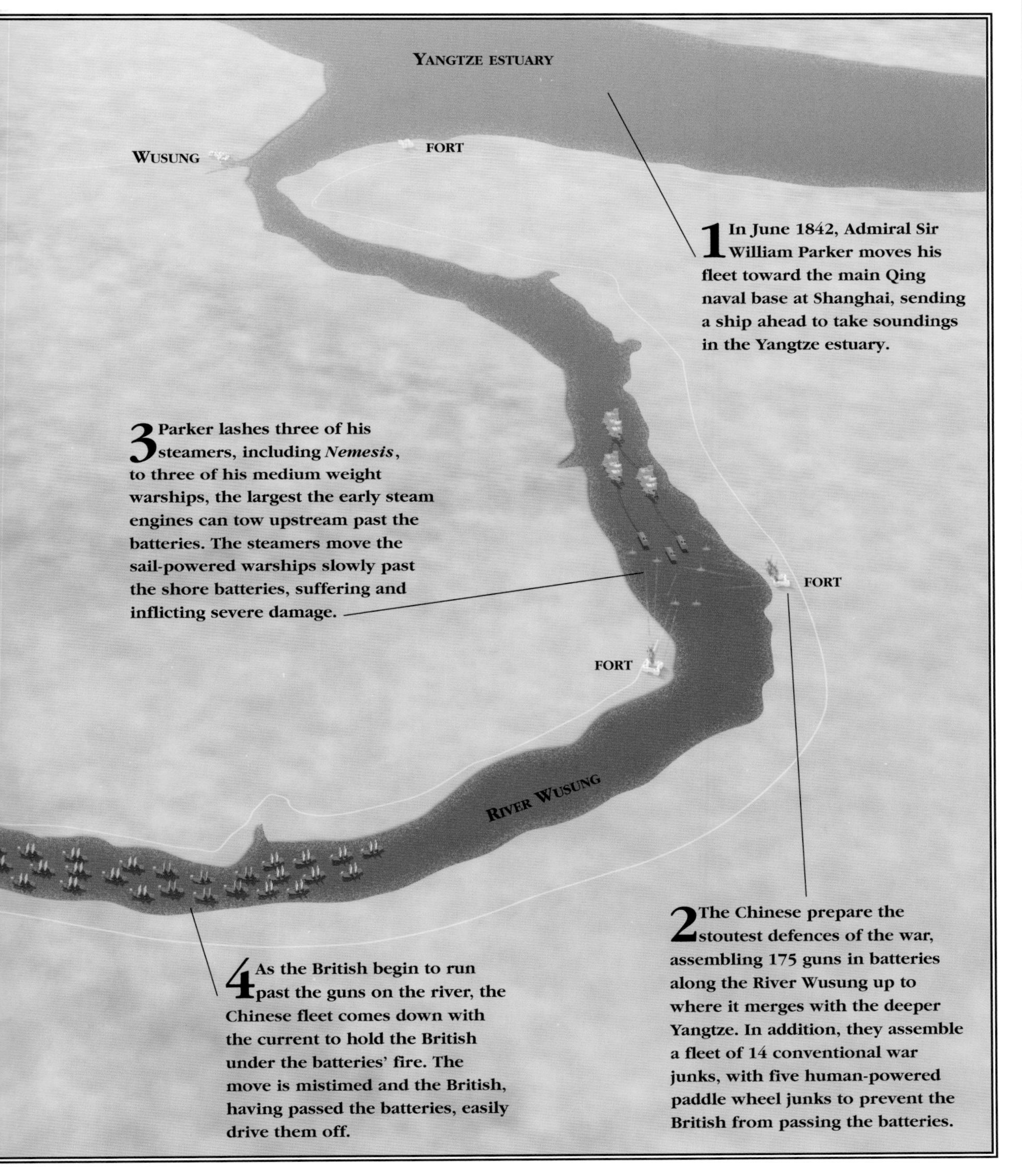
YANGTZE ESTUARY
WUSUNG
FORT
FORT
FORT
RIVER WUSUNG
1 In June 1842, Admiral Sir William Parker moves his fleet toward the main Qing naval base at Shanghai, sending a ship ahead to take soundings in the Yangtze estuary.
2 The Chinese prepare the stoutest defences of the war, assembling 175 guns in batteries along the River Wusung up to where it merges with the deeper Yangtze. In addition, they assemble a fleet of 14 conventional war junks, with five human-powered paddle wheel junks to prevent the British from passing the batteries.
3 Parker lashes three of his steamers, including *Nemesis*, to three of his medium weight warships, the largest the early steam engines can tow upstream past the batteries. The steamers move the sail-powered warships slowly past the shore batteries, suffering and inflicting severe damage.
4 As the British begin to run past the guns on the river, the Chinese fleet comes down with the current to hold the British under the batteries' fire. The move is mistimed and the British, having passed the batteries, easily drive them off.

Bibliography

Anon. 'Seoul', *Korean Cultural Heritage, Vol 2.* Time and Space Tech Ltd, 2002.
Ballard, Vice-Admiral G.A. *The Influence of the Sea on the Political History of Japan.* New York: E. P. Dutton and Co, 1921.
Beeching, J. *The Chinese Opium Wars.* New York: Harcourt Brace Jovanovitch, 1975.
Bond, B., ed. *Victorian Military Campaigns.* New York: Frederick Praeger, 1967.
Chambers, J. *The Devil's Horsemen: The Mongol Invasion of Europe.* New York: Atheneum, 1979.
Davison, M. W., et al. *Everyday Life through the Ages.* Pleasantville, NY: Reader's Digest, 1992.
Fairbank, J. K. *China: A New History.* Cambridge: Harvard University Press, 1991.
Holdsworth, M. and Courtauld, C. *The Forbidden City: The Great Within.* Hong Kong: Odyssey Publications, 1995.
Niderost, E. 'Fighting the Tiger', *Military Heritage* 38 (Vol 4, No. 1), August 2002.
Niderost, E. 'Turtleship Destiny', *Military Heritage* 81 (Vol 2, No. 6), June 2001.
Paludan, A. *Chronicle of the Chinese Emperors.* London: Thames and Hudson, 1998.
Peers, C. J. *Soldiers of the Dragon - Chinese Armies 1500 BC-AD 1840.* Oxford: Osprey Publishing, 2006.
Pow-Key, S., ed. *Nanjung Ilgi, The War Diary of Admiral Yi Sun-sin.* Seoul: Yonsei University Press, 1977.
Rossabi, M. 'All the Khan's Horses', *Natural History,* (October 1994).
Songnyong, Yu. *The Book of Corrections: Reflections in the National Crisis during the Japanese Invasion of Korea, 1592-1598.* Berkeley: University of California Press, 2002.
Sun Tzu, *The Art of War* (trans. Samuel B. Griffith). Oxford: Oxford University Press, 1963.
Sung-jin, Y and Man-hee, L. *Click into the Hermit Kingdom.* Seoul: Dongbang Media, 2000.
Thompson, H., ed. *China.* London: DK Publishing, 2005.
Turnbull, S. *Ashigaru 1467-1649.* Oxford: Osprey Publishing, 2001.
Turnbull, S. *Siege Weapons of the Far East (2), AD 960-1644.* Oxford: Osprey, 2002.
Turnbull, S. *Warriors of Medieval Japan.* Oxford: Osprey Publishing, 2005.
Turnbull, S. *Samurai Invasion: Japan's Korean War, 1592-1598.* London: Cassell & Co, 2002.
Whitlock, P. and Pearce, W. *HMS Victory and Nelson.* Portsmouth: Portsmouth Royal Navy Museum Ltd, 1990.
Wolseley, Lt Col G. *Narrative of the War with China in 1860.* Wilmington, DE: Scholarly Resources Reprint, 1972, of an 1860 work.

Index

Page numbers in *italics* refer to illustrations, those in **bold** type refer to information displays with illustrations and text. Abbreviations are as follows: (B) - battle; (NB) - naval battle; (S) - siege.